Down Along the Haw

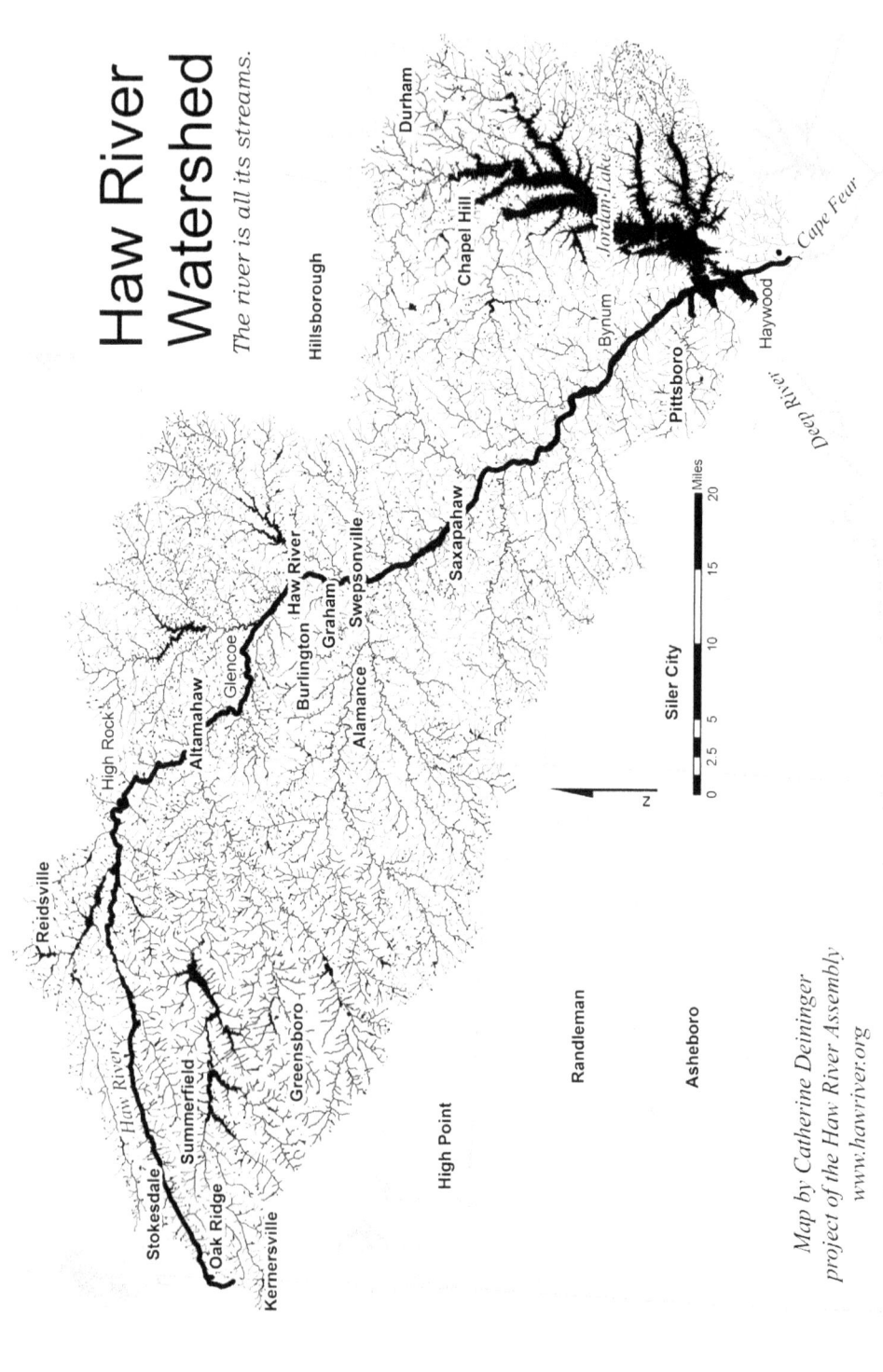

www.ingramcontent.com/pod-product-compliance
Ingram Content Group UK Ltd.
Pitfield, Milton Keynes, MK11 3LW, UK
UKHW050531150426
5217IPUK00026B/1890

Down Along the Haw

The History of a North Carolina River

Anne Melyn Cassebaum

McFarland & Company, Inc., Publishers
Jefferson, North Carolina, and London

Words and music to "Aragon Mill" by Si Kahn.
Copyright Si Kahn/Joe Hill Music (ASCAP).
Used by permission.

"Cotton Mill Colic #3" by David McCarn.
Used by permission of Peer International Corporation.
All rights reserved.

LIBRARY OF CONGRESS CATALOGUING-IN-PUBLICATION DATA

Cassebaum, Anne Melyn, 1943–
Down along the Haw : the history of a North
Carolina river / Anne Melyn Cassebaum.
    p.   cm.
Includes bibliographical references and index.

ISBN 978-0-7864-5948-3
softcover : 50# alkaline paper ∞

1. Haw River Valley (N.C.) — History.
2. Human ecology — North Carolina — Haw River Valley.
3. Stream ecology — North Carolina — Haw River (River).
I. Title.
F262.H34C38   2011   975.6'58 — dc22   2011003519

BRITISH LIBRARY CATALOGUING DATA ARE AVAILABLE

© 2011 Anne Melyn Cassebaum. All rights reserved

*No part of this book may be reproduced or transmitted in any form
or by any means, electronic or mechanical, including photocopying
or recording, or by any information storage and retrieval system,
without permission in writing from the publisher.*

Cover photograph: Lunchstop Rapid (photograph by Sönke Johnsen)

Manufactured in the United States of America

*McFarland & Company, Inc., Publishers
Box 611, Jefferson, North Carolina 28640
www.mcfarlandpub.com*

To the people of the Haw,
ancient, current, future.

Table of Contents

Preface and Acknowledgments   1

Introduction   3

1. From Seep to Swamp — Forsyth, Guilford Counties   5
2. Streaming, Swamping Haw — Guilford, Rockingham Counties   12
3. Cotton Mill River — Alamance County   43
4. Water Power and Whitewater — Chatham County   136
5. Jordan Lake — Chatham County   174
6. Coastal River to the Sea — Chatham, Lee, Harnett, Cumberland, Bladen, Columbus, Pender, Brunswick and New Hanover Counties   192

Appendix: Citizen Organizations Working for the Health of the Haw   205

Chapter Notes   207

Bibliography   223

Index   225

A Negro Speaks of Rivers

I've known rivers:
I've known rivers ancient as the world and older
than the flow of human blood in human veins.

My soul has grown deep like the rivers.

I bathed in the Euphrates when dawns were young.
I built my hut near the Congo and it lulled me to sleep
I looked upon the Nile and raised the pyramids above it.
I heard the singing of the Mississippi when Abe Lincoln went down to
New Orleans, and I've seen its muddy bosom turn all golden in the sunset.

I've known rivers:
Ancient dusky rivers.

My soul has grown deep like the rivers. —*Langston Hughes*

Preface and Acknowledgments

This work explores how the lives of humans, animals and a river run together. Though our technological lives may lead us to ignore these connections, this book argues that not only does the Haw have a fascinating history, but we are so deeply connected to the river, its health is ours.

I, along with my children, was first drawn to the Haw's tributary Reedy Fork in the mid-eighties because it was a fascinating world apart where we felt at home. Paddling under arches of trees, questions would float to mind: why are streams in the South brown, why are wasps' grey nests always on ironwood trees (still unsolved), why was a wastewater plant allowed to take over a river, and who was here before.

This book brings together answers to my questions and the voices of the interesting, often inspiring people I met in my search. It is organized from source to sea, with my narrative of walking and paddling the Haw combined with topics from mill history to algae and interviews with mill owners and workers, archaeologists, environmentalists, farmers, water treatment managers, and many others.

Although interest in the Haw is rising and more books are on their way, *An Historical Atlas of the Haw River* is one of the few out now; it gives a full sweep of the navigable Haw in 20 pages with descriptions of key spots matched to full page typographical maps. Local histories also fill in much of the river's recent past because the river was the story. This is among the first full works to focus on the river and its many facets.

Once asked about his creative sources, my favorite author, William Faulkner, replied that all he did was follow people around and write down what they said, and that turned out to be my method for this river book. It was transmitted from people who live by or work for the Haw who allowed me to question or interview them. I was heartened by people's willingness to share their time, references and information with me, showing great patience when their specialty was far beyond my knowledge base. So I thank all those who answered my questions or gave interviews, for their voices stand for the many people not named here who also care and work for the river.

Any project like this takes time, and without the sabbatical and release time which Elon University granted me, it would not have been possible. I appreciate the support of those in Belk Library and in Technical Support, especially Lynn Melchor, Roger Gant and Rick Palmer, and faculty colleagues, particularly Rosemary Haskell, Brian Crawford, Jim Brown, Helen Walton and Janet MacFall as well as students in my Haw River class.

Special guidance in this research project came from Elaine Chiosso, Michael Holland, Gail Knauff, and Janet MacFall, and generous help came from Mark Chilton, Carole Troxler, Tom Magnuson, Kathy Buck, Catherine Deininger and others at the Haw River Assembly. Kathy Barry and Jerrie Nall of the Textile Heritage Museum also provided important advice and photographs.

Barry Nelson produced an invaluable notebook for me on the geology of the Haw which I hope will soon be published.

The work of local historians was essential to me; they have forwarded knowledge of our counties and our place. Three recent books, *Shuttle and Plow: A History of Alamance County, North Carolina* by Carole Troxler and William Vincent, *An Historical Atlas of the Haw River* by Mark Chilton, and Gail and Bob Knauff's *Fabric of a Community: The Story of Haw River, North Carolina* provided new insight and research.

A loyal group of readers offered important suggestions for revision; I am indebted to Rosemary Haskell, Zo Tanczo, Elaine Chiosso, Cynthia Crossen, Joyce Ahmad, and Janet MacFall.

I also want to thank William Rusch, who provided the six section maps, and Sönke Johnsen, whose photographs are on the cover and throughout the book, for contributing their work so all proceeds from this book can go to the Haw River Assembly.

Family members—Will, Vega, Amber, Sierra, Jamie Lee and Demitri—spurred me on with enthusiasm and good company, and my husband, John Herold, advised me through many drafts and tolerated my late hours even though he is not generally drawn to canoes or immersion.

To all those mentioned here and throughout this book, I am grateful.

Introduction

> Here are as brave Rivers as any in the World, stored with great abundance....
> — Robert Horne, *A Brief Description of the Province of Carolina*, 1666

What is a river? Great rivers leap first to mind: the Colorado bashing canyons, the Mississippi roiling, the Nile glazing mysteries, the Congo luring us to our hidden selves.

The Haw is also a great river though not famous in the world. It is an ancient vein to the sea, fractured and eroded into being after the continents' crunch raised the Appalachians. The Haw was first life, satisfying hunger and thirst for its creatures and for those who came later across oceans, following deer and buffalo trails to water.

When another age of settlers came, the Haw was the force watering bottomlands, driving mills, grinding grains, weaving cloths, carrying goods.

It carried away topsoil when the land was farmed as though European settlers were permanent heirs to the New World's abundance. Its tumbling waters dammed, mussels were silted over. Sturgeon and herring waned. Dye, dirt and sewage changed the color of the waters; in the worst period and places, the Haw was dead. Some could only see a sewer line.

But others could still see the river. The people of the Haw worked to bring the river back, and us back to ourselves and health. In 1972, American rivers were given protection and the Haw's dramatic recovery began. It became again a river to swim and fish in. Yet, the Haw still slides along, carrying too much of our waste, scraped earth, chemicals, tires and ragged plastic bags. Can a way of life that harms the Haw be good for us?

This book on a river's natural and human history takes the reader down the Haw as I explore it. I wanted to discover what I could of the lives lived on its banks and study humans' place in its rich ecology. So I walked or canoed the length of the Haw, reading about the river as I went and talking to people working for the river. Exploring the Haw took me off into areas and subjects I had not imagined and brought me back to see the river as central and everywhere in our lives.

I hope this book will lead others to experience the Haw. This is a crucial time for the river and for us; there is still wildness almost unbroken along its banks, to be protected and opened for walkers and paddlers. The last centuries have given us violent warning of what abuse can do to a river; the last three decades have shown us the Haw's and our capacity to rebound.

Basic Rules of Exploration, some learned the hard way by this author:
Do not trespass on private land.
Even a shallow stream can be hazardous, so do not canoe alone or in high waters.
Dams are treacherous and must be avoided.

1

From Seep to Swamp
Forsyth, Guilford Counties

Source: Stigall Road, Forsyth

Where does a river begin? In rain, falling rain. A river is rain, cycled from earthly evaporation that started long ago from volcanic vapors and icy comets and asteroids melting in our atmosphere.

Where does the Haw begin?

For most, the Haw begins and ends in a glance from the windless capsules of our cars. It flashes below Route 85, under an arch of maples and sycamores, and is gone as fast as a "Cape Fear River Basin" sign flickers green and white in the brain.

For those who slow to walk or fish by the Haw, there is time to wonder about the source. Yet I paddled Reedy Fork Creek for many years before my thoughts drifted up to the Haw's origin. Tracking the source was no arduous climb, just the stirring of my imagination and a drive to where Stigall Road and Route 150 intersect near a woods in Forsyth County.

A scuffed trail leads to where the Haw first seeps in under maple, oak, tulip and sweetgum and tangles of Virginia creeper and grape. The dark soaked gulley is easy to miss and even shifts as rainfall changes the water level in the land. It would be inaccessible, part of someone's back yard, if a veteran kayaker had not orchestrated the public purchase of this land in 2001.[1]

This pool of water, the Haw's official source, trickles on, tripping around pebbles and twigs, obeying the same laws of hydrodynamics as it does surging under and over whitewater paddlers downriver.

The return to the source is, of course, not the primal experience of *Heart of Darkness*, but the lack of drama permits a fuller sense of a river. Most often we think of the Haw or any river writhing across the land like a muddy snake with its source farthest from the sea. But the Haw is more than a line, it is a stream network that starts in all the headwaters of all the streams that tilt to it. The whole river is more like an oak leaf cupped in your hand, its veins netting into the central stem.

This river that sways past Smith Island and into the sea 310 miles away has many humble starts. Three streams immediately join it to the east across Stigall Road on land tilting northward. Starting in the same seeping way, they gather water before they combine on ground so swampy almost all distinctions and channels are lost.

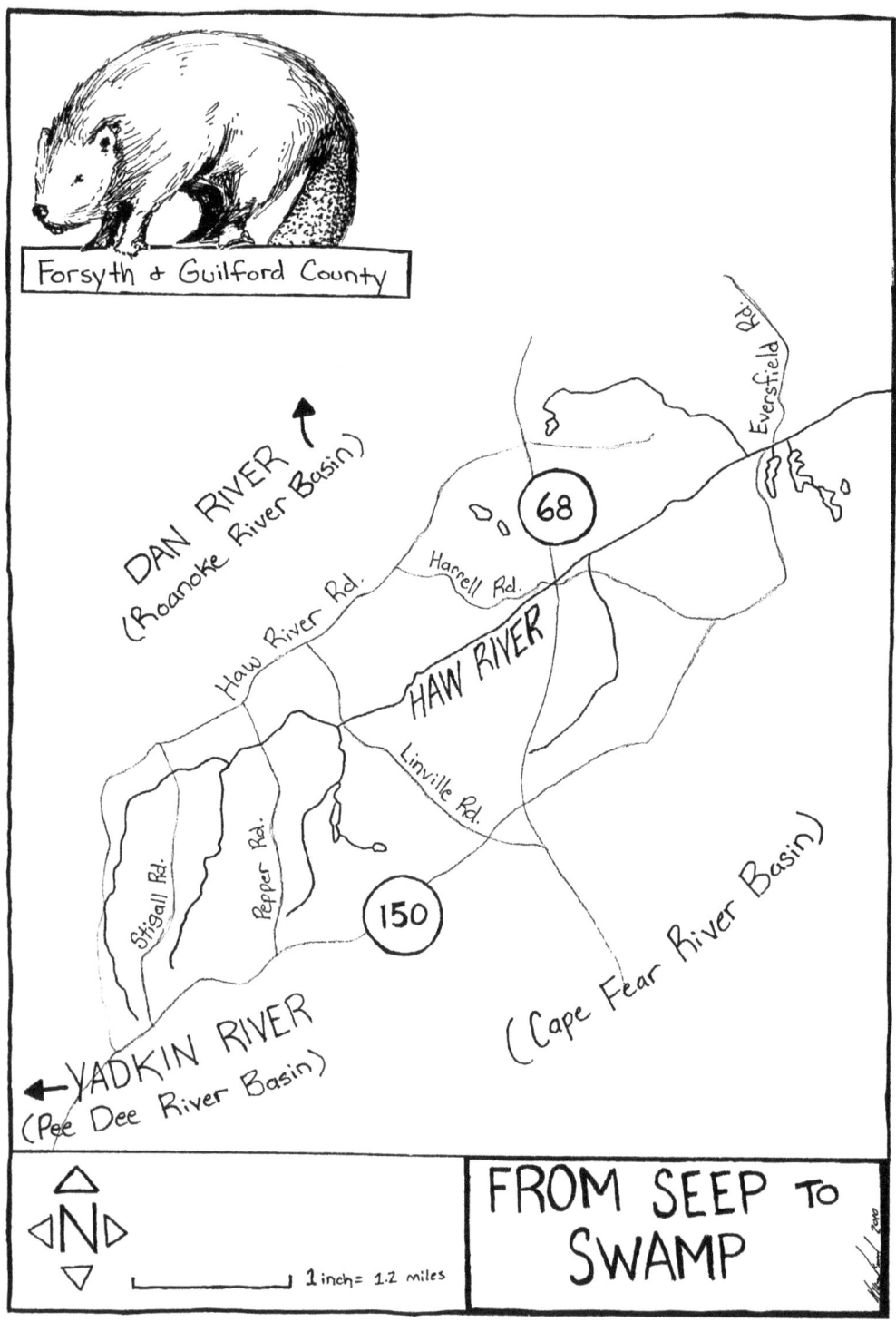

From Seep to Swamp (William Rusch).

1. From Seep to Swamp 7

One source of the Haw as it appeared one February day in 2010; with more rain, the water table rises, and the source shifts headward. Seeps here are under water (Elaine Chiosso, copyright Haw River Assembly).

The whole oak leaf represents the Haw's 1,707 square mile watershed — drainage area, basin, valley. We need the simplicity of one name, and Haw it is, sounding like the calls of crows downstream.

Other streams start nearby, but the land slants them toward the Dan–Roanoke or the Yadkin–PeeDee River Basin. The terrain tips other nearby creeks into the Deep River on course to converge with the Haw and form the Cape Fear River after the Haw has its 110 mile run through wild and urban lands.

If a river seems like a line, but is really a network of streams, there is another illusion to confront at the source. The Haw and all rivers are fed by rain, but much of it oozes to them underground. Sixty times more water is underground than we see streaming across the earth's surface. We are almost always standing on water.[2] In the Piedmont, that water lodges among fractured rocks,[3] so many of the Haw's sources seep or spurt out in a manna of springs, feeding creeks and river, all drawn by gravity to the Atlantic and that "wide water inescapable."[4]

Springs and creeks pulled animals, hunters, settlers to them. Down Stigall Road, a neighbor of the Haw, Danny Wilkes, tells me the Haw passes four springs before the one on his land where water pours from a ten foot rise and early homesteaders built a log cabin.

Swamp Stigmas: Stigall to Pepper Road

"You're not bothering my copperheads, are you?"

What do you say to this graying woman with a notebook peering down into two inches of water—the Haw River at Stigall Road. Danny Wilkes didn't comment further on my quest to walk or canoe the whole Haw; instead he solved my puzzlement about the two five-foot-high drain pipes carrying the trickling Haw under Stigall Road. Though the Haw may be only two miles long here, he has seen it fill both pipes and flood the road. One rain dumped five inches in an hour and sent the Haw past the 100 year flood line, the high water mark likely in a century.

As we talk, I am distracted by darting below—tadpoles, salamanders. Wilkes tells me that there are good bait fish here: horneyhead and crayfish.

With Wilkes's permission, I pursue the Haw east to Pepper Road along deer trails and multiple threads of water, the earliest red of swamp maple blooms sharp against the grey woods. I find the Haw a wetlands, loose, unchanneled, dappled with tufts of grass. This wonderful running together of waters is both swamp and marsh, the difference being that only swamps support trees, but both are wetlands and have a history of being deplored, their atmosphere breeding ill health and evil. Edgar Allan Poe warns us of the ghastly rank miasma in the tarn near the House of Usher. In the 1800s and much of the 1900s, state and federal governments promoted drainage of wetlands for farmland and control of malaria and typhoid; in the 1840s, North Carolina had trouble unloading coastal wetlands in auctions.[5] Swamps lost some of their phobic qualities when Louis Pasteur tracked disease to germs[6]; others discovered the virtues of biodiversity in bog, fen, swamp and marsh.

Prints show that raccoon and deer feel at home here, as well as frogs that thunk into gurgling waters. I splash through hillocks to climb a thick beaver dam. The water's tones deepen as it pours through and around this thirty-foot-long dam, which opens pond and marsh to the sun.

I work to regain my linear sense of river and climb down to follow the drift of waters east, watching the Haw narrow casually into braided channels and then spread at Pepper Road into a wide, cattail-rimmed pond that can accommodate rains. Two five foot pipes are still enough to transmit a larger Haw under Pepper Road, revealing wetlands' power to detain and disperse, the secrets of flood control. These wetlands are also breaking down and releasing nitrogen back to the atmosphere, important work since excess nitrogen is a major problem downriver in Jordan Lake.[7]

One tenth of America may once have been wetlands[8]; we have since lost more than half of that, much of it from advances in machinery in the mid–1900s. Early settlers might applaud the conversion of land so unsuited to travel and farming. My muddy, slurping walk back to Stigall Road never hits a striding cadence, but the territory I cover is vital.

Beavers and Other Trespassers: Pepper Road to Route 68

The possibility of rain hangs over the day and dank February woods when I drive back to explore the Haw from Pepper to Linville Road. A beautiful grey fox lies by the road, its dark eyes facing the spin of black tires. A chorus of dogs at the bridge sends my dog peeing to mark territory. I head down a bank where deer tracks have pressed in three inches.

A cluster of mallards and blue-winged teals startles me until they turn out to be decoys. Past ironwood and tulip trees, the Haw thickens into a swamp, with the dark wine leaves of winter honeysuckle climbing swamp dogwood and bare broken trees. Sometimes there's a coppery film over its olive waters.

I walk in the water which sometimes shapes itself into a stream until Linville Road, acting as a dam, turns the Haw and a stream from the south into a pond. Now pipes aren't enough, and the Haw sails by a beaver hut and the wooden piers of its first bridge.

On the way back, rain comes with evening, hitting the pale beech leaves that hang like snow in the twilight. More copper sandpapery foam and a strange blue oil swirl on the water. I'm puzzled. The tall trees and quiet seem protection from an urbanized world's pollutants. Months later, I learn the blue oil and coppery crust are both byproducts of harmless iron bacteria that grow in chains feeding on themselves by oxidizing ferrous iron to ferric iron. The coppery crust is the ferric iron changed to iron oxide in air and water.[9] Around it, straw grasses are combed flat by high waters, and sand is piled in a bright patch; I wonder how it got there.

I free myself from a blackberry bush enough to admire its grasp and reach. Its successful solar design sends thorny brambles out ten feet and then projects more brambles six feet farther off. Passage is not easy.

The Haw swamps out at Linville Road before going under its first bridge; beaver find protection with an underwater entrance to their hut (center) (photograph by the author).

Finally, I find the beaver chips I was expecting, first a scattering, then a whole patch of saplings reduced to pointed stumps. My dog barks, distracting me.

"Crook, Crook!" He answers my call with more barking. Then I see two men coming toward me, steady men who know just how to dismiss a dog barking on their territory.

I apologize for Crook's commotion. They tell me they live on both sides of the Haw here, and I should have asked for permission. There is no anger. No "Get the hell off my land."

"You should for your own protection."

"I don't know if you notice we come armed." Then I do notice the holster.

"Last week there was a robbery here; the owner had a gun, and he shot the guy. We saw your car by the road and had to check it out." They tell me the landowner's name. "Give him a call next time." I promise I will.

I walk back thinking about this problem of land access. It won't be easy to solve. At Pepper Road, a littering of soda cans gives one reason walkers may not be welcome. But how are people to care for the land and the river if they cannot be on it? In Britain, ramblers' paths allow walkers to crisscross their island. When you buy land with a ramblers' path on it, it is part of the deed and you cannot obstruct it, nor have cattle of an aggressive nature near it. (One British farmer did find a way to discourage walkers by leaving years of manure on his path and keeping it well hosed down.)

The men today were right; I was trespassing since I was not actually in the river, but once reassured, they did not mind my being there. Until we have a public trail, river walkers, like myself, need to secure land owners' permission. I console myself that the Haw is getting deep enough to paddle. That water is public. As American Whitewater explains, "North Carolina's state test of navigability is equivalent to a recreational boating test. If a boater can float the river, it is navigable. The public can use all water determined to be navigable for ... boating, swimming, wading and fishing."[10]

When I return to canoe from Linville Road to Route 68, the February afternoon is dominated by blue sky and sun. I lean over the bridge to savor the spring peepers' ear-throbbing call for a mate, so loud it is a vibration I can feel. Of course, when they sense my presence, the result is silence. Swamp maple blossoms and the clear thick channel below me add a sweet promise to the day. I paddle off sending four mallards into flight.

The channel morphs to a swamp, and I tromp on through inchoate webbings of water, thin and finger deep, braiding around young green shocks of growth, then dropping into a stream too deep to wade. This will go on for three hours of bushwhacking and canoe-dragging until at Route 68, I thrill to the truck rush of the road.

On the way there, I see two men working high in a pasture with Black Angus; I think pistols and wave my son's orange hunting cap, climb barb-wire, cross pastures, to explain myself. Totally unaware of me, yet unstartled, they welcome me as if I were one of the purple martins for whose spring arrival they are erecting houses on antennae.

I learn that the Black Angus grazing these tawny hills belong to the founder of Red Oak Brewery, and dine on the grains left over from beer making then become steaks at the owner's Greensboro restaurant. "It's a complete cycle," they explain to this vegetarian.

We chat on until their work is done; they lumber off in a truck over the hill. I head back to the Haw feeling like Walt Whitman, the sun on my back, walking the land, hailing others, hearing humans' stories in an America as open as these wide fields and sky.

Below in the swamp, my not always happy ramble resumes: mud flats are covered with greenbriars and blackberries, but prints of deer, heron, cottontail, raccoon, and beavers cheer me on. This is beaver territory. Their thick mats of young trees barricade the Haw. Scientist H.C. Pielou puts it mildly when she says: "Their hydrological effects are profound."[11] Beavers compete with us in rearranging earth, flooding bottomland and claiming building sites. "A family of beavers can build a 35-foot-long dam in a week," Alice Outwater, author of the fascinating *Water: A Natural History*, tells us.[12] Some 4,000 foot long dams have been found, twigs hardened to petrification, suggesting hundreds of years of maintenance. All this is done by creatures with small brains, but claws on all feet. They dig tunnels and chambers and dam water, not to catch fish, but to safeguard entrances.[13]

Those seeing land as property may curse, as may a woman who wants an easy paddle, but otherwise "profound" means "profoundly good." Alice Outwater celebrates the beaver; slowing water is the heart of their good work. By creating wetlands and thus flood control, they cause dirt, or sediment, to settle. Their dams keep fast waters from racing through scouring more dirt into streams, suffocating stream life and coating fish eggs. Instead, dirt plasters on a maple's trunk or an outburst of sedges, and, over time, a rich mudflat forms, and rivers are clearer for it.

This slow water is an eco-system where everything from deer to zooplankton feed. The bacteria and fungi clean up a drop of water the way a turkey vulture disposes of road kill, and the algae or phytoplankton grown by sunlight become food for the zooplankton and then insects, fish, us and heron.[14]

River folk hail the return of beavers, a keystone species, now numbering possibly 10 million. It's an impressive comeback though not the estimated 200 million who once reshaped the American landscape and changed the watery dynamics of the Midwest.[15] Beavers were extinct in North Carolina by 1897 and nearly everywhere else. Reintroduced in the state in 1939,[16] they have rebounded; Outwater estimates there are half a million beaver dams in the state.[17]

It wasn't developers who drove them out, it was, Outwater tells us, Europeans needing fur for warmth. Beaver pelts were soft to touch, waterproof, easily shaped. And the castoreum, the musky oil that beavers sleek over their coat and mark territory with, was used to cure ailments from headaches to impotence. Their meat was tasty and counted as fish during Lent because their tails have scales. By the 1300s, the push for furs went into Russia where the demand was satisfied, for about 200 years. Then, with only the fringes of Siberia and Scandinavia able to provide pelts, the New World appeared, and the sad song was played again. Native peoples got metal, beads and smallpox in exchange for the animal, sometimes a pet, that kept waters clean and was builder of the world in Native tales. The slap of its tail was thunder.[18]

Finally, within the rumble of Route 68, I spot a sweet round head sending ripples across water. It dives under its circle of waves. I wait for the tail slap, but this one holds out underwater. The travel back to my car is slowed by the work of beavers, but I admire what one woman's grandfather replied when asked what he was going to do about the beavers on his land.

"Do? Nothing."

"Because," she added, "I don't think it even occurred to him it wasn't their pond too."[19]

2

Streaming, Swamping Haw
Guilford, Rockingham Counties

Moving Waters and Settlers: Route 68 to Sandy Cross Road

In 1778, those journeying along the easy dip of hills at Route 68 could have a pint at Samuel McCrackin's tavern before fording on. Long a crossing of road and river, the Cape Fear Road was just west of Rt. 68 as are remains of an old bridge.

I leave my car by the bridge and head east out of the din of traffic, following a Haw that remains a beaver haven. The river's extensive wetlands are its distinction: narrow floodplains thick with mud, pools, and river strands with pulsing water lights. I tag after deer above the seeping swamp, past mountain laurel coming into bloom and the accordion pleats of beech leaves swelling out of thin tan covers. My dog Crook cools in water the color of his copper belly.

On this glistening stretch, what is remarkable is what I do not see: natural gas lines coming from the Gulf of Mexico travel under the Haw, headed to the Pine Needle LNG Plant on Eversfield Road.

When I phone Bill Parks, plant manager, I learn the plant is by a river of natural gas 10,500 miles long, traveling in three pipes from below Gulf waters up the East Coast. After passing under the Haw here, one pipe branches east and crosses the Haw again under Jordan Lake.

This Pine Needle Plant and one in Newark, N.J., are where the Williams Gas Pipeline Company stores natural gas to meet peak demand in winter. By amazing transformations, the gas arrives as a vapor, is further dried or purified as a liquid, and then frozen to 258 degrees below zero. It takes 200 days to fill both heavily insulated tanks here, but only ten to convert it to vapor and meet demand.

Below the plant on Eversfield Road, in rain pools, spring peepers' trills spiral above the dropping water of a beaver dam. Nearby is a lavish outcropping of skunk cabbage; its young yellow-green leaves glow against the dark swamp. The maroon scathe keeps the temperature of the buds as much as 36 degrees warmer than the outside and heighten the stink, drawing pollinators eager for the first flowers of spring. Hoping the plant would be as potent as its smell, people have used it for contraceptives, rheumatism, headache and more. Also called clumpfoot cabbage and polecat weed, no one tries to glamorize it.[1]

The Haw is still a creek at the new bridge at Eversfield, which spans 48 paces; the thick piles of the old bridge just east are only 11 paces. Crook and I take an old dirt track into

2. Streaming, Swamping Haw

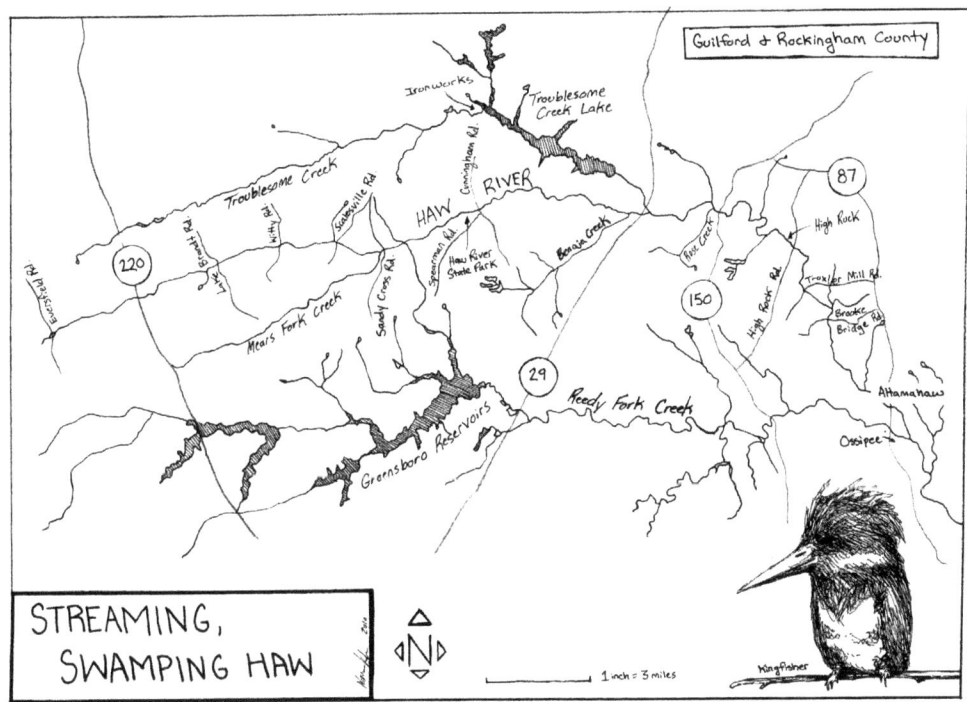

Streaming, Swamping Haw (William Rusch).

the woods as I look for a trace of those who worshipped here centuries ago — the site of the Methodist Church established in 1797.[2]

Leaving the thorny grasp of blackberries and greenbriars, we climb past oak, beech and tulip to a platform east of Rock Branch. No foundation stones rise through the oak leaves, so I have only intuition and rough calculations to tell me this could be the church's site, southfacing, open to the sun, the Haw's healing waters a wavering channel below. A river hymn sounds in my head: "Time like an ever rolling stream bears all its sons away" and then bleeds into the deep chords of the union anthem, "Like a tree planted by a river, I shall not be moved." How did these Methodists sing of rivers?

We walk on with the Haw dropping over logs, pooling, running farther east. On one of the feeder streams, a farmer's family ran a still for bootleg liquor. The bottomlands also concealed their dead cows, illegally left unburied. The woman who told me remembered her grandfather's thrilling explanation that the bones were from dinosaurs.

Farther east, about a mile from Rt. 220, the dark criss-crossed trestles of the Atlantic and Yadkin Railway rise over the Haw. In the early 20th century, this line linked Mt. Airy through Greensboro to Sanford, a six hour trip in 1931.[3] Its predecessor, the Cape Fear and Yadkin Valley Railway was first to provide travel and transport of goods from Mt. Airy to Wilmington in the late 1880s, boosting trade from piedmont to coast.

Farther along, horseback riders at Hardin farms can trot across the Haw's meanders. The Hardins have impressive Watersee Longhorn cattle who are only mildly interested when I drive my car through them. A tidy swaying Haw crosses under Rt. 220, its lapping no match for the automotive hiss above.

When my daughter Vega drops me off to pursue the Haw from Rt. 220 to Sandy Cross Road, the contrast between road and river is great. The force of a log truck's wheeled tonnage blasts me toward a dead raccoon and the bridge railing. Those interested in the river look foolish to those on the highway as an irrelevant Haw carries away their oily runoff. I escape down a deer path and slowly settle into the river world, discovering where I live anew on cloven hoof tracks.

Even Crook pauses in streamside awe to listen and watch the Haw sieving on. When the poet Rilke wrote "happiness falls," he captured the release moving waters evoke. Hindu and Muslim alike find holiness in rivers and sacred sources of life in their confluences. Shiva, destroyer and creator, touched the Milky Way to earth and created the Ganges, and the Sanskrit word for ford, *tirtha*, means a crossing into the world of the gods.[4] Immersion in the Ganges, like Christian baptism in the Haw, purifies.

Why are people drawn to water? Because it is so much of our body that we feel the river in us? Because it held earliest life and holds our earliest memories of floating in the womb? Because we sense the river's water connecting us to far wider worlds?

On its banks, the Haw that invites reverie is endlessly hospitable to bush and thorn and poison ivy tangles. Like the beech, I take to the hills and watch the river from above as it widens and thins in its wooded corridor, gets lost in thick swamp brush and spring leaves, then suddenly reappears dark and sinking at the sharp drop of a bank.

When I see this ramble of the upper Haw across map or land, I don't imagine any of this motion converted into a scientific formula. But, as I learn from hydrology texts, physicists and hydrologists love rivers too — the sweet geometries of their curves and volumes. They plot the rush of waters and the tumble of stones and, like all lovers, have their private language of thalwegs, bed material, hyporheic zones, helical currents, degrees of meandering and v-values of sinuosity.

Einstein, da Vinci, and the Greeks before them sought the laws of tumbling waters. The formulas get smashed and dented a bit because the rock-ridden Haw does not glide through a frictionless plain, but patterns there are.

Take meanders, the course all rivers take in anything softer than rock. From the smallest rills to the broadest river, water sways in half loops back and forth to rhythms captured by a formula: $\Xi = \omega \sin S/M\ 2\pi$, from which emerges that the wavelength of a meander is usually 11 times the channel width.[5]

But why do rivers meander? The answer to this still controversial question starts with gravity's pull. Water seeks the quickest route to the low point of the ocean. This is rarely a straight line and once a bend starts, the force of the water widens and exaggerates it, especially in flatter, softer terrain.

But there's more to meanders. Even the warm waters of the Gulf Stream meander through the colder waters of the Atlantic, so scientists like Luna Leopold claim that meanders are hydrodynamic — water moving by the principle of least action and responding to friction.

When Einstein swirled a cup of tea, he saw in the settling of the leaves the pattern of helical motion[6] — the way rivers roll sideways, actually spiraling downstream. Water near the surface has less friction on it and is faster than water near the bottom slowed by mud, mussels and stone. This uneven speed shapes the river. It digs out the outside curve of the meander where it is faster. Kayakers on the lower Haw anticipate the faster, deeper water

Meanders are the motion of rivers. Tickle Creek hollows, curves, and deposits sand bars as it sways to the Haw (photograph by the author).

will be on the outside of a curve. Slower water on the inside of the curve has time to drop its sand. This explains why a point bar will develop — the sandy bars I saw upstream.[7]

Meanders are how floodplains develop. As the meanders widen their loops, the area flooded by the river extends. Far downstream, the Cape Fear River loops more widely through the malleable soils of the coastal plain. The Haw, by contrast, is striking for its straightness and narrow floodplains, which are due, as local geologist Barry Nelson explains, to the "hard crystalline rocks of the Piedmont" and "a relatively high, steady gradient."[8]

Scientists go on to graph the kinetics of stones in a stream, for when they observe that clumps of rocks in a stream move less frequently than rocks spaced apart,[9] they hear the sweet tumble of numbers in their brains.

I just hear the lap of the Haw, sheltered, moving under a cover of leaves. When power lines cross the Haw and farmers' fields stretch for acres, the openness feels like sudden exposure, and I'm relieved when back in the woods under cover. There the crows' caw sounds a human note of complaint, the cardinals flash their spring wolf whistle, a barred owl hoots who-cooks-for-you, and then there is unidentified bird chatter, the music of the woods.

What few of us can do is smell these woods. Deer always surprise me. One today gazes back for a slow moment before bounding off, the long flick of her white tail rising and falling lightly. These creatures of scent leave messages in their hoof prints I never pick up. They urinate on the mound of their tarsal gland inside their back legs to broadcast availability, life status, dominance, and other messages their long noses and big nasal cavities pick up.[10] One time I did get to watch a fawn as rowdy as a pup until the doe caught my scent.

Otters speak easily to each other through the smell of their anal glands. With an otter's nose, I might enjoy their imperial stare and humping, galloping ways more often. These deeply amusing creatures signal clean water — another reason to wish for more of them.

I hike higher as the Haw splays into channels and pass the rock-strewn path of an old road. Following the river, I am carried along through light, young leaves, soft and yellowy green, lurching, ducking, gliding along deer trails. Even running above the Haw brings to mind the Sufi prayer poem: "We have fallen into a place where everything is music."[11]

This feeling disappears with the asphalt of Lake Brandt Road and the equally hard packed earth of logged land. Huge houses are rising on the Haw's banks above cleared bottomland. Tire treads a yard wide dwarf the narrow prints of deer leading up a vertical bank where an umbrella tree, a mountain magnolia, hangs on. A 30-foot buffer has been maintained, but much of the Haw's protection has been razed to feature it.

Here and elsewhere by the Haw, clearcutting is for houses, twice as large as those of a half century ago, full of isolated worlds — entertainment centers and master bedroom suites. These new river developments with names and stone entrances are a far cry from the homesteads of 250 years ago when people came church by church. Here individual families, likely moving for a job, search for community around work, school, computer, or maybe again church. In a nearby development, one new house is already up for re-sale; it won't be this family's homeplace.

Europeans arrived to settle in the mid–1700s, cutting trees and growing crops, in what was an overwhelming forest. As one historian puts it: "Whatever people of North Carolina accomplished in the name of civilization — the houses they built, the fields they cleared and planted, the towns they made by the river banks — all took shape against the backdrop of seemingly endless forest."[12] The Haw was shaded by giant "oaks, hickories, walnuts, chestnuts, poplar, always woods — dark, foreboding and dense."[13] Breaks in the forests were provided by swamps, beaver dams, burned fields, or wide rivers.

Europeans came through these endless woods, not by river. The Cape Fear would take the Highland Scots up to Cross Creek (Fayetteville), but the Haw was too full of rocks and rapids to invite travel. Settlers came on the Great Wagon Road from Philadelphia, veering off at Roanoke Gap: Baptists, Quakers, Presbyterians, Methodists, Lutherans; their churches bound worship, education, social club, court, and social services together.

In 1761, the McNairys came as newlyweds with relatives and church and settled on Horsepen Creek on Reedy Fork of the Haw.[14] Their house stands today at the Greensboro Historical Museum, its simplicity a tonic to our own consumption-driven age. Like the even smaller Allen House built on Cane Creek now at Alamance Battleground, a few rooms held generations. Harry Watson, a scholar of the American South, finds in them a "premodern concept of family existence" without "specialized spaces" or "compartmentalized lives."[15] Family and community, public and private merged then as they do not now. A house was workshop and home together, the bedroom shared by children, parents, even a guest. So far from the suites and vast closets of homes today, this simplicity was both a necessity and a value.

For a traveler too, accommodations covered only essentials. In the 1700s, ordinaries, or inns, were few and mostly at river crossings, so a lone traveler, almost certainly male, would go to a house and ask to pay for lodging. John F.D. Smyth, a traveler to the Haw Valley and Piedmont in the early 1770s, stayed with the Bailey family, fell in love with 15-

year-old Betsy Bailey and slept with her that night ... along with the parents (in the bed) and on the floor or pallets, "the slaves, young children, grown sons, and three grown daughters."[16] This took place in lower Sauertown, but was typical practice.

Food for settlers was likely plentiful. Through explorer John Lawson and others, we know that buffalo, panther, wolf and elk were along the Haw in the early 1700s. A recently discovered journal of John Powell tells us that as late as April 1737, he encountered buffalo six times; near Mebane, his group came across 12 buffalo.[17] But the buffalo were gone by the 1800s, leaving only their names on the creeks.[18]

The Haw provided plentiful fish and mussels. Great runs of shad were caught in baskets at the end of weirs, V shaped rock walls — a technique of Native peoples. River and stream bottoms provided nuts, for pigs and people: hickory, walnut, chinquapins, hazelnuts, and chestnuts.[19] Streams powered tub mills good for cruder grinding; public mills could take on finer work: "wheat, oats, barley, rye."[20] Their prices were regulated as was their spacing on streams.

Fred Hughes' historical map shows me the land across the river was owned by William Martin in 1756.[21] A half mile southeast on a now disappeared Bruce's Road, the Haw River Baptist Meeting anchored a community with a river identity in 1784.

At Witty Road, where the Haw first crosses into Rockingham County, lands were owned by John Witty (1783), James Campbell (1776), and James Martin (1779). Martin would be coming into more land thanks to his brother, Governor Alexander Martin; graft was part of the new republic as confiscated Tory lands were sold at private auctions at bargain rates to insiders like Martin. Col. James Martin was also accused of selling commissions for 10 pounds each during the war. As Guilford historian Fred Hughes comments, "The Militia was an extension of the political system."[22]

After independence in 1785, this Witty Road crossing was a boundary when Rockingham County was marked off, "beginning at Haw river bridge, near James Martins."[23] The Haw's streams were one reason for dividing Rockingham from Guilford County: "The extent of the county of Guilford and the different water courses therein rendered it inconvenient and troublesome to many of the inhabitants thereof."[24]

At Witty Road, the Haw swerves north into Rockingham County through gently sloping hills, under Scalesville Road, and then drops back into Guilford before looping up into Rockingham again under Sandy Cross Road. The floodplains widen as the Haw cuts deep into raw red clay banks, surrounded by woods, swampy even in drought. Just before Sandy Cross Road, Atamasco lilies on frail stems surprise me in a thin riverside woods. Their white trumpet flowers, bright against the dank earth, turn rosy as they age—one of the New World's surprises for pioneers long ago walking the Haw's banks. The delights of other native plants—spring beauty, pink lady's slipper, cardinal flower, and touch-me-not—would follow in their seasons.

Stream Histories and the Underground Railroad: Upper Haw

Studying maps of the Haw's watershed, I'm struck by how much of the river is not the Haw's main channel. Nine major streams swell just the Upper Haw. One, Reedy Fork, dammed into a string of reservoir lakes north of Greensboro, has its own retinue of creeks: Robinson, Bold Beaver, Cabin, Moore's, Moon's Coffin, Brush, King's Mill, Arvins, Horsepen, Long, Schoolhouse, Richland, Walnut Fork, Beaver, Kenady, Raccoon, Buckhorn, Hunting, Walnut Crooked, Brashear, Silver, not to mention all the creeks of Reedy Forks' major tributaries, North and South Buffalo Creek.

It would be a walk around the world to travel all the Haw's streams, so I detour to only a few here that lead to a mill, to the Underground Railroad, and to a great environmental writer.

Grain mills were centers of social life in the 18th and 19th centuries, and one overshot mill on Beaver Creek is still grinding. In 1764, the Rowan County Court ordered "that Daniel Dillin have licence to build a publick grist mill on the Ready Fork of Haw River at the Mouth of Beavor Creek."[25] A petition for a road to get people to the original tub mill followed in 1769. Twelve years later, the road and mill drew trouble—British soldiers needing supplies, but Dillon, the story goes, had forewarning. "The miller ... had a dream during the night that his toe was burning. When he awoke the next morning, he found British soldiers in possession of the mill. He went to run them off and they began shooting at him. He took cover behind a tree but left his foot sticking out and was hit by a shot in the very toe which he had dreamed was burning."[26]

Dillon survived and his mill would wash away, be rebuilt, change hands, be converted to a roller mill with turbine power in 1913, and lastly be re-converted to water power in

Dillon's Mill, now Old Mill of Guilford, on Beaver Creek, Rt. 68 in Guilford County, grinds flour with millstones turned by an overshot wheel—water falling over the top drives the wheel counter-clockwise. In undershot mills, water flows under the wheel and drives it clockwise. Fellow explorer and grandchild Sierra Herold is in the foreground (photograph by the author).

1977 by a Scotsman, Charles Parnell.[27] You can buy flour there today and smell it grinding.

In early times, streams were addresses. When nearby Oak Ridge Institute opened in 1853, they advertised theirs: "on a knoll overlooking waters of Beaver, Reedy Fork on the South, the Haw to the North."[28]

Streams were always mentioned in sale notices, as in the Moore family's 1826 announcement for "well-watered" land "on Reedy Fork & Beaver Creek two miles below Sanders Mill."[29] The Moores were Quakers, a sect that owned half the mills in Guilford around 1770. They were also part of a Quaker migration leaving the state because they had to pay extra taxes for refusing to fight in the Revolution and because they wanted to live away from slavery banned by the Society of Friends in 1774.[30] By count of the 1850 census, one third of those in Indiana were from North Carolina,[31] and some played a role there in the Underground Railroad coordinating with others here who helped slaves cross Reedy Fork and the Haw en route through northwestern Virginia, across the Ohio River and into Indiana.[32]

Underground Railroad

A historical marker on the Guilford College campus puts the beginnings of the Underground Railroad in 1819 at Horsepen Creek off Reedy Fork where Vestal Coffin and a slave, Sol, helped Jim Dimrey escape. In 1835, Sol himself was sold South and experienced how rivers could mean death or a scentless trail, food, health and salvation. Here is Addison Coffin's 1898 account:

> Solomon ... was sold to a slave dealer, Ike Weatherby, and taken to Southern Georgia ... with a heavy iron collar ... and chained in a coffle (a chain-gang of slaves)....
>
> [Solomon] had carefully ... kept the names of rivers and towns. When he escaped ... he pushed on each night with all his strength, and was making good headway when one day he was startled at the sound of bloodhounds ... so he armed himself with a good club and started to run in the hope of reaching a creek or river. After an hour's run, he reached a large creek with steep banks, and too deep to wade, so he swam across and ran on again with some hope that the horseman in pursuit could not easily cross the creek, and he could have a fair fight with the hounds. It proved as he thought; the hounds came to the creek and swam across, but the horsemen ... could not cross, so went some distance up stream. In the meantime, the hounds came upon him, but he had chosen his position on a large stump about four feet high, from which he defended himself with the energy of despair.[33]

After killing the dogs one by one, Sol, "badly torn and lacerated," rushed by "briars, thorns and bramble vines" into a thicket where he listened to the furious cursing of his pursuers across the creek. He then

> took the back track to the creek, when he plunged into the water and swam and waded down stream several miles, for he feared a relay of hounds would be brought and the pursuit renewed.... [He] pushed on again; his lacerated limbs were very painful and swollen, and but for the bath in the creek, might have been very dangerous.... Weeks later he startled us by suddenly appearing at our house in a deplorable condition.[34]

There Sol recovered and continued north.

Vestal's cousin Levi Coffin, sometimes called president of the Underground Railroad, lived near New Garden Meeting by Horsepen Creek too until he and his wife Katherine migrated to Indiana in 1826.[35] There they worked in the Underground Railroad along with mostly African American conductors whose names are lost.

We can't know the secret routes enslaved people took across the Haw; a likely one was through the Quaker territory of Snow Camp, across Cane Creek to the headwaters of south and north Buffalo Creeks up through the Haw's swamps near Browns Summit, then across Troublesome Creek, and on.[36]

We can't be sure, either, how many found their way by river and North Star to liberty. One estimate is 40,000 of 4,000,000.[37] It's certain the way was harder after the 1850 Fugitive Slave Law raised the fine for aiders and abettors, and Canada became the only safe haven.

Troublesome Creek, which runs north of the Haw until it joins at Rt. 29, was the setting for an extraordinary escape to freedom, as retold by Dr. Bryan Wells, descendant of John Freeman Walls who was born a slave on Troublesome Creek in 1813.

John escaped with his master's wife, but Wells cautions that this is no potboiler romance. Jane's first husband, Daniel Walls, had grown up with John and on his deathbed had freed him and asked him "to take care of Jane and my children and treat them as I would.... It was a Christian relationship founded upon love and they made their ways, literally, through the woods, fighting off wolves and escaping and outwitting slave catchers until they came to Indiana."[38]

They were married in Indiana by Quaker abolitionists before heading out to settle on 200 acres near Puce, Ontario, in 1846. They retrieved Jane's children in a repeat of their incredible journey, but others in the story are not as fortunate. A slave left hanging by Troublesome Creek was rescued but had his ears cut off by slave patrollers who needed proof of his "death" to collect bounty.[39]

Rivers surface often in the spirituals of African Americans. As McKinley Collins, a singer whose voice can sound the depths of Deep River, says: "Rivers have always meant freedom to African Americans. Freedom.... Escape.... Release from life's burdens, yes, that too, that freedom."[40]

Thomas Berry

Down by the south fork of Buffalo Creek, Thomas Berry, Catholic priest, environmentalist, and author of *The Great Work* and *Dream of the Earth*, had the experience that shaped his life.

> My own understanding of the great work began when I was quite young.... Down below was a small creek and there across the creek was a meadow. It was an early afternoon in late May when I first wandered down the incline, crossed the creek, and looked out over the scene.
>
> The field was covered with white lilies rising above the thick grass. A magic moment, this experience gave to my life something to explain my thinking at a more profound level....
>
> This early experience, it seems, has become normative for me.... Whatever preserves and enhances this meadow in the natural cycles of its transformation is good; whatever opposes this meadow or negates it is not good.[41]

Thomas Berry wanted everyone to "hear the voices of the rivers, the mountains or the sea" and feel their "intimate modes of spirit presence."[42] He laments people's loss of connection to the natural world: "Somehow we have become autistic. We don't hear the voices."[43] As environmentalist Christopher Manes adds, "To say nature speaks to me strikes most people as semipsychotic."[44]

Ironically, two decades ago, Buffalo Creek, the source of Berry's inspiration, was designated one of the "two worst water quality problems in the state" because of the inadequate capacity of two Greensboro wastewater treatment plants on Buffalo Creek, as well as leaking septic systems. But people connected to the river are bringing it back; in this case, Piedmont Land Conservancy (PLC), which protects land and streams through easements,[45] and the American Canoe Associations (ACA). As Kalen Kingsbury, the PLC's general counsel, explains:

> The ACA brought a lawsuit against the city of Greensboro for violating certain wastewater treatment discharge regulations, and the settlement of that lawsuit was that whenever they exceeded their permissible discharge quota, they had to pay an amount to Piedmont Land Conservancy and that money would then be used to improve water quality. The ACA tends to do this; they go around and try to bring cities into compliance through lawsuits and we were the lucky beneficiaries of that lawsuit....
>
> Now Greensboro has finally fixed whatever their problem was, so those payments stopped early in 2006, but it accumulated to $379,000.... We are working with a farm that may put an easement on their land that South Buffalo Creek flows through.

Guilford County has built Northeast Park where Reedy Fork and Buffalo join, so the creek and its meadows may again be places where others are transformed as Berry was.

Preservation: Sandy Cross to Cunningham Road

Think about rivers enough and you feel them everywhere. When the land opens up as I drive through Guilford County toward the Haw, I sense streams moving down through fields. Then the drop of Sandy Cross Road signals stream below. And the Haw, enlarged by Mears Fork, divides under two bridges and flows into a wide floodplain.

Two bold signs announce wetlands restoration — humans remaking river land, but there is nothing new in this. We have been doing it for 250 years, as I learn in my walk from Sandy Cross through the new Haw River State Park to Cunningham Road.

Backwater Environment and Restoration Systems have done the wetlands mitigation work here, turning productive farm fields back into productive swamp. Farmers past might shake their heads, for they dynamited and gouged out ditches to have a field where a plow could till and corn and tobacco grow. Now we want something different.

When the Department of Transportation (DOT) destroys wetlands in building roads, the "no net loss of wetlands" policy means that wetlands in the same river basin must be protected or restored to compensate. Restorations Systems spent seven years finding land for mitigation credit, getting conservation easements on it, reversing it to a swamp, and then monitoring the project's success for five more years.

At a midway point, Wes Newell of Backwater Environment arrives with bulldozers and backhoes to fill in the ditches and dig gulleys to allow flooding. After two months of mashing around, the site is raw, but forty acres planted in water-loving saplings of overcup, laurel and willow oak, river birch, American sycamore, and green ash mean a different future. Wildlife counts already indicate what John Pryor of Restoration Systems says he always sees: "Waterfowl and other wildlife return to what was a barren field. We are restoring it to its original sustainable condition. This was a bottomland hardwood and wetland area."

Critics argue that wetlands mitigation is a myth; wetlands are far too complex to be so easily restored. Joe Carroll, who checks up on the Haw's bottomlands here, disagrees.

> We are restoring it with a mix of hardwood trees that attempt to replicate the native mix that was here presumably for thousands of years before the wetland was cleared 80 to 100 years ago.
>
> The federal and state regulators are not going to allow a pristine wetland system to be knocked out, not unless you have a really compelling reason for the greater public good. More often we're mitigating for impacts to lower-value wetlands, such as road-side ditches or wetlands that have been cleared and replanted in row crops or pine for timber harvesting.
>
> Restoration is certainly more complicated than I ever thought, but we can't let the perfect be the enemy of the good. If the measure of success is restored hydrology, improved habitat and increased filtration, then yes, we can do it. Can you create a better wetland than what you lost? Yes, unless it was a pristine wetland."
>
> And don't forget the permanent conservation easements that go with all of our work. The covenants carry with the deed in perpetuity. No one who buys the land, no matter how far in the future, can go back in and cut the trees and destroy the habitat. Over time, nature has an amazing capacity to heal itself.

Wetlands' full dynamics are gone, but land use is now limited to hunting and recreation, so the Haw at Sandy Cross bridge will be troubled only by road run-off.

Past the restored floodplain are the grey woods of the new Haw River State Park, which had a headline victory in 2008. Citizen pressure blocked a golf course and housing development by Florida's Bluegreen Corporation slated for 692 acres on the Haw. The land went to the Haw River State Park instead, for $14 million, tripling its size.[46]

At the park on Spearman Road, Kathryn Royall, director of Environmental Education, is still pleased by the community's response: "The county commissioners said they have never gotten so many letters in support of one issue. The signs for the park were all over, not just in this area, but in downtown Greensboro."[47]

This struggle for the land was short and intense. In June 2007, Bluegreen proposed the development project; August saw it approved over Guilford's planning board's protests and vote, but rallies, editorials, and a letter writing campaign followed, generating state permission to buy the whole property. In January 2008, Tom Powers, Bluegreen's vice president, conceded: "In the light of the considerable controversy surrounding this project, our company believes that this result is in the best interest of the community."[48]

Kathryn Royall explains that the victory owes something to the exceptional drought of 2007-08: "The connections between water and development are finally starting to be made. Bluegreen's plans meant pulling a lot of water out of the Haw; there might not have been much river left.... Some heard concerns that Rockingham and Guilford Counties' water supply would be affected."

The purchase eliminated an on-site sewage treatment plant and heavy pesticide and fertilizer use draining to what Dan Tenaglia, formerly education director here, called the pristine waters of the Haw.

When I mention that Dan Tenaglia gave me a great botanical tour of the Haw when all this was an Episcopal Conference Center, I learn he has been killed, hit by a car, out biking with his wife Karen on a Sunday morning. He was 37.

Stunned, I go into the cold air and woods. The January slopes are thick with pale leaves, young beech everywhere. I try to walk the sadness off. The ridge slopes softly; below is a ruddy leaf cover streaked with wet leaves and the coal sheen of puddles. I find the people's path, too wide to have been made by any deer. This trespassing incursion cheers me, bringing to mind all those who took action to preserve woods and Haw.

As the ridge rises more steeply and the path narrows to deer width, I hang on to beech trunks as big as elephant legs. Then I descend to the floodplain, one of the widest on the Haw, with plenty of wild rose to pull and tear at those who walk here.

Beaver chippings and a deer skeleton send bleached light into the gloom. A wild rose gets a three-way grip on me. Dan Tenaglia would laugh and remind me these briars are a safe haven for wildlife. I retreat through a tunnel of thorns, envisioning in slow motion how a deer would flip through unsnagged.

I recall meeting Dan Tenaglia four years before....

He laughs when I tell him how I missed the turn and found myself in Haw River Plantation on Scarlet Oak Road. "The streets are named for the trees they remove," Dan explains.[49] He has wonderful energy and a focus that does not flag over the next three hours. That enthusiasm peaks when he spots a spongy orange newt or a crab-jawed beetle.

Dan gives me a fast walking tour of the nature education center he and friends Randy and Jonathan head up, which has a tree-house train of cabins for campers, ten feet up. Below, we pass the center's wastewater treatment system where Dillard Wyatt is running a check. Dan has surveyed insects, algae and sediment in the creek often enough to be reassured. Pollution sensitive plants like arrowhead are abundant.

Down the trail, we are surrounded by the green of May, leaves everywhere. "Yes, those are Christmas fern. Named for the stocking shaped leaves along the stem." Dan turns one over and thumbs the fronds on the smaller leaves; the larger ones are sterile. He points out a low and lovely rattlesnake fern and sensitive and column ferns. May apples, everywhere below, are a source of podophyllin, an anti-cancer drug, I learn. Only the two-leaved kind has fruit. I pass on my reading knowledge that the fruit is edible, only to learn it is poisonous until mushy, ripe, and tinged with red. Plants are life and death matters; it's good to have a highly-trained botanist as a guide.

Dan loves living here by a healthy Haw ... and teaching: "I tell people: learn the trees first and get them out of the way because there are only forty or so species you need to learn."

Dan also helps me with shrubs,

Dan Tenaglia, 1969–2007, was nature education director at The Summit, an Episcopal Conference Center, before it became the new Haw River State Park (photograph by the author).

This wetland, seen from a boardwalk at the Haw River State Park, is actually part of the Haw River. Until the Haw nears High Rock in Rockingham County, the river often swamps out of its channel. Frequent wetlands are a distinguishing trait of the upper Haw (photograph by the author).

and I finally get a positive I.D. on the haw, that significantly named shrub I have seen everywhere. "The cherry like leaves come out exactly opposite each other on reddish stems and a tiny leaf that is curled back is at the base of the new stems."

The long-leaved paw-paw produces pears and a no-win situation: "Eat it early and it's like a mouthful of alum and sucks the water right out of your mouth; wait till it is ripe and the raccoons will get to that sweet banana-mango taste first."

We move to swampy ground where maroon dogwood are everywhere. The leaves are like the flowering dogwood with the stem a ruby brown. The boardwalk crosses one waterlogged area after another until ahead, sedately streaming by a deck with benches, is the channeled Haw. Dan scoops out, as he has for hundreds of school children, a strainer of muck. "The Haw River Assembly was here and found a record number of species that are healthy water indicators. But there is an orangey tint to the water; silt is the leading polluter."

As we sift the leaves and magnify the creatures in the Petri dish, damsel flies, mayfly, water fleas and ribbon worm emerge. "We take these up to the microscope and show them to the kids on a big screen. These damselflies attached to grass and shed their hard skeletons. It just splits and sticks to the grass. It needs that grass to hide. You'd better believe it's vulnerable when its shell is gone and it's not flying yet. A perfect meal for birds. The kids got to see one come out...."

By the deck, blue-eyed grass, a star-like flower of the iris family, is surrounded by poison ivy. "Don't say 'it's bad.' If you followed that saying 'leaves of three, let it be,' you'd never walk in the woods or enjoy the hog peanut, that bean plant we looked at earlier. Just say you have to be careful of it if you are allergic." Dan is not. "Birds like goldfinch, butterbutts, yellow warblers, would die in the winter if the poison ivy berries weren't holding on to provide them with food. Touch-me-not [jewel weed] grows in the same spots as poison ivy, and you can use its orange juices to soothe the itching."

At our feet are grasses, so numerous that even botanists struggle to identify them. "I've been to conferences where the top botanists get together and try to puzzle it out. There are only about 4 or 5 people in the world who can identify all sedges or rushes or grasses. For the rest of us, remember, "sedges have edges and rushes are round."

"I am a picker," says Dan, pinching off a sedge's triangular stem. The rush clump next to it has round, hollow stems, just as he said. "Sedges and rushes are both three ranked, meaning the leaves come out of the stems in three different directions, while grasses are two ranked. Look down from above, and you will see what I mean." Sedges have another distinction: "They have no commercial value whatsoever. No one eats them. The Egyptians did find a way to make their papyrus — paper — out of them, but that's the only use I know of."

The air is full of the smell of honeysuckle and wild roses. "Too much.... Yeah, I like the smell too," Dan assures me, "but they're weedy." I am shocked to hear a botanist use that word. "A weed to me is anything that grows in profusion. Botanists are always looking for something new."

"The Japanese honeysuckle [japonica] has beaten out the American honeysuckle for territory so that you rarely see it. The native variety has a rubbery leaf and a glaucous stem." Dan rubs off the white coating to illustrate. And the red or yellow flowers are long tubes, barely open at the end. The exotic is the biggest worry. Introduced after the first botanist arrived on the scene, it has a woody stem that peels and gloriously scented white and yellow flowers.

Dan stoops for a brilliantly yellow-footed centipede. "These guys are great, but they leave a stink behind. They're good detritus feeders. They help rotting vegetable matter break down."

As we climb back up the slope, the topic shifts to human history. "Did you see *The Patriot*? Cornwallis retreated through here. And see this depression. The Troublesome Creek Ironworks dug here for iron. No luck. Now, black cohosh foam flower and bloodroot grow on the rim. A brick factory was here. No sign of it now, but every once in a while you come across a brick."

When I ask Dan how we can protect the Haw, he staggers under the enormity of the answer. "It would take a total change in society, in our way of thinking. It blows my mind. It kills me that someone can own land on the Haw and bulldoze it and have grass growing right next to it. We talked to the developer next door about easements on two lots on the river. And he kept saying, "But that's premium real estate.""

"It's education. Education. Education. But I don't know. You know the saying the last four letters in American spell 'I can.' We want to be able to do whatever we like."

"I gave up. I've given up. I just do this because it is fun. With a chainsaw and a bulldozer, all this can be ruined in a week. And laws are slow and take years to get passed."

"Orchis spectabilis. Be sure to mention this one. It's just a leaf now, but it has a gorgeous pink and white bloom early. How could you not love a flower with that name — spectabilis?"

I follow up: What does he want the kids here to take away? "Don't separate your self.

Nature — it's not something that is over there. Everything is connected. It's not that and me. Whatever you do affects the Haw and nature. When the kids are here for a week, they really get excited about all this.

"I want them to understand how much depends on the river. Our lives depend on water. If the Haw is still clean here, it can be further down. Nature will come back if we let it."

Then Dan moves from macro-problems to micro-events: a glossy crab-jawed beetle he grabs just right. "Two of these guys were fighting it out the other day on a stump that was just powdery with decay. We walked by them in the morning, and you could hear them breathe even though beetles don't make sounds. When they were still at it on our way back, I asked the kids what they were fighting about.

"'Territory,' they keep telling me till finally one said, 'Maybe there's a female in there.'"

"Yup, you got it." Dan laughs.

The Hand of the Farmer

Later I walk on to Cunningham Road; the way is deeply ridged, sharply dropping banks fall thirty feet or more. Beech trees and mountain laurel roots rope the soil in. Below, the Haw swamps and sponges over the land, then rushes on, cutting deep in its banks, twisting so quickly in one section that the fast waters do writhe like a muddy snake.

I climb into and out of deep gullies and hop the small ones without a thought, but a week later, reading Stan Trimble's 1974 study of how European farmers changed Piedmont land and streams, I learn these ruts have a human history.

Native peoples had farmed bottomlands, but their numbers were few and their land use light. Any erosion in those millennia came more from heavy rainfall or animal paths. The Haw ran fuller and clear, its stony bottoms visible. In 1728, William Byrd remarked the neighboring Dan River was "as clear as crystal."[50] Bishop Spargenberg in 1752 reported on the Catawba, you could "see stones at the bottom even where the water is deep." The Yadkin the bishop found "clear and delicious."[51]

Europeans' cash cropping methods, mobility, and slave labor would change that. First, the dark bottomlands along the Haw and other rivers were cleared, tilled, until their fertility was silted into the Haw. Then the uplands were bared and tilled, rain and dirt pouring down the slopes, routing out the gullies I lumbered down and up. Half a foot of topsoil went into the Haw. Reports of this phenomenon came as early as 1818 in Virginia: "Our woods have disappeared, and are succeeded too generally by exhausted fields and gullied hills."[52] A 1853 South Carolina report continues the complaint: "the destroying angel has visited these once fair forests and limpid streams ... the farms, the fields ... are washed and worn into unsightly gullies and barren slopes — everything everywhere betrays improvident and reckless management."[53]

As in the Amazon today, farmers moved on when abandoned land became infertile. Migration patterns west across the Piedmont determined when the erosive land use and turbidity (muddiness of water) peaked. For the Haw Valley, it was likely around 1860; but near slave plantations, the gullies rushed water and dirt into the rivers faster. Slave holders were sometimes called "Land Killers," their often-itinerant overseers "skinning that land" for pay based on crop yield.[54]

After the Civil War, black and white farmers tenanted the land; debt tied them to cash crops[55] and sent more dirt into the Haw. Before long, the land was as exhausted as the farmers; boll weevils added further injury.

Then, the wheel turning, poor soil drove farmers off.[56] From 1880 to 1920, abandoned fields sprouted pine; roots gripped soil; erosion and sediment decreased. When the bottom fell out of cotton prices and stayed low, still more people abandoned farming for the mills that would take a different toll on the Haw.

Fortunately, another change followed in the 1930s. The Soil Conservation Service and the Civilian Conservation Corps brought 50 camps to the Piedmont and planted pines and kudzu(!) to stop gullying in the remaining subsoil. They promoted rotating crops; on sloped fields, this could mean losing eight times less soil. Contour plowing, terracing, and strip cropping led to clearer waters. On the neighboring Dan River, turbidity dropped from 1314 parts per million in 1925–30 to 63 parts per million in 1960–70.[57]

Today the waters of the Haw are clearer, but Stanley Trimble leaves us with a warning. The sudden soil losses caused by modern development may harm streams more than the seeping away of topsoil over 150 years.[58,59]

Now as I step over a gully or canoe low in the Haw's scoured banks, the fluvial morphology of the Haw may come to mind — the changes water and humans brought to land and river.

One more human alteration across these gullies is in the planning: a path connecting the park to North Carolina's 938 mile Mountains-to-Sea Trail from Clingman's Dome to Jockey's Ridge. The corridor could be a river buffer traveling along the Haw from this state park to Cane Creek, river left below Saxapahaw, and then on to Eno River State Park. A secondary loop may run to Jordan Lake. The trail helps realize Dan Tenaglia's and Kathryn Royall's work — connecting people back to the river.

Tributary: Troublesome Creek

At the bridge at Cunningham Road, exploring the Haw pulls me in two directions, east and north. On this watery April day, I choose north and the Old Ironworks and the waters of Troublesome Creek, a branch of the Haw. Massive oaks are fringed in ochre, pale leaves light like moths among the lavender Judas tree blooms, even the pink entrails of an opossum on the road look tender. Whatever sadness and anxiety we carry on an afternoon like this finds a home in the April woods. Rain, tipping from chartreuse leaf to chartreuse leaf, lightens the grey sky, turns a wooded field into creek, then tumbles over the heavy stones of the fallen dam.

Beyond wet violets, Solomon seal, ironwoods and sycamore are the remaining pre–Revolutionary placed stones of the dam, five feet wide. Water was piped across the road to Speedwell Ironworks established in 1770. Its stone foundation in the pit of hills is flanked with lavender, young green, and the first cream of dogwood. Strewn about are gears, too heavy to scavenge, that once turned by Troublesome's power; today they mix memories of industrial force with ephemeral, enduring spring.

Troublesome Creek drove Robert Small's vertical "rip-'em-up and down" sawmill, the way wood was cut until the circular saw 150 years later.[60] A community was nearby. In 1759,

Speedwell Presbyterian Church was organized above on Ironworks Road.[61] Then Joseph Buffington established the Speedwell Furnace, the ironworks, where iron was turned into wrought iron. A belt of titaniferous magnetite iron ore that ran up to six feet wide was at "mine hill," closer to the Haw, but water power for bellows was greater here.[62]

The works were useful to General Nathanael Greene's Revolutionary forces, stationed here in 1781 before and after the decisive Battle of Guilford Courthouse. Recruiting a militia of 4000 had not been easy for Greene, and he had to plead with his inexperienced troops to hold for even two firings.[63] Losses were high. But half of the 1255 American losses were North Carolina's raw recruits who, near home, headed there; the British lost nearly a quarter of their force. Greene could have tried to finish off Cornwallis's badly damaged forces, but retreated to Troublesome Creek and desperately regrouped fearing a further attack. Local historians Blackwell Robinson and Alexander Sloesen picture the scene:

> Rain and a muddy road prevented reaching the iron works until nearly dawn March 16. Here, despite their fatigue and the continuing rain, Greene's weary soldiers threw up breastworks on the bluffs overlooking Troublesome Creek as a protection against Cornwallis's expected resumption of battle. Greene himself, who had not removed his clothing nor enjoyed the comfort of a bed for the past six weeks, fainted before dawn and again the following night.[64]

No attack came. Cornwallis withdrew to Wilmington, giving up North Carolina and marching on to Yorktown and final surrender seven months later.

Of the Battle of Guilford Courthouse, Greene wrote, "The enemy got the ground the other Day, but we the victory ... and tho' it was not glorious for me, it is beneficial to the community."[65] Some in Parliament saw a pyrrhic victory as well. Two centuries later in the film *The Patriot*, Nathanael Greene in the same clothes for six weeks morphs into a swankily outfitted commander whom Mel Gibson, playing a battle-worn coastal plantation farmer, lectures on the heroism of militia men. Never mind, Greene had a city named for him.

A decade later, on June 3, George Washington would stop at the same ironworks on his Southern tour and breakfast with Dr. Benjamin Jones, owner of 35 slaves. At that time, the population of present day Rockingham County was 6219 people, one-sixth of them enslaved.[66]

Alexander Sneed records a gristmill here in his 1810 account.[67] Other mills ran there until the 1940s. The unprofitable ironworks stopped operating after the Revolution, though it was re-opened by Thomas Graham from 1869 to 1871.[68]

The creek's name, Troublesome, raises speculation; was it difficult to ford?[69] Now it is environmentally troublesome—bringing its pollution to the Haw.

Formerly, most of the problem was agricultural runoff—ways of tilling fertilized soil that meant it ended up in the river. Riparian buffers are the cure—letting woods grow streamside so roots grip earth. When the Haw is butterscotch-colored, it is carrying dirt into Jordan Lake and onto the oyster and clam beds of coastal wetlands—a week's trip in fast, high water conditions.

Kevin Moore, Rockingham County district watershed conservationist, reports progress on agricultural runoff. "It's much more forested." Almost all farms are now using buffers and other best management practices: "Most places contour plow 50 feet from the streams, not straight rows right up to the creek."[70]

In some places Troublesome had been straightened to make plowing easier, but this sped up flow and eroded more bank dirt. Heavy machinery could have restored meanders, but Jason Doll, a scientist who worked on the project, favored letting time bring them back: "The Troublesome is too beautiful for that machinery. I saw an elm, six feet in diameter, right beside the creek; the bottomlands are full of mature, old hardwoods." So Troublesome is a creek being restored by itself and simple changes in farming.

However, Kevin Moore won't sound the all-clear for Troublesome; runoff from development is a potential threat. It can be avoided by sustainable development, i.e., conservation subdivision, a win-win solution. If developers buy a 100 acre farm and let 50 acres remain as pasture, woods or buffer and put 100 houses on half acre lots, houses sell for more and faster because of nearby fields and woods to enjoy. Another plus, Moore explains, is that "the cost of infrastructure is cut in half; roads are the biggest cost the developer has." He adds: "Everything humans do has issues that need to be resolved. If we have a high bar, we can still cut timber, farm, or build developments, but we don't want to sacrifice our natural resources."

Troublesome Creek can be a tremendous resource and its fate is important; dammed for Reidsville Lake Reservoir, it becomes drinking water.

Drought: Haw Watershed

> Denial ain't just a river in Egypt.—Mark Twain

There's another detour lurking when I return in early summer 2008 to canoe from Cunningham Road to Rt. 29. Drought. I put my canoe in by the old bridge's stones, but soon find myself mostly walking, experiencing the bottom of the Haw with its dips and pools, a slow irregular staircase to the Atlantic. I tug my canoe along like a dog on a leash until finally I tire of the wet trudge and head home to study the river and drought by other means.

My research begins in a rain storm. Not a few drops hissed away as steam, the parched rain teases we have endured, but a thick, roaring downpour of gusty surges. The Haw will be moving again, and people will be asking: Is the drought over yet? Like turkey vultures swooning above, drought has been circling for a decade; hard to ignore in 2001–2 and by 2007–8, beyond denial. Some cities in the Haw's basin count their water supply in months. The Haw is almost stagnant—a morbid word for a river. How scared you are depends on your water supply. How low is your reservoir? How deep is your well?[71]

Greensboro is buying more water from Reidsville, Winston-Salem and Burlington. Orange Water and Sewer Authority (OWASA) is talking about pumping more water from the Haw, though it already dams Cane Creek.[72] Durham would like to pump more from Jordan, but then so would Raleigh.[73] And Chatham's developers want more pipes in the Haw. Farther downstream, Fayetteville certainly doesn't want less flow.

Some only worry about brown lawns, but for hospitals and fire departments, being out of water is not an option. Farmers may be first to suffer. One drove down my road in July asking if I knew anyone with hay for sale; short grass doesn't last long. Another farmer, worse off, had to bring in water for his cows. Current food prices don't cover this. In drought, homeowners risk having their underground pipes pick up contaminants or even collapse.[74]

The rain today makes it hard to remember the sound of drought searing in. The earth squishes beneath my feet. Is the drought over yet?

A much weaker rain has already lifted restrictions in water-wealthy Burlington. The reprieve was a boon to landscapers and the city water department, which meets its budgets by selling water, not conserving it.

If this is an isolated rain, restrictions, the mother of invention, are needed. Does lawn watering seem trivial? Lawns can use 60 percent of residential water.[75] Old-fashioned rain barrels and cisterns may stage a come-back.

At the state level, the governor's powers to move water from river to river and town to town expand over protests that this policy rewards those who do not have forward-looking conservation plans.

If it keeps raining, we'll deny that this drought is partly of our making. We need rain and a reality check.

Big changes must come. Not all water needs to be drinking quality. Separate pipes and meters for outdoor and indoor use could make this possible. Homes could use gray water — water re-cycled from tubs and sinks — for outdoor use, after switching to biodegradable soaps and keeping toxins away from drains. Good ideas anyway.

Politicians offered other ideas; Robert Orr suggested, "We have to be smarter in the business we recruit. Does it make any sense to subsidize Google, which has enormous water needs, to bring some jobs to the area?" Richard Moore wants attention to leaks, as 35 billion gallons of water a year leak out of underground pipes. Local governments may need help paying for these repairs. Bill Graham suggested tax breaks for water-saving devices.[76] Water consumption in toilets has already been cut in half in two decades as new models shift from 3.4 gallons a flush to 1.6 gallons.

Then there are changes that go deeper into our cultural habits. As super-consumers, we are super-users of water. We run through paper that uses 60,000 to 90,000 gallons of water for every ton; the chips in one personal computer require 2800 gallons more.[77] Just a cotton T-shirt or a hamburger require 700 or 800 gallons to produce.[78] We want endless electricity, but coal and nuclear plants are heavy water users. In cooling Progress Energy's Shearon Harris Nuclear Plant, which dams the Haw's and Cape Fear's Buckhorn Creek, 10,000 gallons of water a minute are lost to evaporation.[79,80]

As Grady McCallie of North Carolina Conservation Network sums up: water supply and use need to govern development. McCallie says we have to decide how water is shared. In some states the Old West notion of "prior approbation" or first-user control holds, and in others it's the capture rule which says whatever you can pump is yours until the courts stop you. However, in North Carolina, the overall pattern is based on the English commons concept of granting reasonable use to all — fair, but complicated.[81] We want the Haw for recreation, food, agriculture, industry, homes — and that's just the humans' list.

Is the drought over yet? Not until we change our ways.

Will we find our way to civilized water use? Eight thousand years ago along the Tigris

*Opposite: **The vast steam cloud of Shearon Harris Nuclear Power Plant's cooling tower is created by the 10,000 gallons of water which evaporate every minute. Limited water supply has been a factor inhibiting plans to expand the plant which dams Buckhorn Creek on the Cape Fear to create Harris Lake** (Sönke Johnsen).*

and Euphrates, civilization began as people came together by rivers to improve their lives.[82] After the 2003 U.S. bombing of Iraq, the Tigris carried raw sewage, human bodies, animal carcasses, oil spills and depleted uranium — war's remnants.

At the same time, Denmark reused 98 percent of its water.[83] There's a standard for civilization.

For the views of someone responsible for the consequences of drought, I call up Allan Williams, director of Greensboro's Water Resources Department, who checks persistently on the Haw and its tributary Reedy Fork, which fills Greensboro's reservoirs.

> Our reservoirs are full now; they were full as early as March, but we were getting down to record low flows between rain events. That flow is generally the water seeping out of the upper saturated soils that basically weep or drain water year round, but in this long drought, there was really no supply to keep that flow up.[84]
>
> Based on our records that go back to 1928, I would say that this drought [2007-2008] is probably the worst drought that we have seen in North Carolina.
>
> We have seen exceptional or number 5 drought before, but the breadth of drought, the geographic breadth of it and the length of time that we were under that exceptional drought exceeded what we were in 2001-2002.
>
> Obviously our stockpile is our reservoir, not bottled water. You cannot sustain a city like Greensboro with bottled water, just because the amount of water that is used for actual human consumption is actually less than .5 percent or .25 percent of water used. It's a doomsday scenario; that's why we are going to do everything we can not to get to that point, and in our case, we buy water, and we go to irrigating restrictions. That's hopefully going to be enough to keep from losing the reservoirs. Now this drought, we got barely below 50 percent on the volume of our lakes, so we are in pretty good shape in Greensboro as compared to Durham and Raleigh.
>
> We are really at the top of the hill, right up at the top of the watershed. There are no major rivers here because this is where they all start. They only drain 105 square miles. Well, 105 square miles is not a tremendously large drainage area. And that's really what our problem is. A lot of people think, "Well, you need to build your lakes bigger." But building our lakes bigger doesn't do us any good if we don't have the recharge basin to fill them up.
>
> Underground reservoirs? They are doing that in California and in Florida, but again it comes back to the geology of this area. You need an aquifer; here there is a fractured rock aquifer, and there is very little storage capacity per acre foot of ground.
>
> I think the answer, whether it's Greensboro, Durham, North Carolina, or, for that matter, the Southeast, we are going to have to realize that water supplies do have limitations, and if we want economic growth, we are going to have to be a lot more efficient in how we use water. Much of our world these days is being constrained as much by water as by oil and energy.
>
> We have managed to continue to grow by shrinking the demand in Greensboro through pricing and by adding these other supplies [including Randleman Dam], and we are using about the same amount of water total today, about 34 million gallons a day, as we did eight years ago with 80,000 customers, and we've got 100,000 customers now. Residents have cut their consumption by 24 percent. In large part because we have raised our price and put in a conservation rate, so the more water you use, the more you pay.
>
> It rewards people who are conservationally-minded and ... basically extracts a lot of money from somebody who is willing to spend $200 a month to water the grass, with every single drop of that drinking water quality.[85]
>
> Americans are going to have to learn that water is more valuable than $2 for 1000 gallons. We are just going to have to look at water as the valuable commodity that it is.
>
> I am fascinated that people want to have green grass in North Carolina in July and August. I moved here from Florida and I watered my grass the first year because in Florida we had St.

Augustine grass that would die if you didn't water in the summer. Since someone told me, "Heck, this stuff just goes dormant in a drought and comes back as soon as it rains," there hasn't been a drop of water put on my grass. It will be brown in August, but I am ok with that.

It goes back to the pricing; we can preach all we want that people ought to drive fuel-efficient cars, but guess what, until gasoline goes up to $3, $4, $5 a gallon, people don't start deciding maybe they don't need to toot around in a great big SUV. It is the same way with water.

You can spend all day discussing whether global warming is changing the drought patterns; regardless we cannot rely on our historic record of stream flows, and they only go back to 1928. Most of the people whom I have talked to about global warming have said, we probably won't see that much of a difference in annual rain fall in this area, but we may see more extremes. Some very dry years followed by some very wet years, well, maybe we'd better be getting ready for those very dry years.

Walkabouts: Cunningham Road to High Rock

Rain does come and spreads across the bottomlands. Such rains in the past would have started up the grinding at the mill John Davis built in 1760, west of the Haw at Cunningham Road. When it later became the Patrick-Cunningham Mill, part of the mill's name stuck to the road.

After the Civil War, one local tradition puts a Ku Klux Klan meeting place by this Cunningham Mill on Mill Creek. In the night, pebbles thrown against a klansman's window were a call to a nightride.[86]

All this is well past as I start my walk up on the high banks along the Haw's loop through Rockingham County. Yellow and orange tulip tree flowers litter the ground. Black snakes undulate away, snapping like gum. The Haw spills out and channels, still as apt to be swamp as river. When Crook and I drop down close to a strand, we startle a Canada goose nesting by water's edge. An oblivious Crook ambles closer, and the goose lifts her winged self into a statuesque hiss. A gosling looks out from a buoyant nest of radiant yellow down. I rush us on to the pleasures of blue-eyed flax and blue-tailed skink before any harm is done on either side.

The way gets less certain as the Haw spreads at Benaja Creek and creates a swamp of significant size, but the river bank always cradles you back. That's a comfort of river travel; going down will lead you back. Circling Benaja Swamp, I range widely and pass a woman gardening whom my calls do not alert.

"Oh, I thought I heard something," she says mildly and points me to the short route around the swamp past the thickest clump of American honeysuckle I have ever seen, 20 long red trumpets at each blooming.

I head down onto train tracks that will lead me, like the river, to where the Haw spreads under Route 29.

When my husband drives me back to continue to High Rock, my canoe is strapped resolutely to the car though the road map does not show the Haw, only a misplaced Reedy Fork. "They only care about highways," John sums up the reverse perspective of 200 years ago.

I have talked to sane people who have canoed this part of the Haw, so I am hopeful though the put-in is depressing. Obscenities swirl on pillars holding Rt. 29 aloft, the image of a white car splats in the water, then a tanker truck rockets horizontally. A walk-about soaks one boot before I even leave the shadow of the bridge. The Haw splays messily to the left as Troublesome Creek joins in; I take the right channel and begin a day of second-guessing about the main channel and minute differences in water level. There are compensations; the thick body of an owl flaps away, a turtle drops from a log, and before long the sounds of tires on asphalt and metal on air are gone.

Then endless walk-abouts begin; maybe they number 30, maybe 50. I climb over and around beaver dams; scramble up and down fallen trees the stream lacks the force to move aside. Thirty to fifty crawlings and perches on rotten branches, clutchings at thorny briars, leave me swearing in hisses and shouting four letter words into the gentle woods.

In another frame of mind I might care that these tree blockades are health to the river. They slow the water, create stiller pools and trap leaves and twigs, the wood's organic litter, whose rotting feeds a water community.[87]

So on I scramble toward Rose Creek, past Kinady, now Candy Creek on maps. This area between the creeks is thick with landholders' names on Hughes' Revolutionary Map.[88] William Churton, 1755, was North Carolina's surveyor general and creator of a map of North Carolina that became "Collett's map." John Collett simply replaced Churton's name and added some errors.[89] James Wright had a mill on the Haw near Rose Creek in 1783 when he petitioned to turn a room of his family's house into a tavern.[90]

I paddle, climb and claw on toward Rose Creek where the Haw River Presbyterian Church was in 1770, near a "Haw River Bridge" now gone and on to the high and modern three-pier bridge at Rt. 150. Here Little Troublesome Creek puddles in, carting urban runoff, but not Reidsville's treated wastewater, which pours out of a pipe six miles long on the other side of the bridge — a most remarkable detour.

The story of the pipe begins when Reidsville's wastewater treatment plant's discharges into Little Troublesome filled it with fecal coliform. These microorganisms live in humans' and other mammals' intestines and are not dangerous but suggest the presence of bacteria that are. In dry times, treated sewage was 95 percent of its flowing water. The creek was on the state's list of badly polluted waters (the 303[d] list); it was also carrying way too much alga-fertilizing nitrogen and phosphorous — the problem in Jordan Lake, delivered from its upstream creeks.

So the proposed solution to pollution was dilution. And in 1998, the treated wastewater was piped six miles to the Haw where it didn't send the fecal coliform readings off the chart. Outrageous, I think, but none of the state's environmentalists I consult agree. Kevin Moore reminds me, "It was making it to the Haw anyway, and Little Troublesome was brown and foul, and I mean foul, when it went directly there."[91]

Patrick Beggs of Ecosystems Enhancement Program (EEP) stresses that runoff from roads, lawns, and development is a worse threat to the Haw from Little Troublesome.[92]

Jason Doll, who worked with the Division of Water Quality (DWQ) when the piping issue came up, tells what happened. The city of Reidsville wanted permission to pipe directly into the Haw as well as higher limits on fecal coliform, ammonia and biochemical oxygen demand. The DWQ opposed these — no surprise. The matter was settled in court when a judge gave the city the right to pipe treated wastewater to the Haw but sent them back to the DWQ for a permit. So the DWQ sat on the permit, just as sheriffs do in Southern

A bankful Haw in early spring at Rt. 150 in Rockingham County has the dimensions of a river. The six-mile pipe from Reidsville Wastewater Treatment Plant comes out at left — an example of regulated point source pollution. Non-point sources are general runoff to the river, like car oils that will drip in off the bridge (photograph by the author).

novels, and Reidsville had to issue a moratorium on new connections, stopping development.[93]

The solution came in a new and, what Jason Doll calls state-of-the-art, three-stage wastewater treatment system. Doll may be right about the 6-mile pipe being less important than an excellent treatment system. Before that, Little Troublesome "had no appreciable aquatic life."[94] In holding a line here on water quality, the local creek, the Haw, and Jordan Lake all got some protection, pending a new approach to wastewater that does not include sending it into the river.

I still eye the pipe with suspicion as I canoe near the opposite bank, but the result is not to be sniffed at.

The beauty of the Haw soon distracts me. The way is easier. What are a few walkabouts when the beech are showy with late winter leaves; holly and mountain laurel grow among steep rocky outcroppings, and a beaver shuttles to his bank home. Kingfishers, who nest in river banks, sail back and forth always as on a high wind, red-wing blackbirds flash their bright shoulders and crows argue in the sky. I watch them above pale patches of sycamore and a blue sky, intense with cold. I veer toward feeling this is a great run. Canada geese slide off in pairs. Grey wasps' nests hang, from ironwoods, like lanterns over the Haw.

When I straddle a fallen river birch to pass my canoe under it, I am in another world, atilt and transcendent. I climb along the peeling bark like a forest creature moving in the territory of trees, rising through branches shining back in a sky-filled creek.

Henry David Thoreau gazed into a stream a century and a half ago and found reflections of his transcendental philosophy: "Time is but the stream I go afishing in. I drink at it; but while I drink I see the sandy bottom and detect how shallow it is. It's thin current slides away, but eternity remains. I would drink deeper; fish in the sky, whose bottom is pebbly with stars."[95]

Twenty-five centuries ago, Taoists saw their philosophy in streams; "humble water's yielding, yet relentless flow that wore down all hard and strong obstacles [was] the essence of nature and ... an exemplary model for human conduct."[96] Water seeks the lowest place.

And now ahead, the river stays open. The last quarter mile to High Rock is as smooth as it comes. Trees that seem to block the creek are illusions and open wide enough to careen through. Finally, the river is movement; its philosophy — keep going around the rocks of life; keep the faith; you will get to the sea. My river day ends in a flourish of ripples at High Rock's tumbled-down dam. A busy mill and stage coach stop in former times,[97] but today there is only John leaning on the bridge railing, waiting for me.

Fords: High Rock

High Rock was long a ford. Animal tracks led Native peoples across the Haw here. Europeans followed and found support for their wagon wheels on its rocky ledge and named it for the granite mass on its north bank.

I come back in the truck of Tom Magnuson, scholar of fords and head of the Trading Path Association who did graduate work at Duke University, but looks and sounds like a Marine. He approaches his mission with practicality and wit: to find all fords and thereby old roads that point to the Trading Path and lead on to prime archaeological sites. At the bridge, Tom Magnuson sees a different landscape "Mmm, pack-horse road coming down," he says, twisting and pointing at thin woods. He is homing in on a fording spot east of the bridge. "I bet there was a town there. You know where there was a ford there was a tavern, a farmhouse. People had to do something when they were stuck by high water."[98] He estimates where the old road is by the way the present road swings west of a straight line crossing the Haw.

"Conditions for bridges are opposite those needed for fords. If it was less than fifteen feet wide, you had a bridge; over that and you would ford it. Is it sandy there? If it is, the ford's not there, no good for wagon wheels."

We go on and turn right where Elon University students await instructions on finding the approach road. "Who knows how to GPS this? Do both sides and try and get down in there? The depression is running straight to the Haw. Looks like it.... See the berm, too straight for a gully, the bottom is flat, and it looks like it's been filled with debris. I tell people look for appliances.

"These were hand dug. Kick back that mold to see if there's any cobbling. They brought in cobbles to carry water out; they wouldn't want a quagmire. These roads were court ordered and assigned to someone to build."

Prof. Magnuson asks two students to check the county records. "If there was a road, it will be there; roads made a property more valuable, and a surveyor got a percentage of the value, so he put it in."

This team underway, we head back toward Weitzel's Mill on Reedy Fork as Magnuson tells me:

> People coming into this region from the north had 26 major streams to cross and 148 minor ones. Each one could stop you, either with not enough water to float a leaf or violent enough to tear the *Titanic* apart.
>
> The fords were where people stopped — the geo-political choke points. And there are nine major fords on the Haw: High Rock, Ossipee, Shallowford, Haw River, Gilbreath's, Swepsonville, Cedar Cliff, Saxapahaw, Gunter Island. The trails went from ford to ford not town to town. If there was a major obstacle, a town would be there. Look at all the old roads, Routes 158, 29, 49, 62, every 20 miles a town. They didn't travel at night. And they'd go to the branch head, the source for the best water.[99]
>
> High Rock was one of the best fords. Shallow with a wide floodplain, so you could ford it even when it was flooded. If you were going down to Catawba or Waxasaw, you could cross here and then cross again at Weitzels's Mill or upstream a bit where Buffalo Creek joins the Haw.

We slow on the bridge at High Rock again to see the gap in the rocks, where Tom thinks a wooden sluice ran from a mill, river left.

> There were a complex of fords here; one about a half mile past where the river right angles southeast and then this one.
>
> The rock was less of a marker than the river bend. Without a compass, you navigated by ridge and fork. They all traveled on ridges, better visibility, no canes and briars to mess with. William Byrd was only making two miles a day, and his horses were starving when he had to cut through bottom land. I got in some briars the other day and had to get down on my belly and crawl out on a beaver track.
>
> What you look for in a ford is a good way down and a good way up and a solid shallow spot that can carry the weight of pack animals. There's an Irish saying, "May your climb to the ridge be an easy one."

People who have made a ford are almost vanished. "I'd be driving the cars Elon students do if I had $5 for every time I've heard, 'You should have talked to so and so, but he died last week.' I did get to meet one man whose school bus forded the Eno."

We stop at Doug Sockwell's farm on Reedy Fork, in his family for seven generations. In the sharp January sun, another group of Elon students are tromping through winter wheat, where Cornwallis had a staging area for the Battle of Weitzel's Mill at Reedy Fork Creek, March 6, 1781. Cornwallis expected to pick up supplies at the mill only to find Greene's men loading the last grain. A brief, running battle ensued. The British didn't pursue to High Rock; later Col. Tarleton speculated that decision cost the war: "An immediate movement of the King's troops across High-rock ford might, at this period, have produced various and decisive events."[100]

There are no battle cries now; the day is soundless save for metal detectors students wield over red clay that holds wagon bolts and horse shoes. Orange pennants mark their finds and trace the road down to Weitzel's Mill.

Mills were crucial, if unwilling, supply centers in the Revolution, and High Rock where Aaron Pinson built a mill by 1753 was one of General Greene's base camps, as the marker says[101]:

> Gen. Nathanael Greene
> Maintained headquarters here,
> Feb. 28–Mar. 12, 1781
> Before meeting Cornwallis
> at Guilford Courthouse
> Ford is 100 feet west.[102]

General Greene rode here in rain after his victory at the Battle of the Cowpens, covering 150 miles in three days on roads heavy with red clay and across swollen rivers. As he pushed on, Greene had to beg to recruit an army and develop strategies, one of them was to choose "the small triangle between Troublesome Creek and Reedy Fork for his operations."[103]

By the beginning of the next century, High Rock Ford lay in newly formed Rockingham County, named for the Marquis of Rockingham, the popular British prime minister who had championed the repeal of the Stamp Act and negotiated the Revolution's end. Just above High Rock, when most homes were log cabins, an impressive Federal style house, still standing, was built before 1807 by Joseph McCain, Jr., for his bride Polly Scales.[104] It became a tavern frequented by wealthy men wanting a reprieve from the nearby mineral springs health resort of Lenox Castle where they vacationed with their wives. Here by the Haw they could enjoy "cock-fighting, card-playing and horse-racing, and ... the liquor flowed more freely.... The outlines of the race track are still visible."[105]

Archibald Murphey, North Carolina's visionary entrepreneur who figures prominently in the Haw at Swepsonville, had plans to make the Haw navigable up to High Rock. Alas, they were fantasies; Lenox Castle, which he owned, failed, and the revelries at High Rock faded. The mill here kept running though. By the mid-1800s, "clustered about the gristmill at High Rock on the Haw River were a sawmill, an oil mill, a cotton gin and a cooper's shop."[106] A store, post office, still, ordinary room, and stage coach stop were also recorded here.[107]

As Grimsley T. Hobbs wrote in *Exploring the Old Mills of North Carolina*: "Mills quickly became community centers where farmers lingered...."[108] At High Rock, people could visit, catch up on news, debate, relax, or fish in the mill pond. In such settings, the miller was often influential, as well as relatively wealthy.[109] Mills may seem quaint today, but once they were economic centers with the allure some now find at malls.

As the century went on, technologies changed, and the High Rock community faded. In 1912, the dam was dynamited after a Reidsville doctor traced an outbreak of malaria to the stagnant waters of its millpond.[110]

River Geology: High Rock to Altamahaw

From High Rock to Altamahaw, the Haw passes through the northeast corner of Guilford County into Alamance, with only a few farmers' fields, several houses, two quiet bridges, and a winery along its wildness. "In wildness is the preservation of the world," wrote Thoreau. Sometimes wilderness is substituted in error, but wilderness, a day's journey without any human signs, is rare in the Piedmont. Wildness is plentiful along the Haw.

I start my paddle to Altamahaw watching a rabbit's wild race across the road. It drops safely below the bridge as a hawk lifts away. Downstream, a river birch leans over a beaver hut. Farther on, I surprise land-owner Ronnie Beall who says he sees few canoers. Beall has found a Savannah arrowhead here and reports that "Greene and his men were running all up and down the Haw." Standing high on the bank, he points 20 feet up to where waters have flooded.

Just below, two men with rifles startle me and then reminisce about when mink were here. Grey wasps' nests hang all along the smooth olive stream. Four mallards rush to the banks and away. It is another spring season as distant peepers throb, maple flowers litter waters, and a kingfisher swoops and chatters.

With the sun to the southwest, there is no denying the Haw's direction has shifted 90 degrees. Maps show two violently-arced meanders between Rt. 150 and High Rock and then the sudden descent. What makes a river headed northeast suddenly turn to Wilmington? This question drifts into deeper questions about how the Haw and all this circling of river, sea and cloud began.

These are the questions geologists take on. It is their work, as Barry Nelson, chief geologist for Northwest GeoScience, puts it, "to make the rocks speak. Geologists like to speak of the fabric of the earth," he explains, reminding me that to understand the Haw's origins, we must realize that the land we stand on, even the rock we call "our firm foundation," is over eons almost as fluid as water.[111]

The story of the Haw and its earthly home begins in a contracting "cloud of gas and dust"[112] four and a half billion years ago. Our molten earth pulled together and the chain of events began which created land and water and shaped our very recent Haw. First, the cooled surface hardened into crust which broke into plates; these, driven by convection currents in the mantle, jarred together unleashing volcanic eruptions and vapors which produced our land, atmosphere and water. Bombardments from space by icy comets and asteroids added more water. Heavy elements of iron and magnesium, which sank in the crust as it cooled, were ages later thrown up by volcanoes, leaving mafic rocks below Chicken Bridge for future kayakers to careen around.[113]

But that is too far ahead in our chronicle. First rivers played their role in creating land. Rock and dirt laden, they poured off the land mass that had formed from Wyoming to Labrador, moving particles of earth toward the present day Piedmont, extending land to Western Carolina.[114]

Around 1.3 billion years ago, crustal plates crunched together to form Rodinia, "Motherland" in Russian. In this vast amalgam of continents, the western curve of South America was pressed into western Carolina, then the continent's edge, and they ground together for 200 million years, raising mountains that spiked the sky and stretched 600 miles wide.[115] It was the slow erosion of these Grenville Mountains, by rivers depositing earth further eastward, that formed the land we stand on and the Haw runs through.[116]

Rodinia would rift apart as the earth mantle, under the crust, rebounded, like infinitely slow piano strings, and pushed the continental plates apart.[117] When South America swung off, an island-arc of crust called Avalon Terrane, separated. This Avalon Terrane we will follow closely as it forms a layer under the Haw watershed. As seas filled in around Avalon, the continental plates stretched farther apart. There was yet no Atlantic Ocean for the Haw to flow to; that would come.

First, the plates of Africa and North America reversed and drifted toward each other, putting Avalon Terrane, like the Hawaii islands today, at the crossing of continental plates.[118]

When the African plate jammed under Avalon into the earth's mantle, it created cracks for magma release. Fire rock poured out, sending down flakes of ash to become mud and, millions of years later, smooth slate. When you touch the slate stones of Duke Chapel or the cemetery wall in Haw River, you are connected back to that lost — no transformed — island arc of Avalon. The volcanic landscape became the Carolina Slate Belt the Haw moves over after it enters Alamance County.[119] These eruptions also spawned rhyolite, a fine-grain, felsic rock sought for arrowheads, produced when granitic magma hardens exposed or near the surface and cools rapidly.[120]

The volcanoes of Avalon collapsed into sea-filled caldera, where 650 million years ago, molten eruptions hit water, cooling suddenly into the pillow rocks of New Hope Creek swinging north of Chapel Hill on its way to Jordan Lake.[121]

Under Avalon Terrane, water "sweated out of rocks ... moved upward"[122] forcing quartz, gold, silver, and copper to the surface. This would make North Carolina the largest producer of U.S. gold before the California Gold Rush.[123] A bit of this gold showed up on Back Creek, but it was sought in vain elsewhere along the Haw.

The continental plates moved on to grind together, Avalon Terrane crunching in, the Appalachian mountain building in full progress. Pangaea, the last "all earth" assembly of the continents, was forming, folding the earth, metamorphosing rock so that igneous rock, born of fire, and sedimentary rock, dropped and layered, are twisted and swirled into shapes only geologists can decipher. The mountains that formed were as high as the Alps (16,000 feet) or perhaps as the Himalayas (29,000 feet).[124]

In hundreds of millions of years, these mountains too would flatten under rain and river, and another rifting apart would begin as the mantle under its lightening mountain weight rebounded, splitting the Appalachian chain and West Africa apart. The waters that filled in became our Atlantic Ocean, still widening today. The rift fractures of this breakup charted the course of the Haw, the Catawba, Dan and Yadkin Rivers alike, along stress lines northeast toward the Potomac River Basin.[125]

Much later, after another 200 million years had passed, something — uplift of the land or headward erosion of a stream moving above its source — caused these rivers to be captured by another stream and diverted to drain southeast. This capture, or stream piracy, which geologists like Barry Nelson talk about with excitement, occurred when a stream west of High Rock swelled, moved headward above its source, and turned a tributary of the Dan toward Wilmington. A look at a stream map shows how close the Haw was to the tributaries of the Dan; before its capture, it was likely one of them.[126]

When North America and Africa tore apart and great blocks of stone between fracture lines fell, the future path of the lower Haw was also changed. One such break dropped the Durham Triassic Basin, which the Haw crosses just below Rt. 64.[127] Kayakers love the rapids this descent created. Those in cars can see the basin's drop on I-85, after mile 168, going from Chapel Hill to Durham.[128]

Some river, let's call it the ancestral Haw, would fill this dropped basin with "red mudstones and sandstones."[129] In Triassic and Jurassic times, 200 to 150 million years ago, the basin was a hot swamp, just right for dinosaurs and their reptilian predecessors, as North Carolina is thought to have then been riding on crust only 10 degrees above the equator (not 34–36 degrees as today.)[130]

As rain and rivers wore the Appalachians down, soils that collected in the basin's lakes and swamps would alter the Deep River's path much later. These softer soils enabled its capture only 4–5 million years ago by a tributary of the Haw. This capture struck Barry Nelson with force one day as he focused on the Deep's right angle turn at High Falls where the tributary eroded headward across the Rocky to the Deep.

Waters' wearing away of the Appalachians carried off land, perhaps eight miles in thickness, to bare the ground we live on. This vast erosion means that the Appalachians and the Cane Creek Mountains we see today are not, strictly speaking, mountains at all, but water-carved basement rock, worn down into mountain shapes by rivers sandpapering rock away.[131]

We must look up to see the past. The tops of Grandfather Mountain and Pilot Mountain represent the Schooley Surface of 23 million years ago.[132] The Haw's Cane Creek Mountains by Saxapahaw reach to the Harrisburg Surface of 5 million years ago. We stand down on the New Harrisburg plain, on land that has been raised up and eroded by river many times and has even been ocean floor.[133]

Rivers, a predecessor of the Haw among them, built the coastal plain as they carried sediment farther and farther east, fanning out new deltas.[134] As Barry Nelson says, "Flowing water is the greatest geologic agent at work in forming the landscapes on the earth's surface."[135]

Another change occurred 100 million years ago when the Cape Fear Arch slowly lifted, driven by a force under the surface. The arch, which runs parallel to and just south of the Cape Fear, makes the Haw's fall zone, the drop from piedmont to coastal plain, the highest and widest on the East Coast. At Washington, the fall zone narrows to a line. In North Carolina, it widens to almost the length of the Haw's 110 miles and has a drop of roughly 820 feet, creating room for many rapids and mills[136] but making river transport difficult.

How long has the Haw been here? Barry Nelson tells me it was here near its present course at least 10,000 years ago. Probably the Haw River drainage system was running here much longer. Forty million years ago when the first monkeys appeared, some ancestral Haw flowed northeast. It was captured and turned to a southeast course about 5–10 million years ago. And when the first early hominid, australopithecine, appeared 4–5 million years ago, a tributary of the Haw brought the Rocky and Deep Rivers to the Haw.[137] The rocks speak to us, but not in dates more exact than that.

To grasp this vast time (and the Haw's place in it), John McPhee, author of *Annals of a Former World*, suggests you stretch your arms wide to represent the 4½ billion years of our earth's existence. Now imagine, it takes the span from your left fingertips across your left arm and over to the middle of your right palm before the Appalachians rise and Pangaea masses together. The space of human history is so slight, it could be rubbed out by one filing of a fingernail. The Haw could have found its groove any time in the last joint of our right middle finger.[138] That's a time long ago, but not timeless.

On that scale of time, it's just a nano-second since Troxler Mill was built in 1855 near a fish dam below Mill Creek on Troxler Road.[139] Now the mill has collapsed, and the Haw takes a narrow passage through the old dam's fallen rocks, a short, sweet run past a sycamore's massive roots.

Brooks Bridge Road comes up next by a dam built for Greensboro's last intake.[140] A winery is nearby. Definitely avoid the dam with a portage, and enjoy just one glass of

Norton, a native American grape, at Grove Winery. Sobriety is essential on water, especially as just downstream, a table, then a chair hit my canoe as I make notes about the dam and a rock slide marking a vanished bridge. After three walkovers within 100 yards, I watch a heron lift off and the Haw becomes its most beautiful with clumps of hollies rising fifty feet and rocky ledges thick with mountain laurel. The Alamance Alluvial Forest is worthy of its special designation.

When I slow to sketch a red-eyed duck with a black back to its head, still unidentified, I realize the river has also slowed. It is impounded, turned into a lake by the dam in Altamahaw. This town's name and Saxapahaw's have made it several people's best guess that *haw* means river.

Coming to the dam, you look off into space above treetops. Under this sky, the Haw will drop, gathering force to drive the mills of Alamance County into a major 19th century textile center.

Looking into space can pull you right off the planet and over the dam, so the reverie mustn't take too deep a hold. Paddlers must avoid the fifteen foot dam drop by taking the channel behind the small island, river left.[141]

3

Cotton Mill River
Alamance County

On Safety and Dams: Haw Watershed

The dam at Altamahaw is a far drop for a human body, and the waters of the Haw are worth a warning. First, this book is no guide to canoeing safely on the Haw. Bob Brenner and Paul Ferguson, pioneers of the whitewater Haw, have both written great ones. River guides and paddling classes are available.[1]

Secondly, this warning should have come at the start. High waters make even small creeks forces to be reckoned with. In flood, a swamp maple lying over a stream can catch and drag you down. You will not experience uplift; rivers flow to the sea by gravity. We only have a minute or two to come round right if trapped in sieves (underwater tree branches) or entangled in a sweeper (overhanging branches or vines).

Dams are always a risk because of the vortex of current and commanding hydraulic spin that extends out from them. In *Paddling Eastern North Carolina with Paul Ferguson*, the author warns that the more regular the dam surface, the more deadly the spin, with no irregularities or breaks as escape routes. The dam at Brooks Bridge has a hydraulic spin that extends six feet out though the dam is only three feet high.[2]

Being alone, over-estimating your abilities, and not wearing a life preserver are avoidable causes of disaster. This book is not full of stories of those who have played too carelessly in the Haw, but fatal stories there are.

A fatality nearly occurred at Altamahaw on September 29, 2004 ... to an expert. Because of his life-long work helping people safely enjoy rivers, Paul Ferguson put his story on the web.

> Pete goes first and paddles carefully into the narrow channel leading to the left bank. I follow with an uneasy feeling about how close this path is taking me to the edge of the dam. There is little flow affecting my boat as I paddle past the upstream side of the land in the narrow channel. Everything is going well. Paddling to beach my boat near Pete's canoe, I cannot make my bow stick to ground.... I make another attempt to land my boat and think the bow sticks. I get up to walk forward in a crouch, but the effort sends my boat back from the bank. I paddle again but do not immediately notice a slight current catching my stern and swinging it toward the dam. I paddle harder and harder as Pete runs down the bank because he sees that I am getting into trouble. Suddenly there is a moment of horror. I realize the current at the lip of the dam is rapidly accelerating my stern. I am 10 feet away from Pete. We lock eyes and each knows that nothing can be done. There seems to be a bright light illuminating the

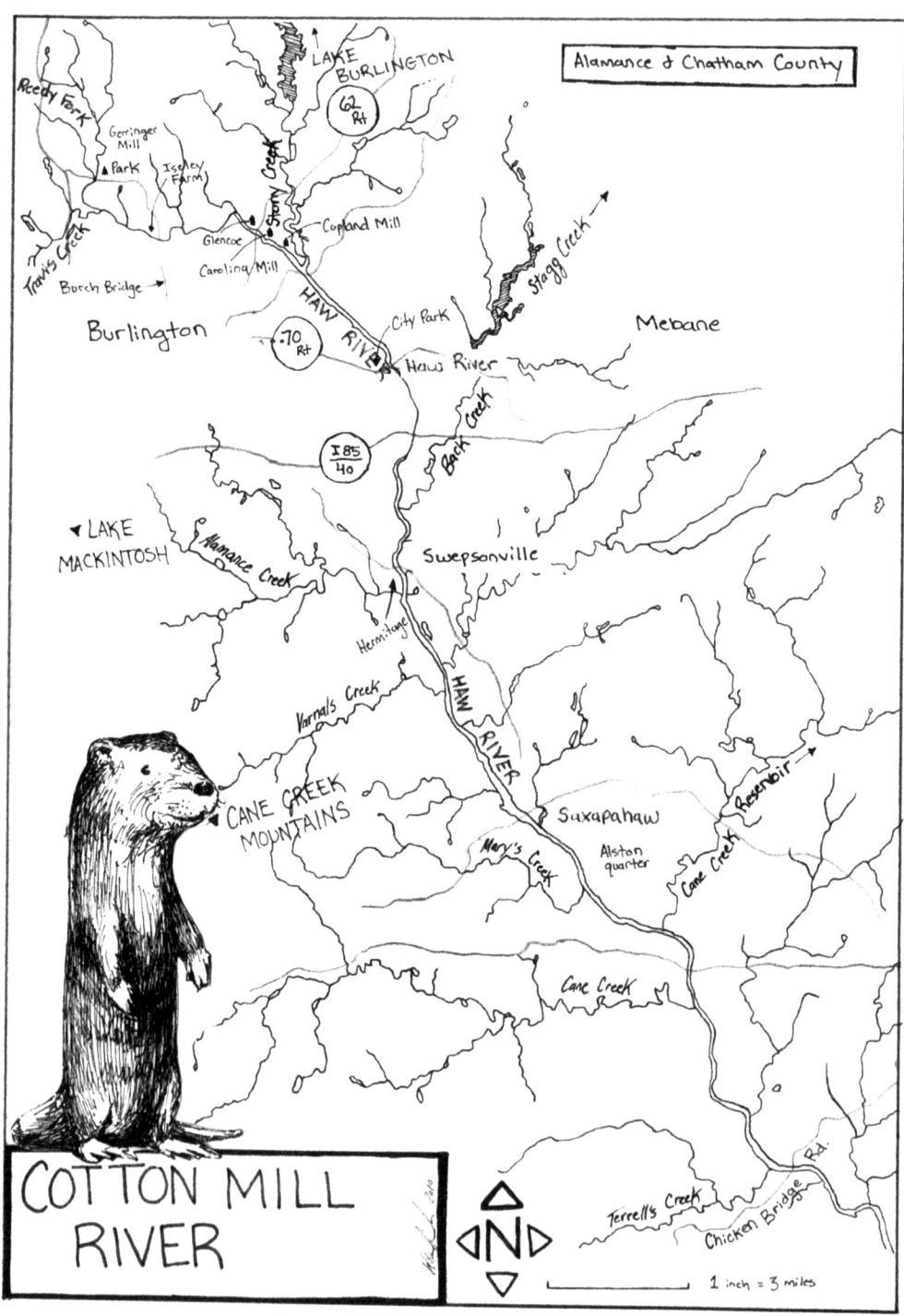

Cotton Mill River (William Rusch).

scene. I'm sure the bright light was in my mind from the focus of all my mental faculties on dire consequences of the situation.... I am being swept backwards over the dam. It all happens in just a few seconds. I feel the canoe surge and start to drop. I feel my body hitting something as I am thrown out of the canoe and go underwater. I do not know what or how I hit. My body is thrashed as violent forces play themselves out. I come up from underwater next to my canoe and grab a gunnel while standing on the bottom in about four feet of water. The canoe is about 10 feet out from the dam.... I try to walk but something is wrong. There seems to be a log between my legs preventing them from moving. Using my hand, I push down to free it. The log is my left leg, broken somewhere above the knee, and now dangling in the current. Pete is scrambling down the bank. He is relieved to see me alive after expecting to be on a body recovery mission.[3]

Pete Peterson pulls him to the bank where Paul Ferguson is steady enough to advise on his own rescue and even photographs the dam with the camera still around his neck. Later, while recovering from damage that just missed being a snapped spinal cord, he reflects:

It would be a lot easier on my mind if I could point to some external cause of the accident — equipment failure, a lightning strike, a squall. There was none of that. I was simply working too close to the dam. Why was I doing it? I can only guess that good judgment was overcome by overconfidence. In hindsight the risk was enormous — loss of life. The reward almost nonexistent — a few extra minutes crossing farther upstream and perhaps some muddy shoes....

Hear him.

On this day the Haw below the dam threads thinly around stones, yet it can tear through the woods on both sides, full flood.

Dams cause other problems as they generate the energy we need. As Alice Outwater says, "When you look at a map of the Haw River watershed, one of the things that jumps out at you is that the Haw has a heck of a lot of dams."[4] The drop that made the mills flourish brought the dams which the National Inventory of Dams totals at 165 in the whole watershed, though the Haw's main channel is blocked only at Brooks Bridge, Altamahaw, Glencoe, Swepsonville, Saxapahaw, Bynum and Jordan Lake.[5]

The number 165 is not impressive next to the 47,655 dams worldwide[6] that together collect so much water that they affect the earth's rotation and may even contribute to earthquakes. But 165 makes an impact that some call "intelligent use," others "destructive."

"Muncipal water supplies, flood control, electrical generation ... irrigation and transportation and ... human control" are on Outwater's short list of major benefits.[7]

John Jordan, whose father's name is on the B. Everett Jordan Dam, argues eloquently for dams: "They are one of our few sources of renewable power. You can't have hydroelectricity without a dam. We have trade offs on everything. First you have to determine costs and then someone has to pay for them. Then you've got to decide, 'Do you want the lights on or off?'"[8]

Hydro power is a strong point; it supplies only 4 percent of our state's energy.[9] Water supply is another, especially in the densely populated Haw Valley.

Jordan Dam's original purpose was flood control. The project was spawned when the military base at Fayetteville was under water at the end of World War II (Sept. 18, 1945).

Eagles and osprey certainly benefit from Jordan Lake. We might also add family recreation to the list; the safest paddling on the Haw is in the lake-like sections above dams.

Yet, those impoundments represent a loss to the many paddlers who enjoy a moving river and rapids.

So what are the other down-sides of dams? In sum, they have a negative effect on water quality and aquatic life. The Haw's flow mixes temperature and adds oxygen. Dams do the reverse; oxygen stratifies in their pools and water temperature varies only one degree from night to day.[10] Temperatures in streams may drop as much as 7 degrees in the winter and 20 in the summer—crucial triggers that some aquatic life, like Mayflies, need to hatch. Dams also prevent surging flood waters from clearing stony bottoms for fish eggs.[11] They separate and weaken populations of animals as well. When an old dam at Carbonton on the Deep River came down, the endangered Cape Fear shiner appeared in a new river section within two years.[12]

A huge supply of shad and sturgeon fed the people of the Haw until dams sealed them out. River herring too, alewife and blueback herring could stage a miraculous comeback if able to return to their spawning grounds. Herring, along with cornbread and black yaupon tea, helped many a family survive, and they are also food for striped bass, osprey, bald eagles and blue heron.[13] Dams' effect on fishing was reported in 1884 after a large shad run near Moore's Mill below present day Rt. 64: "This is the first we have heard of shad being caught so high up that stream since the navigation works were constructed over thirty years ago."[14]

Before 1850, shad were recorded up the Haw as far as Burlington. People complained of losing a great, free food source. An 1881 law mandated that all dams have fishways, but an 1884 report stated that hadn't happened.[15]

The Haw and the Southeast are home to more mussel species than elsewhere in the world, but 49 are endangered, threatened, or of special concern. Mussels are invaluable water filterers, but they need fish to carry their larvae, so blocking fish means fewer mussels.[16] While there are innovations that help some fish pass dams, none compare to a free-flowing river.

Everyone agrees we don't want dams giving way without warning. In 1999, 35 state dams buckled under the flood weight of hurricanes Floyd and Dennis. One in five are large enough to be life-threatening if they burst.[17]

And if they are blasted out, any heavy metals deposited from dyes and industry output will be dislodged and head downstream. It's best if dams come out a meter at a time, so trees and shrubs planted in their sediment can hold dirt and toxins.[18]

"Removal gets more complicated in the Midwest and Southeast because we don't have this big sexy fish [salmon] that people love to latch onto," says Martin Doyle, associate professor of geography at UNC–Chapel Hill. "We have pigs, and a lot of pollution."[19] That may be one reason why dam removal on the Haw is stalled while 430 U.S. dams of all sizes have been taken out since 1999.[20]

Frontier to Mill Towns: Altamahaw and Ossipee

Altamahaw, where the Haw and Reedy Fork join, was frontier in the early 1700s. Tom Magnuson writes with relish of the anarchy of frontier life, which he dates 200 years out from the settlement of Roanoke in 1585 when places like Altamahaw were home to "religious and political refugees, escaped indentured servants and slaves and assorted other folk [who]

Altamahaw dam's drop is fifteen feet. Though perilous for paddlers, the steep fall meant that only a short mill race provided enough drop onto the mill's turbines to power the mill's gears. A new canoe access and park at the dam opens up one of the Haw's most beautiful stretches. The author is with her grandson Demitri who pursues a turtle (John Herold).

fled or were taken into the back country...." They blended with the Indian peoples and created a culture and society far "different from anything imagined by enlightenment theorists and imperialists in the Old World."[21]

For Magnuson, backcountry could mean progressive: "Frontiers are stateless places ... lawless, but not ungoverned."[22] Carole Troxler, co-author of *Shuttle and Plow: A History of Alamance County*, maintains, "Especially before county government was organized in the back country, the area was appealing to fugitives from indentured servitude and slavery."[23]

Troxler highlights one case in particular to show that that freedom and equality could extend widely; one man of African descent, William Chavis (or Chavers), born about 1706 to a prosperous father, came to own roughly 2,000 acres. Two tracts were here near the Trading Path's crossing of Reedy Fork. He sold both after improving them — by the labor of his slaves.[24] Chavis was also able to use the court system to bring justice. At the Orange County's courthouse in 1754, then on a stream east of the town of Haw River, he argued for the safe return of a kidnapped young black man, John "Busby" Scott, to his mother.[25]

This more elemental frontier culture did not endure in the 1800s; as settlers increased and powerful slave interests gained influence, a more racist culture took hold.

Altamahaw has the "greatest fall ... of any place on the river, there being seventeen feet natural fall in less than a mile," reported the *Alamance Gleaner* in 1886.[26] Though the fall is possibly greater in Chatham County, the enthusiasm is justified. Without the power and drop of the Haw, textiles would have taken off elsewhere. Later, mill owners worked together to keep textiles booming in Alamance County even when the raw power of the running river was replaced with electricity.

There were actually two textile mills here. Captain James Nathaniel Williamson built one on Reedy Fork Creek about 1880 when John A. Gant and Berry Davidson were setting up another a quarter mile north on the Haw. Williamson purchased the Kernodle Grist Mill and built a state of the art facility with "automatic sprinklers, fire hydrants, and running water for toilets."[27] One June, students in my Haw River class wait with me inside Ossipee Mill in cool relief after our humid trekkings. The workers who struck here in May 1887, for an eleven hour day, would not be impressed with our stamina.[28]

Our guide has exceptional credentials: Roger Gant joined the firm his grandfather founded and served as president of Glen Raven.[29] Today, Glen Raven produces performance fabrics of worldwide reputation, like Sunbrella awnings and marine textiles and is even innovating in geo-textiles that reinforce soils. Ossipee Mill has become one of their warehouses.

Roger Gant's keenness and energy are worthy of the best of the mill owners. He soon whips in, slows for a warm greeting, and then, standing, revives languid students with off the cuff textile history — family history, in his case. As Roger Gant traces the enterprises of the Holts, Gants, Williamsons, Whites, Davidsons and Trollingers, he describes the aristocracy of textile mills that developed along the Haw when water was power.

First, tribute is paid to E.M. Holt, whom he calls "the textile pioneer of this area and the whole South." His warmth reminds me of how cooperative mill owners could be. John Q. Gant learned the trade from E.M. Holt and was even cared for by Holt's wife, Emily, when typhoid struck him in his first months of employment there.[30] Roger Gant describes E.M. Holt as "a farmer who sold yarn to women to weave, then brought in looms that wove it, so women didn't have to make cloth." It was E.M. who broadened trade of this cloth "to Philadelphia and Baltimore, sending up cloth and bringing back corn whisky — a lot of corn in a very small package."

"From those beginnings," Roger Gant reminds us, "a whole technology and industry was spawned from Lexington to Raleigh.... E.M. Holt's sons made a whole lot more money than he was making, and that was already considerable."

Outside in the heat, Gant points out the handmade brick that James N. Williamson used for the mill, his first independent project. E.M. Holt helped establish Williamson after he married his daughter, Mary Elizabeth Holt, having won her admiration by helping bury the machines of Alamance Mill from a feared "Stoneman's Raid" in 1865.[31] "[Williamson] ran it for thirty years until it went broke in the Depression. The early 1920s brought a terrible depression in the textile industry which had been good during the First World War, then the 1929 crash came. A lot of mills went broke in those periods."

Though creekside, Ossipee Mill's "water power was [only] for the fire pump. Water never ran the mill. In the early days, steam was generated from wood fires; we still have plots of land [woodlots] scattered around."

Ossipee was "not a big mill, maybe employing 100 workers at the most. It ran 12 hours in winter and 14 hours in summer if they had the business. Now," Gant tells us, "you'd go under pretty fast if you are not using your investment 24 hours a day. You can't survive."

We head up Hwy. 87 to the Altamahaw plant in full blaze of the sun; Roger Gant shifts into a smooth run leaving students trailing dizzily behind.

At the bridge, the Haw is more rock than water. "Did I swim in the Haw? My father would not have allowed that. Some did of course."

Just past the bridge, Roger Gant points out the mill race, the dug channel that directs water to the turbine. The drop is so great here that the race is not long.[32] "And when we cleared out the mill race, men would come from all around to get fish. There'd be thousands of fish. That was a big day for boys around the neighborhood." Under the turbine is the dark drop to the tail race that channels water away. "You have to get it away fast or the power is lost."

A few steps farther along, the original office building, dark and impressive, looms above us. In 1880, John Q. Gant, in partnership with Berry Davidson, built the mill on 53.5 acres Davidson owned. A saw and grist mill were at the dam site before this mill; a company store and about 40 houses were built for workers who came from the mountains as well as nearby.[33, 34]

Roger Gant tells us modestly that this mill was "quite profitable at the turn of the 20th century." Some wanted that wealth distributed differently. W.W. Oakes, president of the National Union of Textile Workers at Altamahaw, wrote in 1901 that wages were "too low for the average man, woman and child who are toiling day by day for their daily bread" and long hours were turning women to "skin and bones" and children to "dwarfs."[35]

In the twenties, the Gants, showing necessary adaptability, sold out to the Holts, then bought back the physical plant, and in 1933 turned it into Glen Raven Silk Mills, producing among other things rayons known as Canton crepe and sharkskin.[36] In 1939, Gant tells us, the mill converted to electricity with a new turbine and generator. The Haw could still generate 350 hp — enough for running the lights and providing free power and some income: Duke Power had to buy back power at 75 percent of their price.

By then, the center of operations had shifted to Glen Raven; John Q. Gant's weaving

Raising and lowering a gate at the start of a race regulated power and protected the turbines and mill from high waters. This mill gate is at Glencoe (photograph by the author).

empire took hold. It still provides jobs for 475 people in Alamance County and 2300 others around the world.

As we head back to our van, Will Holliday, the military buff in our class, asks about the mill's work in World War II. This is another success story that Roger Gant recalls with pride. "We made parachute cloth. Every yard passed inspection—not a single yard was returned."

Cows in the River: Altamahaw to Glencoe Mill

It's a May day when the smell of honeysuckle on the Haw can stop your paddling. My son Will, and Neil, a student on break, join me for the ten miles from Altamahaw to Glencoe, the next mill town.

We start out scraping, launching a pair of squawking heron, most elegant and awkward of birds. The banks are in bloom: elderberry, dogwood, honeysuckle, wild rose, blackberry, poison ivy, and cascading seed pods of ash and ironwood. Neil spots an 18 inch black carp and mussels on the bottom. Will catches four palm-size bream in ten minutes and sends them on their way as a red-tailed hawk studies him from a sycamore.

Native peoples were drawn to this spot. Their artifacts from 8,000 years ago have surfaced on the triangle of land between the Haw and Reedy Fork that has enough rise to grow the berries of summer and nuts of fall. In the 1900s, tenant farmers lived in a row of cabins here.

River cooters, Pseudemys concinna, usually fall into the river as a group if a canoe passes. The flared plates at the back of the carapace or shell are one identifying trait. Males are distinguished by long front claws used to stroke females' faces during courtship (Sönke Johnsen).

With the confluence of the two streams, the Haw reaches dimensions fit for a river. Reedy Fork is larger to the eye, but the Haw's name carries. We paddle on past the site of old Gerringer Grist Mill, river left, where Shallowford Natural Area, a new park with trails and campsites, has opened. A white-winged scoter splashes, faking a broken wing to distract us from her ducklings. A heavy short grey snake with orange markings, most likely a mud snake, drops into the water. Three river cooters soon follow.

A half mile downriver, we reach boulders that formed a dam, partly natural, for a grist mill. Foundation work for Iola Mill and deep pits are there, river right, and it was on William L. Spoon's 1893 Alamance County map, but Roger Gant thinks it was never completed. At the dam site, Will scrunches my Kevlar canoe over plentiful rock, and I pull Neil until we glide again, past Shallow Ford, on the Trading Path, once connecting to Shallowford Church Road. An 1893 map puts it a third of the way to the first big bend after Altamahaw.

Then Traverse or Travers, now Travis Creek, comes in. Some of Cornwallis's and Greene's men moved along this creek and by the German settlement of Frieden's Church[37]; long before this Native peoples foraged and camped its banks.

The Haw turns just where cows graze on Jane Iseley's farm. They remind me of watching a cow amble into the Haw while paddling with Janet MacFall, professor of Environmental Studies at Elon University.

"I love the way cows look in the water," I said, the gold tones of bucolic English paintings in my mind.

"Mmmm," said Janet. "The problem is that they're not great for the river."
I rolled my eyes.
"I didn't believe it either until I saw the numbers."

I was soon to learn she was right; cows need to be out of streams. And their tread leaves wide paths of erodible dirt. The splash and plop of their excrement adds another problem. Don't believe it? Alice Outwater did the calculations: a lightweight heifer of 750 pounds puts out 65 pounds of manure a day. One dump in the river adds 10–15 pounds packed with nitrogen and phosphorus. And it's "not just 40,000 cows. It's the cumulative effect of every acre of [fertilized] plowed cropland and every animal."[38]

When Alice Outwater compared 1850 and 2002 census data for the Haw Watershed, she found we had far fewer horses, sheep and pigs than in 1850, but almost as many cows — 43,000 now, only 4,000 less. Add 40 million chickens which were only a scattering at each farm and not even counted before, and you have 7,227,928 pounds of manure a day which can find its way to the Haw.[39] Then factor in chemical fertilizers to grow to full production 420,000 pounds of silage corn, just one crop in the Haw watershed.[40] Our farms are a factor in the Haw's health.

Jane Iseley, whose fields we paddle by for the next two miles, has found ways for both the Haw and the farm in her family since 1780 to prosper.[41] Her father, a school principal, raised tobacco here until he became seriously ill. When he did, Jane Iseley left her globe-spanning career as a photographer and came to his room at Duke Medical Center to ask him if he thought he could teach her to farm. He did. So she learned to pace off and make up tobacco while her father instructed, sometimes leaning on his walker in the field. He lived another seven years to see the farm prosper and Jane Iseley start to diversify into tomatoes, then strawberries, pumpkins, mums, corn, cantaloupes and organic tobacco. She also set up a store in the barn, to sell her and neighbors' produce, and ran hayrides with scavenger hunts to teach kids about farming. The Iseley farm is now basically pesticide free; good stewardship of land and river is important to this woman who remembers the Haw of the fifties:

> You could hardly stay in the house it smelled so bad; it was just so terrible. Through the years that, of course, has improved although now the municipalities want to claim that farmers are the polluters of the river. Farmers do own most of the land adjacent to the river, so you can manipulate your figures to say they are the majority of polluters, but in actuality you've got a lot of trees and grass that are stopping all those nutrients from getting in.
>
> In almost two miles of river frontage, there are only two places where the cattle can get in the river; they don't like to and it's very unusual. We are right now putting in waterers. Most of our cattle get their water out of the creeks, but we are working with the DOT/EEP folks now [to change that]. There are advantages for us and disadvantages. They will pay for fencing out the cattle, but someone has to maintain that fence, one on each side of the creek now. But the advantage is ... you get as much as a twenty five percent weight gain in cattle drinking better water.
>
> This farm only has two places, very limited distances in the woods where the creeks haven't already been protected. We never cut the timber. You know farmers are the best stewards of their own land that you are going to run across because it's our livelihood.
>
> In the field I worked in yesterday sowing oats and grain, you could see across where there are new houses practically right on the river.... Then municipalities say runoff's all the farmers and the cows ... and I see all the red mud and runoff coming off the other side down into the river, it's hard to convince me. It's not what I see or smell.

I think you will find all up and down the river that most farmers are very concerned about their land.

One cloudless October day in 2007, over a hundred people gathered at the Iseley Farm for a hay wagon tour, through titled fields. "You know farmers always name their fields," Jane tells us. The roast on the banks of the Haw was Piedmont Land Conservancy's celebration of Jane Iseley's easements, a guarantee her family land can be farmed but not developed in the years ahead.

The Iseley farm extends to Burch Bridge, which was once a covered bridge, then Ireland's Bridge and before that a ford.[42] When we pass the bridge, I realize there has hardly been a house the whole way: some cows and a deer, otter slides, and now rocky banks covered with dogwood. When thick honeysuckle hits at Indian Valley Golf Course, Neil says, "Let's just drift." So we do, past the steep knoll and chimney of Fort Snug, river left. This is where the bachelor owner of Glencoe Mill entertained out-of-town guests in the early 1900s, far from the highly regulated world of Glencoe village. Fort Snug's vista on the Haw could inspire reverie. As a boy, Billie Phillips found its 15 rooms, wrap around porch, and view a wonder:

> A spiral staircase located inside led to a tower [with] windows on all four sides.... This facility was the most elegant structure in the entire county at that time.... We were not permitted to go near the clubhouse when I was a small boy. Many years later, when it had stopped being used, we could go up the river to the club house and pretend that we were having parties.[43]

Below Fort Snug, granite supports mark the 1909 cement dam which replaced one of wood and stone; portage around it, river right. The high concrete dam below creates calm waters for family paddling, but beware the dam itself. That take-out is river left. It puts you on the path by the old mill race and the island where a water wheel once turned a grist and saw mill and leads on to the restored, vibrant world of Glencoe village.

Fort Snug, an elegantly furnished hunting lodge on the Haw, above and separate from Glencoe, was where mill owners brought out of town guests and entertained. The chimney can still be seen from the fairways of Indian Valley Golf Course and the river (courtesy Textile Heritage Museum).

Renewed Community: Glencoe

> Now the mill has shut down
> It's the only life I know

> Tell me where will I go
> Tell me where will I go
>
> > And the only tune I hear
> > Is the sound of the wind
> > As it blows through the town
> > Weave and spin, weave and spin
> > —"Aragon Mill," Si Kahn

When Si Kahn sings "Aragon Mill," there are people who hear their personal history in it — how community, way of life, and work were lost as mill towns closed down all over the South, as Glencoe did in 1954.

Ghost mill towns are everywhere, but at Glencoe, the mill buildings, offices and mill houses are being preserved: the efforts of many to build a new community in the structures of the old. Preservationists hope the museum and village will lead visitors to imagine the lives of workers in isolated river mill towns.

The Holts set the scene here with the last water-powered mill on the Haw. In the early 1880s, E.M. Holt and his sons, James and William, added to the grist and saw mills, once owned by Joseph Shaw and Levi Vincent, a cotton factory and a town.[44] They needed to be by the Haw for power, and workers, without cars, needed housing, but the design gave the mill owners control of much more than a worker's job. Your house, your job and your community could be lost in one stroke. He, and it was always a man, could see that you did not find work in any other local mill, as owners were a tight elite.

The people who came were struggling in the post-bellum South. They gave up their independence to do "public work" when two dollars a day was top pay for skilled male labor.[45] One man remembers as a boy wailing home to his mother when he saw his earnings amounted to 12.5 cents a day.[46] The twelve hour days were not unlike farm hours, but the change wasn't easy as Dave McCarn's "Cotton Mill Colic No. 3" reflects:

> Lots of people with a good free will
> Sold their homes and moved to the mill.
> We'll have lots of money they said,
> But everyone got hell instead.
> It was fun in the mountains rolling logs,
> But now when the whistle blows, we run like dogs.[47]

Those who want to preserve mill history, its pains and pleasures, are at work here. Helen Walton, Elon University math professor, board member of Preservation North Carolina, and my first guide to Glencoe, is one of them. In early spring, we swing through Burlington on a tour that only someone who feels history can give. Our first stop is Ireland Street, the site of the house that Robert Holt, one of Glencoe's owners, built for his fiancée in 1899. They never lived in it. Just before the wedding when the house was full of gifts and new linens, the bride died of yellow fever. Robert Holt could not bear to live there and sold it. Almost a century later, Sam Powell, Helen Walton, and others saved it from being torched for fire practice. It now stands at Glencoe at the site of the superintendent's house, a spot marked by enduring daffodils. Its presence there completes all gradations of class in the village. The owner's house stands above it, highest of all, for Robert Holt did go on to build a house at Glencoe plus Fort Snug, though he never married.

As we cross the Haw on Rt. 62, the river moves Helen Walton's mind to all the cotton

The control and order of life in a mill town is reflected in these mill houses in Glencoe. The higher status and seniority of workers would help them rent these two story dwellings closest to the mill. The ell or back addition, seen on the second house from right, allowed for an indoor kitchen (Sönke Johnsen).

mills it turned: Altamahaw, Glencoe, Hopedale, Haw River, Swepsonville, Saxapahaw, Bynum. "Where there's a town, there was a mill."[48]

We circle by the owner's house, now restored in lavish Victorian style by George and Jerrie Nall. Outside, porches, sunrooms and porticos stretch wide. Knowing my labor sympathies, Helen reminds me, "[Mill owners] really saved the South after the Civil War. Lots of the very poor were able to find jobs; a church and school were here. They had plots for growing vegetables; some mill villages did not allow that.

"There was no union here or any record of employee uprising," she adds.

At the superintendent's house, Helen Walton points out that both this and the owner's house offer a view of the whole village, a vantage point of control. "The water towers were used to put out fires in the mill. Cotton is very flammable." We round the corner to Front Street, now renamed Glencoe, with company store, offices and houses lined straight up the street in punctuated rhythm. An old barber shop adds one variation and is now a bed and breakfast. In Glencoe's heyday, about 250 lived here.

"It was a self-contained community," Helen Walton says, "from the mill on the river up to the Methodist and Baptist churches." Your house told your status. "New workers start at the top of the hill and move down from one room cabins and duplexes with shared kitchens to four room houses at the bottom nearest the mill."

We look at a house for sale; fixer-upper doesn't begin to describe it. Copal trees and sky come through the walls, but imagination and practicality have rebuilt this place. "A tree fell on it, and it's still standing," Walton notes brightly as she points out the "post and beam construction and the brick nogging insulation that fills the wall ... not seen elsewhere in mill houses." She's proud too of the kitchen ells which Robert Holt added to all the houses, so that cooking no longer had to be done outside. "All the outbuildings are still in place." The houses sell for around $35,000, plus the commitment to restore them in mill house style and tones.

Now for the mill work. We peer at the vats inside the dye house walls where chemicals ate away at the mortar. "Can you imagine what it did to their lungs?" Walton grimaces, heading on to the main mill itself. She would like to see the front addition torn off so that the three-story mill and tower could stand as it did in 1882 in stark and square authority. We drop down to the dark metallic core — where the head race shot water onto the turbine which turned a universe of belts and spindles and set the looms whirling. I can almost hear the thunder of river and mill working together. An 1899 report on waterpower records a claim of 160 horsepower here, 90 percent of the time, to drive 186 looms and 3500 spindles and roll out soft brushed cotton plaids.[49] Turbines, under restoration, may turn again to the Haw's power.

In the mill itself and up the tower stairs, we come to the top floor, which is recovering from a society ball; plastic sheets block off a center space and machines have been shoved aside. Still, this is a beautiful open space. Tall paned windows line the long south wall; the sun reverberates in through bare trees. For decades workers stood here amid the din of machines weaving plaids and striped cotton, bringing their young children with them. In the cotton dust of the spinning room that was "like a fog," some workers made it a habit "to talk to the machinery, to cajole it, to give it a whack on bad days and to praise it on good."[50] Looking out a window, I wonder if the glinting Haw skimming below cheered them or stirred feelings of entrapment.

The mess at the top floor does not daunt Helen. She has watched as houses were reclaimed from weeds, and she was on the board of Preservation North Carolina when it went way out on a limb to buy this 103-acre property for $200,000. The choice risked bankruptcy. From 1993 to 2000, 138 textile mills were vacated, offering 10,000,000 square feet of abandoned space for preservation. At Glencoe, sales were slow at first, but all 32 houses plus building lots have sold; Glencoe has the hope of being a real community. And a combination of public and private forces are at work to put 32 acres and land along the river here in the public commons.[51] There are plans to have office and artist space in the mill and restart the hydropower. "Glencoe can bring $10 million back to the community," Helen concludes. She is after all a math professor.

I'm introduced to Lynn Cowan, Glencoe site manager at its rebirth, who grew up in Alamance County — "textiles and tobacco, that's what we knew." She had her own high school experiences here in the seventies: "[Deserted] Glencoe was where your mother told you not to go. There was stuff down here that you didn't need to be involved with, and, of course, it was where everybody from high school came."[52] Decades later, Cowan is back to study its history:

> Glencoe started as a very separate community. As transportation changed, it became more connected — with cars you could leave the village more readily and do other things. That would be the change with the most impact. It was a very close community. Everyone shared

everything they had. At the top end of the street, as we understand it, each family had their garden, but it was treated almost like a collective garden. These were mini farms; they would have had animals and raised a lot of food.

As I understand it, everybody took care of everybody else. If you were the child of a mill villager, you didn't misbehave just because you were out of Mama's eyesight because you had a village full of mothers. Anybody was liable to call you down. You know there was this feeling of a large family. If someone was sick, they were taken care of.

All the stories we have heard are related to a common goal. I don't think there was any attempt to soften class lines, more a sense that we are in this together. There are stories about Holt Green, nephew of Robert Holt, who ran the mill until World War II. He was very gregarious, one of the guys. When the dam broke, there are stories about his going into the water, right in there beside them. He was very liked. He was in the secret service under cover and was killed in Czechoslovakia. Holt [Green] would actually bring flowers to some of the people — a woman had a child and he'd show up to congratulate her — and was very attentive to sick mill workers, getting the doctor to come in the middle of the night and things like that.

Unions? We don't have any stories about a great deal of unrest here. Obviously you know what was going on in the textile mills here in the '20s and '30s throughout the whole Piedmont area. What I have heard from people who worked here is that up until World War II, there was very much a sense of camaraderie. Then Walter Green took the mill over and ran it until it closed in 1954; he was more introverted though and had a law practice elsewhere.

A lot depended on the particular owner and his personality. Some functioned in a paternal role that was accepted and supported by the community and some the community just didn't like.

Personally? I would have liked the community, people really did interact and were part of a bigger thing. I would not have liked working 12 hours a day, 6 days a week. The cotton dust, the heat and the noise were incredible back then. So it was just a hard way to make a living.

Summer, the river would get so low that typically [the mill] would shut down for about five weeks and there would be picnics down there by the river, and the mill owner would hold picnics. No, we have the time books; they actually were only paid for time recorded, so they weren't paid then. They were self-sufficient, and they had credit at the store, but it meant you went in debt.

To capture the spirit of the place, Lynn Cowan takes me on a tour of Ann Hobgood's house, vibrant with creativity. This artist and carpenter was the first arrival at New Glencoe, and her house was often shown as encouragement to others. On the slats of her porch swing, she has painted the words of Don and Paul Faucette, two mill workers:

> If you need it and we got it, it's yours. We'd kill our hogs this time, and a month later we'd kill yours. Well, you can give us some and we can give you some. Down in the church basement, the women have a quilting bee and they'd go down and they'd all quilt. One of them would have a crop of cabbage and they'd get together and all make kraut. And up in the mill company's barn they'd have a corn shucking. They'd just visit around and work voluntarily. They all done it and nobody owed nobody nothing."[53]

Lynn Cowan sends me on to talk to the women responsible for Glencoe's Textile Heritage Museum. In the lavish owner's house, Kathy Barry and Jerrie Nall tell me the idea was sparked when Kathy Barry visited a mill museum in Lowell, Massachusetts. Neither had worked in a museum, but they felt how much heritage was being ignored here. "Just sitting rotting here" is the way Jerrie Nall puts it, adding, "I think every piece of paper they ever worked on over here remains."[54]

In their museum, a supplement in the *Charlotte Observer* in the early 1920s bolsters mill owners in the tense labor atmosphere of the times. Article titles herald the bias: "Ideal Working Conditions," "Most Modern Mills," "Religious and Educated Employees...," "Ossipee and Hopedale Mills Run by Man Who Puts Milk of human [sic] Kindness into Policies of Management."

The banner for Glencoe reads: "Compulsory Education Inaugurated at the Glencoe Mills Almost 40 Years Ago — Clean Community of Splendid People — This Alamance Mill Another Plant Which Does Much For Its Employees."[55]

The Holts did start compulsory education here in the 1880s well before it became mandatory statewide in 1907. "Of course," Kathy Barry comments, "they'd call them out of school if they were needed in the mill." But children could have a few years of schooling before mill work began.

The article on Glencoe emphasizes the lack of crime, the beauty of the setting, and the morality — all are credited to mill owners' fulfilling a mission to establish good communities; "Fish and game abound," we are told. The whole is "a city set upon a hill, or to be correct, two hills and these hills overlook the winding waters of the Haw River."[56]

In *The Quest for Progress*, Sydney Nathan illuminates the power of this paternalism:

> In 1896, seeking to make their wages go further, Glencoe's inhabitants deserted the community's company store in favor of merchants miles away who charged a third less for flour, meat and coal. The mill owner responded with a masterful appeal to company loyalty. Flyers posted all over the village read: "I cannot live without you, or some others like you, and you cannot live without me, or someone like me. Our interests are therefore mutual and the more I make, the better wage I can pay you. You should do what you can to help me."[57]

The fact that the textile profits were as high as 60 percent in 1899 reveals why workers would want a larger share.[58] Museum records provide information on wages per day:

    1889   11 hour days — 6 days a week
               Men $1–2
               Women $.50–1
               Children $.40
    1905   10.5 hour days — 6 days a week
               Men $.75–2.75
               Women $.60–1.
               Children $.40
    1925   Men $2.10–6.60
               Women $2.10–2.38[59]

The community may have felt "like a family," but, aside from low rents of $1.50 to $2 a month (1889), the economics did not. Paternalism was a limited answer to questions about fair wages and conditions raised by socialism in the North, where workers made 40 percent more for shorter days.[60]

What was it like to live in Old Glencoe? In *Habits of Industry: White Culture and the Transformation of the Carolina Piedmont*, Allen Tullos captures different views. Ethel Faucette, born in 1897, started out at 85 cents a day twisting and drawing and ended earning $1.69 an hour in 1954.[61] She recalls the pleasures of summer drought.

> There's a big old rock out there in Haw River they call Lily and I forget the other one's name, but there's two of them. When you begin to see them two rocks, you'd know we was going to get a rest. Cause the water was getting low.

We run by water then, had water wheels. That was the power that run the mill. And when the water'd get low, maybe they'd stop off for an hour or two. Well, there was a crowd that would get their instruments and get out there in front of the mill. They would sing and pick the guitar and the banjo. Different kinds of string music. And maybe they'd stand an hour or two and the water'd gain up, and they'd start the mill back up.[62]

Don Faucette, Ethel's son, remembers the river meaning money and lights and food (until pollution):

Water was low and sometimes the mill would stand. Once when I carried the dinner up there, Daddy was standing out there up to his hips, getting the mud out from the back of the dam so the water would come down the race so they could go back to work....

You'd be sitting there studying and the lights would go down. One of us would say, "An eel has hit the wheel." In a minute, that eel would grind up and the lights would go back on. There used to be some good fishing too in the Haw River.[63]

His brother Paul adds: "There was carps up there that weighed twenty-five pounds. There was catfish, bream, bass, white perch. White perch is called crappie now. Blue gills.... We used to fish there until they started turning that dye loose up the river. After that, we'd have to go out there and rake the fish off the gates to let the water keep the mill running."[64]

Some of Ethel Faucette's memories cast a shadow on life in this Haw village; her father, Man Marshall, was superintendent at Glencoe Mill for forty years: "We didn't sit at the table and talk about what the other fellow done down there — if we did we got our mouths mashed. He says, 'You leave the mill out of your conversations.' He didn't allow us to say a word about it."[65]

Union talk was out of the question: "Nobody wouldn't vote for the union. My dad didn't pay a bit more attention to it than he would a dog barking. I just remember them talking about it. We didn't never ask Daddy nothing about the mill because that one thing he didn't allow. He said it wasn't none of our business — that's just what he'd say."

"And you better do just what he said, too," adds George Faucette, her husband.

"And we knowed it," says Ethel Marshall Faucette.[66]

This authority and control that stifled unions kept the status quo in Glencoe going; perhaps it felt like a common fate in a tight community.

Finally, I meet with Garland, Thelma, and Donnie Massey, two brothers and a sister from a large family who lived in Glencoe from the late 1930s until 1977, when they were one of the last families to leave. When I asked them to recall their childhood, Garland mentioned lots of broken bones. These siblings were born with soft bones or rickets, caused by vitamin D and other nutrition deficiencies, and only Garland was ever able to walk; he worked in a print shop in Greensboro.

GARLAND: Typical childhood, I guess. The parents worked in the mill. Our sister babysat at night while our parents worked, and then if one of us got hurt, a neighbor would go into the mill and get 'em. Sometimes two of us got broken bones at the same time. There was one phone in the office then, and when someone needed a doctor, they had to go to the office and call.

THELMA: They said it was hard work. You know Momma would come home, Daddy too, all sweaty; their clothes would be all wet. They worked hard.

DONNIE: They'd be drenched.

GARLAND: My Momma and Daddy used to say they worked from 7 every morning to 6 or 7 every night.

THELMA: And they worked at Carolina [Mill] too; Carolina to Glencoe and back to Carolina then back to Glencoe. It's wherever the work was available.

GARLAND: They just went to a better job or they lost their job. Then Glencoe would take them back. They lived in company houses.

THELMA: It wasn't cause they couldn't get along with people. It was mainly cause the work was slow.

GARLAND: Well, our Dad didn't go to school.

THELMA: He couldn't write his name. That was pretty common then.

GARLAND: He could spell it though. He was a section hand in the mill, and he could do anything they needed from one floor to another.

Unions? I don't think it was accepted with the owners of the mill.

THELMA: Mr. Roosevelt. When he came in, he said it would be different, and it was. Wages went up.... I remember one day Mama was so pleased and happy. She'd gotten a raise and she was making $13 a week, and Daddy was making $15 a week. We thought we had already arrived at riches. We got along better then too, didn't we?

GARLAND: When our parents would get paid, we would get a nickel a week, didn't we, Thelma? And they had something called a three-center, sort of like a Coca-Cola, and we'd get a three center and two cakes, cookies of our choice, whatever we wanted.

Donnie can tell you something about the river cause about every Sunday morning they'd go down there to eat breakfast.

DONNIE: Oh yes, my friends used to pull me around in the wagon a lot. We'd ride up there in a boat and fish. It was mostly posted property, but we didn't let that bother us. You know where the river is close to the road. They used to call that the race. There's an island there. Two buddies of mine took me over there on the other side of the island behind the dam to fish. They hitched me in the wagon, but it got so rough one of them would have to carry me; the other one would carry the wagon. We had a good time.

Course if I missed church, they would fuss at me. I didn't care. We'd go up there early Sunday morning and ride a boat and eat breakfast, catch yellow bellies, catfish, and some bream.

GARLAND: They killed the fish.

DONNIE: After they stopped dumping dye in, it's come back some.

GARLAND: Walter put in a new dam; water was too low, they had to cut the lights off at a certain time of night, at 9.

THELMA: We got lights in the early forties. Donnie was born then. Before that we used lamps, kerosene lamps.

GARLAND: And they had baptism in the river, I was baptized in a church, but Thelma was baptized in the river, at the swimming hole, the Dongo. Two men took her out there in a straight chair and pulled her under.

THELMA: I trusted the Lord as my savior and wanted to be baptized. They had a song they sang at my church, "Shall We Gather at the River."

GARLAND: The river could flood up to the road, to the first buildings.... Our cousin was drowned up there. He was skating on the ice one day after Christmas, and he went under.

Sometimes the water would get a little high and the windows would start to vibrate. You would have to put something in them to stop it.

THELMA: Rattle, rattle, rattle, rattle.

GARLAND: Something about the water being heavy, it would make the house vibrate a little bit. You could feel it.

DONNIE: I used to lay in bed and listen to the water running over the dam. It was sort of like a lullaby. From the back porch or any where you went, you could hear the water running over the dam.[67]

A baptism in a serene Haw River in 1913. Young women, likely of the Haw River Baptist Church, are submerged while young men wait near the bank. This took place above the dam in the town of Haw River (Haw River Historical Association).

> Shall we gather at the river,
> Where bright angel feet have trod,
> With its crystal tide forever
> Flowing by the throne of God?
>
> Yes, we'll gather at the river,
> The beautiful, the beautiful river;
> Gather with the saints at the river
> That flows by the throne of God.

Copland Mill Generations: Stony Creek and the Haw

Before I canoe on from Glencoe, I explore by car what this wider river has fostered on its banks: Copland Fabrics, Goat Island, East Burlington Wastewater Treatment Plant, and Haw River, the town.

At Copland Fabrics, Stony Creek joins the Haw in a lush confluence that was home to a number of grist mills[68] before John Trollinger and his partners built the High Falls Manufacturing Company in 1832 — the first cotton mill.[69] After the Civil War, it changed hands and became Juanita Mill until bankruptcy in 1904[70] turned it over to James N. Williamson (of Ossipee Mill) who changed the village name to Hopedale.[71] Consolidated Textile Corporation bought it in 1921 and sold it two decades later to J.R. Copland, whose great-grandson Jason Copland runs Copland Fabrics today.

Copland Fabrics, now run by the fourth generation of Coplands, stands at the confluence of Stony Creek (left) and Haw River (background right) near where the first cotton mill on the Haw was established by John Trollinger in 1832 (photograph by the author).

My tour starts under the water tank's Crafted with Pride sign. Is it price or loyalty to America that makes a difference in sales? Mark Andrews, my guide, gives the nod to the bottom line.

Andrews explains how the company has held on: "Jim [Copland III, former president] decided that if we were going to be competitive in the market we were going to have to modernize our plant. We spent a lot of money bringing all new equipment into all our departments. We went from 2000 looms in 1980 to 150 today. The second thing we did in the late 1990s was diversify our markets."[72] Sheer curtain fabric is still the mainstay, but they also make upholstery, bedspreads and curtains for chains and resorts like Disney World which remodel every three years.

I am led through acres of rooms, some with a roar worthy of old Glencoe, even with ear plugs in. Other rooms hum vibrations, mere mechanical murmurings. No voices. No dust. Tubes as thick as elephant trunks circulate on overhead tracks to vacuum any specks. I count only 15 people in our travels by rows of spotless ivory and olive machines, munching out computerized reports and the translucent rayon found in windows all over America. Mark Andrews tells me 300 workers now produce what 1200 did in 1980; however, the downsizing was by attrition rather than massive layoffs.

The river has changed too. Outside, Andrews sketches a line through the parking lot where the river used to run before 50 to 75 feet of land was filled in for new buildings. During Hurricane Fran in 1996, the Haw tore by with such force that the waters of Stoney Creek were blocked and backed into the basement. "People said it was the river trying to take the land back," Andrews tells me.

The mill pulls in about one million gallons of water each week from Stoney Creek, not for power, but for steaming, boilers, and wash ranges.[73] We walk by a 150-foot long washer. Copland Fabrics treats the outflow with sulfuric acid to reduce its high ph before sending it on, not to the Haw, but East Burlington Wastewater Treatment Plant.

The details of the trade amaze me. Fabric is dyed in 800 shades of white. Yarn that is 95 percent polyester is twisted 19 times an inch to give it strength. One $100,000 machine can draw 5,000 threads onto a frame in one hour, something that it would have taken 19th century workers two days to do.

The man who started Copland Fabrics, Jim, "Old Man" Copland, was known as a "slave driver" to some and a savior to others, by accounts in *Like a Family: The Making of a Southern Cotton Mill World*. Some said "Copland would 'just pitch a fit' when something went wrong in the mill."[74] But when Icy Norman started mill work in 1929 and cried with frustration, it was Jim Copland who got others to teach her the weaver's knot and checked on her with a "How's my little girl doing?" Copland had been a friend to Icy's dying father and her mother, who desperately sought work in the Depression: "And Mama says, 'Yes, we've been everywhere hunting a job.' And he [Copland] says, 'You don't have to hunt no further. You've got a job. We're tearing the cotton out and putting in all rayon.'"[75] All agreed Copland was a "ball of fire," a rare one to make the transition from mill worker to owner.

James Copland III was the third generation to run the mill. A Phi Beta Kappa Morehead Scholar at UNC–Chapel Hill, Copland can talk venison barbeque and hog killing with workers and argue eloquently for the textile industry he is passionate about.

> My granddaddy didn't have enough money to eat; he didn't have anything and worked in the mills, at Swift Manufacturing, there in Columbus, Georgia, and in his early twenties he was the superintendent, believe it or not, and went on to be general manager for Dan River mills in Danville, Virginia, which at that time was the largest mill operation in the country. Think about that; my granddaddy only went through school to the first grade, and he college-educated every one of his seven kids. He came from Dan River to Burlington, worked for Spenser Love who started Burlington Industries, and then went to Virginia Mills [on the Haw in Swepsonville] and was general manager. There's a story.[76]
>
> He started Copland Mills when he was 66 years old. It was a closed down mill [Consolidated Textiles], and he had some people in town he knew who invested in it and the people who owned the property got stock in lieu of cash. He started in 1941, had 105 box looms, and then war broke out. My daddy came to work here six months after it started. They did not make government goods; they made curtains, rayon lightweight curtains for the mass market.
>
> When I was a little kid we lived here, right on the river in the mill village. During the war, we had a three room plank house with a pot bellied stove. River? No, no, no, we never played in it because it was too swift and too deep and too dangerous. We weren't allowed to go close to the river. I don't remember that much about the river back then. There was an old bridge; we used to go out there and look at the river. Daddy and Mommy raised chickens and had an old rooster that used to chase me when I was a kid, and my daddy went out there and put him in a sack and walked down there to the river with me with that old rooster in that sack. I watched him throw the sack in the river, in the water, and saw the rooster going down the river, and Daddy said, "Son, that rooster is never going to bother you again." I was about four or five.
>
> The river got pretty doggone polluted back in the late '50s and '60s. Now I am not going to put the blame all on the mills. I think the biggest polluter was the city of Greensboro really ... got so there weren't a lot of fish in it, but the river is clean now. The fishing is good in the river; there are ducks on the river, there are heron on the river and plenty of deer running up and down the river. The river's right pretty now.

Did you always want to run this mill?

Yep. Yep. I never wanted to do anything other than that, looked forward to it, was excited about it. I would have done it for nothing and thought it was the most exciting thing in the world. Now I can't say that today. Today it is extremely tough because of the trade policies of the United States.

It is so sad what has happened to the textile industry, but the problem goes way beyond the textile business, it is all manufacturing. We are paying a terrible price ... 650,000 people have lost their jobs in North and South Carolina. I hunt in South Carolina, and I go down through the smaller towns. You see mills closed, stores closed, houses abandoned, even the churches closed. I mean it is so sad, and those mills were well run, but you cannot compete with foreign government-owned businesses and foreign government subsidies.

Other countries subsidize, and the reason they do this is they want U.S. currency and jobs for their people. They get money through loans, through the World Bank, and they mostly don't have to pay them back.

Then you've got the currency manipulation. This is the worst situation of all. Let's take Turkey. They devalued their currency 25 to 1 versus the U.S. dollar. What did cost one dollar to buy from Turkey now costs four cents. It's impossible to compete. It's the same thing with Chinese currency, Indonesian and Pakistani. So those things killed U.S. textiles. And the fact is the U.S. textile industry is the most efficient, has the best quality, has the lowest cost per man hour in the world — bar none. Government currency manipulation, undervalued foreign currency, and govern subsidies should never be allowed in world trade; a legitimate manufacturer has no chance of competing against these foreign government practices.

Now you mentioned labor laws. Please know if the labor were free over there, we would still win hands down if they did not have subsidies and currency manipulation. Look at Pakistan's working conditions where they may pay 12 cents an hour or $560 will buy you a six, seven year old boy for five years' work. They live in compounds with walls all around, a latrine here, a place to sleep there, and work for 19 hours a day. They don't have any workmen compensation costs, health benefits, social security, labors laws. Our government's trade policies do away with good jobs in America and send those jobs over there with those working conditions. It's just awful.

Free trade works only if things are equal, but you cannot compete with foreign governments' subsidies. U.S. trade policy favors multinationals. Our government just doesn't connect the dots; they just don't get it.

We are going to have to say to foreign governments, "you can send all you want over here, but we are going to put a tariff on it." It's going to equalize the playing field. But the government is not going to do that. They are just going to go on until they can't go on anymore. Then it's too late, too late 'cause I will tell you, and I know something about these mill owners, we have been burnt so bad, our government has destroyed everything we have worked for all our lives, for generations, in one fell swoop.

We won't build 'em back.

It started under Reagan, Bush, under Clinton the same, but under George W. Bush it accelerated, and look at the job situation now. Americans need second jobs just to make ends meet. They can't earn a living wage because subsidized imports are destroying their earning power.

Trade policy ... it's a dagger in the heart of America.

The argument ranges onto the pages of *The New York Times* where James Copland is quoted next to a dissenting World Bank economist who says: "Textiles moved from England to New England to the Carolinas. It's hard to blame China for using the same market forces."[77] Jim Copland makes this reply to me: "Blame the U.S. government, their policies destroy U.S. manufacturing."

Others outside the Beltway, like John Emrich, CEO of Guilford Mills, side with Jim

Copland III. "All I hear from Washington is that [free] trade is a win-win proposition. Then I look at our growing trade deficit and think about the 3,400 good people in our good factories that we had to let go, and I want someone in Washington to show me where we have won."[78]

Jason Copland, taking over in the tradition of his father, was quoted locally: "We will not build a plant in Asia. We are Alamance County or bust."[79]

River Horror Stories: Goat Island

Goat Island waits in the river at the end of Sellars Mill Road, generations forded here, some taking their flour to be ground at the mill, river left. The island is now a trip to the dark side.

If you are not brave enough to canoe to Goat Island at night, summon the courage to walk the rusty grated bridge the Haw tears below on a February day that smells of sleet. Goat Island is the creepiest place on the Haw. A stripped abandoned car and a "No Anything" sign put me in the mood. Why? What's happened here? Listen to what I have heard.

A man lived here who raised goats and tortured children. He had been cut in half, and at night became a goat man. If you doubt me, you can find on the island the oven where kids were cooked; it's near the remains of a house with bullet holes fired by drug dealers and mafia men. After Hurricane Floyd flooded it, there was a terrible smell, some said rotten fish; others knew different.

A pond here never goes dry, but the reason is too horrible to tell. Some have seen a black horseman, one of Satan's.

Amanda and her husband are my serendipitous guides and source; she grew up here. Undaunted by stories to keep her away, Amanda played Ouija board at night and hid playing hooky days, sometimes near a school bus, so old the steering wheel is in the center.

"Listen, be open. You will hear the spirits," Amanda speaks. Together the three of us explore the island past grim and greyed beech, holly, iron wood, oak, maple, river birch, cedar. Sycamores look naked, and myrtle and ivy, those sweet plants of domestication, sinister.

A dog howls mournfully as we escape back over the grated bridge. "Don't look down," Amanda warns, "or you will feel like you are floating away."

"Is it any wonder a woman committed suicide here driving off into the Haw near the bridge?" Amanda asks.

Safely on Sellar's Road Extension, I can almost imagine a beautiful island pasture of goats, but today looking past the stony tree bark at sky and river, I see a Goat Island that belongs to the demon world.

Haw in Pipes: East Burlington Wastewater Treatment Plant

If we follow the Haw in all its streamings, sooner or later we find ourselves in the tanks and pipes of the urban Haw. Except for well-users, these pipes are how the river fills bathtubs and water glasses and then drains back to the river. We take from the Haw, and we give back to the Haw — after treatment.

Today, I am following the trail of human and industrial wastewater as it goes back to the Haw through one of the 90 treatment plants in its basin. The domain of pipes' headquarters is an old mill building off Webb Avenue, where I stand amazed before a huge map of the hidden world of the Haw. Red and green parallel lines mark its strictly segregated pipes — water and wastewater — that run down every street, on opposite sides. Today to understand the health of the river, I follow the red lines that flow out to the East Burlington Wastewater Treatment Plant.

As Eric Davis, water and sewer operations manager, drives me to the plant, I sense the hidden world of twin pipes on opposite sides running under us in reverse directions. Davis, a former science teacher, has since wrestled with the physical challenges of cleaning up wastewater and now takes on the harder task of finding the funds to maintain and improve an aging system.[80]

Past a soaring American flag and into a long brick building, I meet Chief Operator Clarence Sell who has the hearty humor needed to run a wastewater plant. "No one thinks anything about us as long as it keeps happening," says Sell, echoing Davis's worries of maintaining, never mind upgrading, a system with hundreds of miles of underground pipe.[81] A bad crack or a clogged pipe raises a fast, furious outcry.

However, all's well this blue March day of sharp breezes that only occasionally carry a mustiness, certainly not the rigorous nasal experience I expect at the place where half the wastewater in Burlington drains.

My tour begins at the intake a few hundred feet above the river where the waters will return, after a week or two of treatment. First I hear the incoming waters, ten feet below a grate, pulsing in with the heavy rhythm of a beast's heaving breath. Next, I see them down a cement shaft, gray and smooth pouring through the first filter which strains off branches and large objects. What pours by amounts to 5.2 million gallons an average day, one million more rushes through in the morning than afternoon. It's much more if a heavy rain and pipe leaks add storm water.

The water goes on to more filters, one a tank with a rotating arm to remove equipment-destroying grit. Then the grey stream is pumped underground up the hillside, and we follow it. Around us spreads a virtual waterpark of pipes, pools and holding tanks which reclaim the Haw's waters by filtering, separating, providing attack organisms, dousing with chlorine, de-chlorinating, and releasing it down eight aerating steps to rejoin the river.

This water park is not cheap, Eric Davis reminds me. The chemicals cost $80,000 to 90,000 a year and then there's the electricity — $500,000 more. It takes a lot of energy to lift water uphill and fill it with an endless stream of oxygen bubbles so the organisms that break down the organic matter, i.e., human manure, can thrive. Except for the chlorine, the processes are biological and the same as the Haw's, which would need more water and vast stretches to do the cleaning and oxygen-adding work done here.

And like the river, Eric Davis tells me, at almost every stop, it is all about flow. Blowers keep air moving through the water and pumps keep water circulating. Davis remembers one morning suddenly noticing an eerie silence. He located the missing hum of engines and knew they had to act fast. When water stops moving and air stops flowing through it, bioactivity drops to zero in 30 to 60 minutes; treatment ends; aerobic bacteria die, and the air becomes foul indeed.

The pipes all over Burlington are sized to ensure water moves: too big and it just sits; too small and flow is blocked. Keep the water moving; that's the river imperative.

3. Cotton Mill River

An overview of East Burlington Wastewater Treatment Plant shows the rectangular aeration pools and behind them the circular tanks where bio-solids settle out. To the right and outside the picture are the initial straining operations and the Haw River, where water is piped after treatment. Rt. 70 is in the foreground (courtesy Utilities Department, City of Burlington).

Up the hill, the primary treatment of separating continues; in vast circular tanks, water swirls slowly. The heavy organic solids settle out and pour thickly over to digesters. We climb to a rooftop and peer into wide silos where the human waste is broken down by organisms into bio-solids. Call it *sludge* and an eyebrow goes up; there is heated controversy here. Davis feels that bio-solids are safe: laws govern its separation from the human food chain — no cattle can graze on the land for a month, and the fields can only grow animal crops. Farmers want bio-solids for its water content and high phosphates and nitrogen: the very chemicals needing reduction in the Haw.

Others in the community don't want any toxins and heavy metals in the sludge spread around and say some of its nutrients run back to the overloaded Haw. They want proof it does not cause health problems. In 1988, a law banned sludge dumping in the ocean.

Terri Buckner, one participant in a local "sludge" listserv, offered a solution, pending crucial changes in policy and standards to protect us from toxins:

> Sludge is the result of the post–World War II love affair with chemicals. If it weren't for the chemicals, we'd call it compost and gladly apply it to our farmland and gardens. It's unfortunate that as a society, we so often fail to connect the dots between inputs and outputs.... The ultimate change has to be personal behavior — breaking our love affair with chemicals requires education and awareness....
>
> If you stop buying household products such as toxic drain cleaner, and pesticides and personal care products that are laden with perfumes, you help.[82]

The biosolids vs. sludge issue is important. Twenty-nine million gallons are spread over the fields of Alamance and four adjoining counties a year. Some nearby claim adverse, even deadly, health effects. If the EPA and FDA were more cautious about the stream of unregulated chemicals they permit, there would be ways to use this "bio-solid" as the valuable resource it is for fertilizing agricultural fields and supplying energy. And these two uses avoid dumping more nitrogen in our rivers.[83]

But that is the subject for another book, and today Eric Davis and I get back on the trail to clearer waters and head to an immense square pool stripped with ribbons of oxygen-fixed bubbles that keep the organisms growing and then multiplying as they feed their way down the pool breaking down organic matter in what is usually a five day cycle.

At the top of the hill here, the pumping is over; gravity will move the water down the hill and back to the Haw.

The third stage is a chemical one of chlorination and de-chlorination. Eric is interested in the results of an experiment to reduce chlorine by using it later in the process. The heavy metals once dumped into the Haw are not taken out here but are now the responsibility of industries themselves and the EPA, which sets standards some find dangerously slack. What are not even regulated are the hormones from birth control pills or medicines flushed down the toilet. Eric Davis agrees that that is a hot-button issue and has read these hormones are disruptive at very low levels.

Clear water sprays knock out the bubbles of harmless foam that could give a bad impression. Finally, the water tumbles down steps, gaining oxygen, before a pipe shoots it back to the Haw. How clean is it on release? Eric Davis says he wouldn't drink it, but he would swim in it.

Inside Clarence Sell talks about the whole cycle: "It's 99 percent clean when we get it. We just work on the 1 percent." And it's only recently that this work has begun; Sell explains: "It used to be much simpler; you just dumped your waste into the Haw."[84] Sell believes the first system which used trickling filters and chemical clarifiers came about in 1959 and was stopped "because it didn't work well enough."

"We haven't grown in quantum leaps since then; the technology is evolving, but it's still in its adolescent stage." But progress has been made. This plant was discharging 3 milligrams of phosphate per liter, now it discharges .3m/l, ¹⁄₁₀ of that, well below current state limits. New regulations for nitrogen are on the near horizon; swimming, fishing and drinking water in Jordan Lake waters are at risk due to excess phosphorous and nitrogen. As Sell acknowledges, "Stormwater regulation will come too."

What's a better treatment system? Davis answers as Sell nods, "More pipes, tanks, pumps, more circulating through [the equipment] we have now."

And they would like to see a reverse in the loss of industry and revenue from the decline of mills. Sell adds, "We have 25 percent less flow now than in 1981. We've gone from 8 million gallons a day to 5.2 million. That's less revenue.

"The new nitrogen limits are going to be expensive, in the millions. There's already a $2.2 million budget just for this plant and $10–12 million for water and wastewater treatment in Burlington.

"It's all about how much bang you get for your buck," Eric Davis continues. The energy that this treatment takes comes at an environmental cost. That needs to be factored in."

Clarence Sell agrees. "Oh, just give me the money; we can discharge distilled water." And this sets Sell rumbling around in a closet until he plunks a bottle of Burlington Blue in front of me — a promotion to make a point: "Premium quality drinking water from the

City of Burlington's water system. "When empty, simply refill from the tap." The label also sets some water treatment history straight; the first public well came in 1888, first water treatment plant in 1919.

Do they drink bottled water? "They say a bottle is equal in price to about 400 gallons of tap water. Triton starts as Burlington tap water; they tell you so on the label. And though Triton provides some further treatment, tap water is certainly more regulated than bottled water, which has no regulations on it," Davis says.

Davis and Sell want to know if I am going to drink my bottle of Burlington Blue. My emphatic "no" draws a laugh until I explain that I want to keep it, but the reaction unleashes some public relations frustrations.

Davis complains, "They see us as damn polluters."

Sell adds, "And we're working environmentalists."

"Pollution isn't just coming from the point [pipe] sources, but we're easy to find."

"It takes a long time to change people's opinion after the facts have changed. In 1972, point sources were the main polluters."

"I'm all for what the Haw River Assembly is doing; we just want them to see our situation."

Sell would like to see a tax on fertilizers to reduce runoff from lawns and fields to the Haw. "Address the problem at the source. Don't just look at the point where it hits the river." He never puts fertilizer on his lawn. Sell has another solution too: "Every city should have to discharge upstream of their [drinking water] intake. We do need to take care of what we are putting out."

Streaming Chemicals: Haw Watershed

Back at the old mill office, I get a chance to speak with Steve Shoaf, director of utilities, about excess nitrogen and phosphorus in the Haw. His eyebrow goes up or is it a steadier look, a signal I am missing something? When he asks if I have read *Our Stolen Future* or *Living Downstream*, I know I will be.

What is in water beyond too much algae-boosting nitrogen and phosphorus? Worldwide, water-borne diseases such as cholera, amoebic dysentery, and typhoid take a heavy toll.[85] Ninety percent of the 1.7 million who die each year from these are children. That is one low estimate of deaths that our water treatment prevents compared to lands where 70 percent of industrial waste and sewage pass untreated into the world's waters.[86]

However, the books Shoaf recommends focus, not on waterborne disease, but on new chemicals, circulating through water and air, spreading to the Haw and everywhere. Read on, but beware the nocebo effect — the ability to get real physiological effects from just learning and worrying about something.[87]

Sandra Steingraber's *Living Downstream: A Scientist's Personal Investigation of Cancer and the Environment* traces the effects of some of the 75,000 synthetic chemicals introduced in the last half century, less than 3 percent of which have been tested. Like Rachel Carson, author of *Silent Spring*, Steingraber wrote as she battled her own cancer. She grew up on the Illinois River, once the most productive of inland rivers supporting 2,000 fishermen, until in 1900, the Chicago Sanitary and Ship Canal raised the river level with industrial and sewer wastes. After World War II, chemical pollution took out mussels, diving ducks, scaups; herbicides finished dabbling ducks. Twenty species of fish disappeared. Bladder cancer was persistent in her family. When people suggest genetic predisposition, Sandra Steingraber lets them know she is adopted.[88]

The number of new untested synthetic chemicals is as upsetting to Steingraber as their persistence. What she calls "vintage" ones, DDT and PCBs, are found in high concentrations in the dirt of New Hampshire's mountain peaks and in the bodies of Siberia's Lake Baikal seals. Synthetic chemical toxins are routinely used by one in five American families on their lawns.[89] Chlorophenoxy herbicide or 2,4-D, invented in 1946, is found in Weed Bgone, Scotts Turf Builder and others though it has been linked to non–Hodgkin's Lymphoma.[90] And the chemicals can be chillingly random. Dioxins and furans develop from a mix "as banal as newspaper plus plastic wrap plus fire." Steingraber calls these agents of cancer "the unplanned unwanted offspring of modern chlorine chemistry."[91]

This information has local relevance. Dioxin is a by-product of the medical incineration at Stericycle in Haw River. Sandra Steingraber cites further evidence from the Haw, downstream:

> In North Carolina, a cancer cluster in the rural community of Bynum was linked to consumption of river water contaminated upstream with both agricultural and industrial chemicals. This study is particularly compelling because the sudden increase in cancer deaths that emerged in the 1980s corresponds closely with the time of peak exposure to known carcinogens in the river (1947 to 1976) once the normal latency period for cancer is factored in.[92]

To prevent more cancer deaths, Steingraber advocates tighter control of these synthetic chemicals. Less than 1 percent of our food is tested for them. She lists 40 possible cancer-causing chemicals in our drinking water, 60 in our air and 66 sprayed on our food crops (though only 0.1 percent of pesticides actually reach the pest).[93]

The makers of 2,4-D can claim "no scientifically documented health risks ... exist from the approved uses."[94] Sandra Steingraber would like to see those who produce and import the chemicals carry the burden of proving their safety. This is more the practice in Europe.[95] In the U.S., we wait for fatalities to make the case for toxicity.[96]

Pesticides come under heavy scrutiny in *Our Stolen Future*, which reports use has multiplied by 30 times since 1945 despite increased potency. The authors, Theo Colborn, Dianne Dumanoski and John Peterson Myers, agree with Steingraber that those who produce the 8.8 pounds of pesticide consumed per person each year should have to demonstrate their harmlessness to us and our waters.[97]

Their focus, though, is on xenoestrogens, synthetic chemicals that change plant, animal and human hormone messages and can unintentionally block reproduction or sexual development. Sexual mutations, decline in animal species, and worldwide drop in human sperm count led them to look into those chemicals. Endocrine disruptors can be present in the chemical itself, as in birth control pills excreted into waters, or in one chemical's chance contact with another. Varieties of dioxin, DES, DDT, furan, PCBs and many others can all act as endocrine disruptors.[98] Xenoestrogens can operate at parts per billion and are rarely tested for in drinking water.

Medicines that make their way to the Haw represent another problem. The life-saving antibiotics from the 1940s are now dangerously over used. Two million pounds manufactured in 1954 became over 50 million pounds by 2000. Forty percent of antibiotics are used on livestock to promote growth and pass on up the food chain. They are even sprayed on fruit. And it is estimated that half those given to people are of no use[99]; viral infections and colds are not caused by bacteria.

Our overdose of antibiotics means good, as well as bad, bacteria are knocked out. This explains why resistant strains of TB bacteria are increasing.

At Elon University, Janet MacFall's environmental studies students wanted to see how pervasive antibiotics were in the Haw, so they set out to find the extent of resistant bacteria in the river. In pairs, they picked two sites, one "contaminated" and one "pristine." They tested samples for resistance to four common antibiotics; if no halo formed, it meant resistance — the antibiotic was without power.

Students were amazed to find the results were the same all over, immediately downstream of wastewater treatment or on a well-buffered rural creek. Almost half the samples show a resistance to tetracycline; the results for kanamycin and streptomycin were worse (75 percent resistant). In the case of erythromycin, widely used for lung, ear, urinary and intestinal infections, 93 percent of samples were resistant. As one student said, "It doesn't even start; it is just everywhere."[100] Another commented the results "made me mad at every single person on earth." He has a point; the problem is cultural. As Dr. Stuart Levy of Tufts University Medical School suggests, "The resistance problem may be reversible, but only if society begins to consider how drugs affect 'good' bacteria as well as 'bad.'"[101] Cutting out the routine use of antibacterial soap and antibiotics is one place to start.

As a culture, we believe in "away": we will wash it away, throw it away, flush it away, move away. Soap will go down the drain. Pills disappear down our throat. Garbage trucks vanish down the road. This belief in "away" that hurts the Haw and us is ironically the promise we feel in its streaming waters — to take us and everything away — even as water and air currents bring antibiotics and chemicals here and everywhere.

Winged Town: Haw River

> A river touching the back of a town is like a wing.... With its rapid current, it is a slightly fluttering wing. River towns are winged towns. — Henry David Thoreau[102]

"Haw River history is river history," says Gail Knauff, summing up this town that took the river's name; she and her late husband Bob Knauff chronicled that history. In *Fabric of a Community*, we learn George Carteret was given all of the Haw River watershed and more for his help restoring Charles II to the throne in 1661 after Cromwell, then it passed on to his grandson John Carteret, later Earl of Granville, who was the lone lord proprietor not to sell his land back to the crown in 1729 — transactions of little relevance to a Siouan people who fished and forded here.

The arrival of Europeans did change things. Only a few Native people continued to camp at the ford, on the Haw's east side,[103] their numbers had been seriously depleted. Pine Ford or Piney Ford was also known as Trollinger's Ford after Adam Trollinger, who homesteaded here in 1745, building a stockade to deter wild animals. The Trollingers came from the Rhine valley down the Shenandoah valley,[104] likely singing German equivalents of "one more river to cross," as they forded five river systems of Virginia.[105] Two years after the Trollingers settled, Adam's son Jacob and two enslaved workers built a grist mill, an amazing feat in that wilderness.

In 1752, only 1113 "taxable persons" were recorded in Orange County, which included modern Alamance.[106] Slaves were not noted though Mark Chilton surmises they built all the mills on the Haw.[107] East of the Haw, so many Lutheran Germans were on the creeks that in the early 1770s, traveler J.F.D. Smyth was pressed to find an English speaker.[108]

When Moravian Brethren stopped at Pine Ford on Halloween 1755, ignorant of local custom, they camped on the Haw's west bank, which was not for public use; they preserved Trollinger's response.

> Oct. 31 / The road was miserable, but we crossed the Haw River and camped near Drollinger's. He was not at home when we arrived but returned late, somewhat intoxicated. He made a great stir when he saw that a fire had been built on his land, but when he learned who we were he excused himself, — [saying] "he was ashamed that he had drunk so much, — we should not think ill of him, — we were heartily welcome, — he was a poor fellow who could not help himself, — but he was at our service." He soon went to his house, and we were well content.
>
> Nov. 1 / Drollinger attended our morning prayers, and gave our drivers some hay, and went with us to a road leading to a mill. He was much ashamed of last night, and wished that we could spend the day with him so that he might kill a cow, and share it with us as a peace offering. We felt sorry for the poor man, for he seems to love the Brethren, and the Saviour will not let his willingness to serve us go unrewarded.[109]

In 1781, Adam's son Jacob Trollinger would go into a rage when Cornwallis and his soldiers seized the grain at the mill. For this, he was bridled and tied to a tree. His fury later turned to revenge, and he sent both sons John and Henry, and the enslaved Thomas Husk, to fight the British.

All survived, and Henry went on to build the first bridge at Haw River — a toll bridge on the stage coach route from Tillsboro [Hillsborough] to New Garden [Greensboro]. In the next generation, John Trollinger started his cotton mill at Hopedale (1832). In the fourth generation, Ben, William and John Trollinger built Granite Mill (1844), a hundred years after the first mill here, using the power of the Haw and the strength of its rock bank.[110] The granite was formed when molten magma jolted out of a volcano about 600 million years ago and cooled slowly in cracks in the earth.[111]

Ben Trollinger was active in the major technological revolution of his time, the railroad, diverting its tracks from Pittsboro to Haw River, by serving on the railroad's board of directors and himself financing the 260-foot bridge across the Haw. Ben Trollinger's public energies and capitalist spirit did not always bring rewards. He donated 631 acres of woods and fields for Company Shops (now Burlington), named for the maintenance stop railroad engines needed every 100 miles or so. Trollinger understood his would be the only hotel and staged a dramatic opening overlooking the Haw, but then a hotel was built in Company Shops that put his out of business and led to his bankruptcy. But Trollinger, true and resourceful, started a turpentine operation on the coastal plain and repaid his debts.[112]

E.M. and Thomas Holt were able to buy Granite Mill in 1858 and added Cora Mill in 1881 and the T.M. Holt Mill in 1893. They had insisted on payment in gold during the war and were even accused of profiteering "by state officials and some rivals"; in any case, they were posed to expand.[113] Thomas Holt moved into a mansion in Haw River and created an estate on 400 spectacular acres river right, just south of the bridge. His passion for gardening and control are portrayed in a popular story about his hiring a gardener: Each applicant was required to move the same pile of rocks between the same two spots all day long. All quit, some sooner than others, until one man finally kept doing just as he was told and earned a house and a job creating a showplace on the Haw with 200 North Carolina trees which was open to the public as a park.[114]

A pre–1890 photograph shows Granite Mill (built in 1844) in the background and, in the foreground, a horse and buggy on a very provisional looking early bridge over the Haw (Swain, et al., Waterpower, 153; courtesy Haw River Historical Association).

With the river and railroad station, the town was a center of trade. Goods from Swepsonville came by barge and plaids from Alamance Creek by wagon. Two Holt-owned grist mills were running—one for community needs and the other, the state's biggest, produced a hundred barrels a day of "Morning Star" flour that was traded in state and in Virginia.[115]

Like other mill towns, Haw River was a sharply segregated world; Old Town, Johnstown, Sugar Hill, Terrapin Slide and Pinetop, Over the River, and Red Slide were European American neighborhoods. Ruby Doo (now Green Level) was where African Americans lived; none worked in the mill they had helped to build.[15] Before the Kirk-Holden War, the Ku Klux Klan was active here; Adolphus Moore, Thomas Holt's brother-in-law and partner, was "among the first of the 101 persons arrested and charged with being klansmen."[117]

The Trollingers and Holts did support public education, albeit segregated. Thomas Holt was elected lieutenant governor in 1889 as the "workers' friend" after trying in vain to get workers' hours cut from 72 to 66 a week.[118] Though he made that reduction in his own mill, it did not quell labor tensions after he died in 1896.

In September of 1900, mill worker Anna Whitesell's firing sparked a major confrontation about workers' right to unionize. The igniting incident is described by Bill Vincent, co-author of *Shuttle and Plow*:

> An overseer at Thomas M. Holt Manufacturing Company in Haw River dismissed a weaver, Anna Whitesell, for making unnecessary trips to get filling for her looms. Whitesell "flew into a passion, denying [the overseer's] charges and scorning his threat." The overseer then made derogatory statements about Whitesell's union affiliation, and he gave Whitesell's looms to an orphan girl who was not a union member. After hearing the circumstances of Whitesell's dismissal, the girl refused the assignment and asked to speak to union representatives.[119]

When management refused to meet with union members and replace the man who fired her, workers reported to work but would not operate machinery. The response was to shut down Thomas M. Holt Manufacturing Company, Granite, and Cora Mills as well. This affected 300 union members and over 900 non-union workers. Alamance County textile mill owners met and let it be known: "We do not hire help through committees but individually, and we certainly will not treat them in any other way than as individuals,"[120] and added, "We are determined to run with non-union labor or not run at all."[121] Workers had until October 15 to renounce membership in the National Union of Textile Workers (NUTW) or move out. Organizers articulated the dilemma:

> "Is it right," asked a union committee, "that mill owners shall organize, meet together, fix prices on their yarns, their loom product, and regulate wages, and then say to the laborer: 'You take what we give you, do what we tell, and say nothing.'" TA Allen concluded that the attacks on the NUTW showed "that freedom of thought and action is not allowed among the common people. When a body of citizens are denied the right and privilege to unite for the common good into a lawful organization, we may then say our liberty is not longer our own, but is to be utilized for the benefit of some one who controls the purse strings of hundreds of men and women."[122]

On October 15, 1900, only 154 workers showed up for work, and 3,000, union and non-union workers in nearby mills stayed out in solidarity.[123]

With business typically slow in the fall, the mill owners held out and used their power to evict. "The great majority," reported the *Alamance Gleaner*, "remain firm and are moving away."[124] In this lose-lose scenario, the Knauffs report, "Many mills operated with a severe shortage of help for many years afterward."[125]

Some would add that "the defeat broke the union in North Carolina."[126] Allen Tullos, author of *Habits of Industry*, puts it this way: "The visible defeat of the NUTW in Alamance affirmed the authority of the Holts and gave mill owners a precedent in the industrializing region."[127]

Charlie Poole, the renowned banjo player, would keep at least the mood of rebellion going. He was a pioneering musician who started in mills here at age eight or nine when the union fight was astir and he was already picking on a five string banjo. Later, he would skip work to serenade co-workers from the bridge; they would take a break and wave their thanks. Poole went on to organize the North Carolina Ramblers and have a best-selling Columbia Records hit that sold 102,000 copies.[128] This man "who came to epitomize the hard-drinking 'rambling man' of musical tradition"[129] once slammed his banjo over a policeman's head after being commanded, "Consider yourself under arrest." "Consider, hell!" was Poole's reply.[130] The Haw's profound influence on his work is evident in these lines Poole sang:

> If the river was whiskey and I was a duck.
> I'd dive to the bottom and never come up.

Gang behavior was another response to mill life, especially in the Charles Holt years; the bridge over the river was a boundary Haw River boys defended from intruders. While Thomas Holt and Alfred Haywood kept a tight rein on this behavior, Charles Holt, who took over in 1896, was their childhood friend and would bail them out of jail on Sunday so they could work on Monday.[131]

Through all this, the Haw continued to be a part of recreation and industry. When Max Lieberman started a hosiery plant in 1911, the popular bachelor kept his canopied motorboat at the Haw River dock. Gasoline tugs were chugging goods in from Swepsonville until the 1920s when Cone Mill took over from the Holts and began producing moleskin then the popular corduroy. In 1911, a trolley drew in more folks who rode from Burlington for 10 cents to walk the wooden trestle bridge in style or take pictures by the Granite Mill dam.[132]

Sometimes the horror of wrecks was the lure; in 1911, 1928, 1936 and 1960, train cars crashed down into the Haw. Conductor L.D. Rochelle, who survived in 1911 and again in 1936, thought of Haw River as a lucky place.[133]

Workers at Holt Granite Mill in Haw River in 1912 included Charlie Poole (top row, first on right), who gave up mill work to play "You Ain't Talkin' to Me" on the banjo and record hit records. Another worker, Pearl Spoon, is top row sixth from right (courtesy Pearl Spoon and Kinney Rorrer).

The Haw River survived, too. It endured a toxic period when declared dead. Dr. Donald Francisco of UNC–Chapel Hill, who studied Haw mud samples in 1970 taken below Granite Finishing, found "the only piece of sterile nature I had ever seen. I found nothing alive in it." In 2001, almost 30 years after the Clean Water Act, he described the river as in "fantastically better shape."[134]

Richard Jarrett (1933–2003): Coming Up When the Haw Was a River

When I tour the Haw River Museum, Gail Knauff makes sure I meet Richard Jarrett, who is installing a barber shop to set off an old barber's chair. Gail Knauff suggests a lather bowl and brush. "I'm already on that," he tells her. "And I have a leather strap for that rusty razor as well."[135] They are co-conspirators for the Haw River Museum, keepers of past connections.

"It's a shame that we have to guess at so much about the past." Richard says, "I think everyone's story is interesting, and everyone's story is sad because you know where they are going."

> I lived a bit off the river in the group of houses they called Johnstown. Not right on it. I think it was 1948, I went down the river. It was one dam after another. You know what portaging is. We timed it for July and August, crops were in all along the run, fresh corn, watermelon; we ate catfish and pork and beans. I was a junior in high school. We went two to a boat, six of us. They were flat bottom boats. Swepsonville, Saxapahaw, Bynum. There was no Jordan dam then. Further down you had to paddle, it was stiller water. At Elizabethtown, it was dead still water, and then tidal waters. Most of us gave up about there. Some who had motors later got to the beach, to Wilmington. We must have called home for a truck to come. We liked the fast rapids; they'd shoot you along.

A 1930s aerial photograph of Haw River after Moses and Ceasar Cone's company bought Granite Mill, removed the tower and built a large square building by the road. Old Town (upper center), was where the first Granite mill houses were built in the 1840s, and has since been turned into parking lots. Over the Haw in the foreground are the bridges for Old Rt. 70 and the railroad; a new bridge now runs in between them and over the old grist mill location (Haw River Historical Association).

There were muskrat then. You don't see them now, not on the river. The pollution started with Cone Mills. Raw dye just poured in. Pollution came in at Ossipee, Buffalo Creek off of Greensboro. From here down it was terrible. When they first turned the dye in, we lost a lot of fish. You couldn't stand to be near the river, the smell of dead fish was so bad.

I know 'cause I used to work for the Wildlife Commission and would go count dead fish and weigh 'em. Back then there were no such thing as controls. Prior to the pollution, you could drink out of the river. There was frogs, snakes, bream, catfish, crappy, carp and suckers. About all there was to do when I was coming up was to play in the river, swim, fish. People would swipe a boat and take it out. I had mine swiped several times. When I come along, only two or three had boats.

We used to catch snapping turtle. It was a rite of passage for boys. We'd go out with a pole and tap the river bed. Tap on the shell to tell the front from the back, then grab it by the tail. Then you'd put a stick in its mouth and tie the stick to the tail.

I don't know where the gong was or the piling for the dock. They'd pole barges up the river with mules on tow paths. When the barge came up loaded with cotton [goods] from Swepsonville, they'd hit the gong and mill folks would come down and help them unload. [Raw] cotton came by railroad cars.

The power company put two cables three miles up from the bridge, right under where the big power line is now. You walked on one and held onto the higher one. They were taut. You weren't bouncing around.

What was the river to me? I did not think I was fortunate to have it here. I'd go with my daddy south and see some of those rivers, the big ones, and I didn't think much about this one. Then I came to learn the history and why people settled here.

Everything was on the river. People would come from Burlington and all around to look at the river. Come to look at the governor's house; the grounds were open to the public.

Gail Knauff and Richard Jarrett of the Haw River Museum stand between the Haw and the museum alongside one of the old gristmill's millstones (photograph by the author).

In the fifties, this was a bustling town, had four grocery stores and two taxi stands. You might go to Burlington for a movie or Christmas shopping maybe. I was swimming less. I had girls on my mind, not fishing. I took a four day trip to Elizabethtown in 1954, too. No women could go. We never took a change of clothes.

In the sixties and seventies, everyone started talking about pollution. Now we've got more pollution from cities and from population. Then it was industry. We would say I wonder what color the river will be today. Two to three colors would come out in a few hours. You could sit on the bank and watch it. People called it an open cesspool.

[Before that people in] houses near the river used to throw trash down into the river. Course, we didn't have much garbage then. You had a slop bucket for pigs. Charlie Dickey would pick it up, or you would have a neighbor with pigs who would come for it.

Lots of local people used to go camping on the river then and set for catfish. We didn't have all the equipment then. No tents. You know what we used for tarps? Corduroy. You can get a piece of corduroy before it is cut wet. We didn't carry a cooler full of drinks. We'd camp close to a spring.

Trollingers Today

Mike and Spencer Trollinger, two descendants of the first settlers at Haw River who live in Burlington, are ordering lunch when I catch up with them at the Blue Ribbon Diner in 2003. The brothers are well known around town as owners of Duncan Service Station and as kayakers.

MIKE: We grew up here; we never went on the river.
SPENCER: Not when I was small; he got a kayak about five or six...
MIKE: ... years ago. When we were growing up it was polluted. So about six years ago when

we got our kayaks, we started looking for places to go around here, creeks, whatever. You have to know where to put in and take out and don't go at low water. You'll be dragging a lot.

SPENCER: We took the commissioners down the river.

MIKE: With Mike Holland.... What drew me to the river? Just looking at it. That first time. I loved it. I wished I had started earlier.

SPENCER: Remember that bald eagle we saw that day, in that Saxapahaw stretch?

MIKE: Yep. There's bald eagle in Alamance County. You can go down the river on the weekend, and you won't see anybody.

SPENCER: You see something different every time ... deer ... beaver, you can see where they make their home. A raccoon fell out of a tree one time. I guess he was twenty feet up. We watched him drag himself out of the river. Pretty funny. We must have startled him.

MIKE: A good spot for beginners is the put-in just below 54, but I would not do it alone.

SPENCER: No, I wouldn't go out on the river alone.

MIKE: Never. You never know what's going to happen. One time, it was winter, and we met up with a couple and they fell out. They would have lost the canoe if we hadn't been with them. It was full of water. It's kinda hard to stop a full canoe.

MIKE: The Haw's got some rapids in it. It did have class III's but the lake covered them over. Two still are there, Gabriel's Bend and Lunch Stop. The first time we went through in some big long kayaks and the river was kind of up, we met these kayakers at the turn in this big old boat. We made it through there, but they weren't the boats to take. We took in some water, but I didn't lose it. I've lost it twice on the river. Fell out. Flipped twice.

Spencer: Told you, he's an expert.... Our family goes back. We're Trollingers.

MIKE: It's kind of neat going through that section at Haw River knowing that your ancestors — what was it, 1750s? — were there with a grist mill. It was at Trollinger ford; the name was changed to the town of Haw River.

MIKE: Goat Island used to be in our family. I didn't know that until the other day. Mike Holland wanted us to clean up the river. Before I left, my Dad told me when you're on the island remember our great, great uncle used to take his mules and plow there.

SPENCER: And raise corn because the soil was so rich, and I had no idea. The Trollingers owned land over here around Fountain Place and Davis Street, all that was farm land, very poor farm land, so they sold it. They owned half the darn county.

MIKE: Got the railroad across there. Our family built the bridge. Actually, Trollingers built the hotel and the bridge across the river then Company Shops was built.

SPENCER: We did our part in the Revolution too [after] General Cornwallis came through and confiscated goods.

MIKE: That was something. They tied [Jacob Trollinger] up to a tree and put a bridle in his mouth. It's a wonder they didn't kill him.

SPENCER: I guess it's a good thing they didn't kill him. We wouldn't be sitting here right now talking about history.

MIKE: We found one of those Indian fishing weirs one time. Fishing is pretty good in the Haw, catfish and bream, bass. I've heard stories handed down about how the fish were so thick in there, big huge fish.

SPENCER: You know what would be nice — if the dams were gone. We'd probably get shad, stripers and all, this far up. One day when we have time, we're gonna go clear to the coast.

MIKE: I'd love to have gone Saturday. It was moving quick. One time we were on it like that. We had a lot of fun. We put in at 62 and went all the way to Swepsonville in under three hours. Last Friday it would have taken you 30 minutes.

As Mike and Spencer rush back to work, I ask who else I should talk to. Spencer calls back, "Just talk to the river."[136]

Yee Haw Paddle Day: Glencoe to Swepsonville

After investigating the river's banks and chemicals, I come back to the Haw in sunny April cheer for the first Yee Haw Paddle Day in 2008 — a mass canoeing venture from Glencoe to Swepsonville. It's not quite Coney Island, but Nall Park is paved in canoes; yellow and red kayaks flash in the Haw. The broad gravel steps make this put-in deluxe and a community-minded memorial to the Nalls' son, Dr. Steven Keith Nall.

Brian Baker oversees the day. A former lawyer, he now works for the Haw River Trail and Alamance County to protect the river with buffers and bring "our community together around the resource that has been at the center of social and economic life for centuries: The Haw River."[137] We could be old Glencoe today, on holiday by the river.

Once on the water, however, it is clear we are mainly strangers, trying to figure out whom to follow down a scrapingly low Haw. The talented and the bold emerge; we follow their threads. People meet and re-greet, funneling through thin rapids. Not kingfishers' and herons' cries, but human voices and laughter dominate today.

We pass below the Rt. 62 bridge and by the land-anchored remains of another dam, Carolina Mill's. Water diverted here into the trough of a long mill race which can still be seen by those walking the Haw River Trail to the mill. The race's length was needed on this gently sloping stretch of river to achieve sufficient drop and falling force for full power.

I follow a couple straying between two islands where a wide waterfall's dark waters froth. Down this channel, fallen trees jam into a blockade and soda bottles glint emerald and diamond amid other refuse; it's a river's dustpile. Across the water, kayakers drag down a rocky ledge. We are not moving on tongues of torrent today. After I scrape down, one man shouts over: "Put this down in your notes, a kayaker went through without touching a rock."

Then Carolina Mill itself emerges river left; its arched windows and spare design rising three stories above us, bricks reverberating sun. The concrete structure at river's edge houses a fire pump. The mill was built in 1869 on land purchased from the Trollingers and officially owned by J.H. and W.E. Holt and Company.[138,139] It ran for over five decades.[140]

Behind it a cottaged street of the mill town mounts the hill. Cars descend it rimming the river, en route to Graham-Hopedale Bridge where Copland Fabrics spreads massively over 15 acres on the Haw's left bank.

Stony Creek and its tributaries—Jordan, Owens,

On an early 1900s postcard of Hopedale Cotton Mills (once Carolina Mill), the cottages of the mill village rise in background left. The Haw River Trail now goes by the mill (North Carolina Collection, University of North Carolina at Chapel Hill).

Buttermilk — pour in over Burlington, carrying so much water that the creek has been dammed twice for the first reservoirs: City Lake (1927) and Lake Cammack (1961). This water power led to a succession of mills, and town name changes: High Falls, Big Falls, Stony Creek, Hopedale.

Jesse Gant, father of John Q. Gant of Altamahaw and Glen Raven renown, had his homeplace on 500 acres by Stony Creek and had shares in the saw and grist mill and Trollinger's cotton mill. As his son records, his beloved father, born in 1893, "took an active & prominent part in effecting the division of the county."[141] Indeed, in 1842, a convention took place here to debate the separation of Alamance County from Orange County. Most west of the Haw wanted to avoid fording its many streams and paying taxes for a new Hillsborough courthouse. They prevailed by a referendum in 1848.[142]

During Reconstruction, Jesse Gant was elected county commissioner, only to be denied office. He also lost 100 enslaved workers and saw his son, among other prominent men, on trial for being Night Riders.[143] Alamance County was a Klan stronghold, and its actions would fuel the Kirk-Holden War.

Wyatt Outlaw, the African American man whose lynching started that war, likely grew up on a branch of Stony Creek, called Jordan Creek. Wyatt's murder was enough for Gov. W.W. Holden to declare Alamance in a state of insurrection on March 7, 1870, and fight to bring the Klan under control.[144]

Eighty years later, the long march to equality continued on Stony Creek with Tom Long's struggle for basic rights, as related by his friend Alan Sanders:

> There was a regulation or an unwritten law that African Americans could not fish from a boat up on the lake where Stony Creek is dammed. If you wanted to, you had to get a white friend to go along, and you would have to row until you were well out of sight, then you could sit there and fish from a boat. African Americans were supposed to fish below the dam which was nothing but a pile of rocks.
>
> One day, when Tom was fishing there, he was either bitten by a copperhead or one struck and missed, but he was mad and got out of there. The next time he wanted to go fishing, he went out in a boat. The police hauled him in. They couldn't charge him, but they let him know he wasn't supposed to be doing that. But he kept on going out in a boat and getting yelled at. However, others started going out too, and this is how one of the old rules got rubbed out.[145]

Today the rapids above Stony Creek's convergence are a staircase of rocks. I appeal to a couple near me to check behind; my Kevlar boat was not made for this, and its scrapings promise "crack coming."

Safely below, we chat and identify ourselves as flatwater folks. "I'm Ruth," "Sam, hello." That wonderful moment when you latch on to a river buddy, whether you paddle just that day together or many more, you go down the river talking about your work, your family, the rocky snags of your lives and then lean back so the river and green flush of woods runs through you.

At the natural dam above Goat Island, once part of Sellars Mill, the Yee Haw paddlers again rub together like ants on a mission. I remember this one — extreme river right. Some bump down; others flip and lug boats heavy with water over to a bank full of curious dogs, river dwellers and goats. But we all paddle on by the infamous Goat Island, an inviting place when shining in the sun.

On this day of river luxury, refreshments await us at the new put-in at Town and

Country Nature Park, river right. Tony Laws, director of Burlington Recreation and Parks, is even there to pull us out of our boats. This burly man, all hearty efficiency, gets everyone running toward the same goal — river and trail access. He is enjoying this day and reminisces about the long road to it:

> When I first came here back in the sixties, the Haw River was pretty much a dumping ground so to speak. People all just put it off, "Oh, the Haw River that's a nasty place. You can't fish in there." It did look bad, lots of foam in places ... and a very unpleasant smell, like rotten garbage and acrid like some chemicals were in the water. But we have come a long way. The river that was looked down upon as a cesspool has shifted to a really important asset to our community.
>
> Our hope is to have trail down both sides of the river. Now that's going to be a long time in coming. We have a lot of public property; you'd be surprised how much, but there are also a lot of private owners. So we are putting in pieces, one at a time. It's probably not going to happen in my life time unfortunately, but yes, eventually.[146]

Tony Laws launches Ruth, Sam and me downstream where shadows fall on us from the bridge at Route 70, now renamed Three Governors to celebrate three men's ties to the town of Haw River, the highest state office, and the river — they all lived on its banks. Thomas Holt's (1891–1893) grand house and gardens were at Haw River, river right, and W. Kerr Scott's (1949–53) and Robert Scott's (1969–73) homestead was river left below Rt. 54 in Hawfields.

"CONE FAB" emerges on a smoke stack; "RICS" will fill in when we see its circular and rectangular shapes. In 1899, Granite Cotton Mills was the largest on the Haw. Its 436 looms surpassed 186 at Glencoe, 200 at Virginia Mill in Swepsonville and 300 at Ossipee Mill.[147] Now the dam has collapsed into a rubble slope where Ruth's kayak lodges. "If there's a rock, I will find it." Today, she speaks for many.

The family part of the Yee Haw Paddle starts at Haw River, and the climb through cane stalks, the old take-out, is plugged with standing and squatting paddlers and canoes. Forty signed up and 180 showed up.[148] It's a triumph that has Brian Baker in the water, with other volunteers, sending canoe after canoe past two bridges' pilings while calling into his cell. "We need two more people here."

A train rattles over the Haw, just as Ben Trollinger planned a century and a half ago.[149] Across, Red Slide Park opens a link in the river's walking and paddle trails. Haw River, with its neighborhoods of mill houses, is poised for gentrification. Paul Kron's students at A&T University have drawn plans for a 21st century town center with roof gardens on Granite Mill and terraces and trails down the steep bank. Investors may soon convert Cora and Holt mills, below the bridge, to condos, offices, and stores. The train could become a commuter line, and the river the center of life.[150] It won't be a mill town anymore, but the Haw River Museum, at the bridge, captures that history.

Paddling continues on a smoother Haw that offers several passages through rock outcroppings rather than none. Then the roar of I-85 comes down the river. Frayed plastic bags waiver like ghost birds in overhanging branches ... 1,000 years to degrade ... 12 million barrels of oil a year to produce ... fatal food to ocean life.... When will we ever learn? In Europe, people take their own bags to the store.

In the colorful stream of trucks and cars ahead, some passengers may notice red and yellow dashes in the Haw's leafy green waters.

On the other side of I-85, highway sounds vanish as Alamance Community College comes up river left. We are soon at the 54 put-in, the Haw's first, championed by Grahams's Melody Wiggins and others, where trail mix and water await us. Below, the town of Graham's frothing wastewater rushes in, river right, white over dark rip-rap; by now paddlers are thinking—whatever floats our boats. Back Creek enters river right. The site of fords above it and ferries below and bridges, too.[151] The rusty iron piers from an old bridge rise like sculptures on each bank; Spoon's 1893 map records Galbreath Bridge here, most likely a toll bridge since it bears a name.[152] River left is the Scott Homestead, farmed since the mid–1700s and homeplace of governors. Another Scott, State Senator Ralph Scott, with State Senator Wade Paschal, passed a bill in 1955 to end sewage and industrial dumping in North Carolina rivers. That April, a caravan full of Alamance County supporters travelled Rt. 70 to secure its passage.[153] Ralph Scott knew first hand what the 11 industries and 10 city sewage pipes were doing to the Haw.[154] However, the bill resisted by industry and municipalities was not enforced, records of water quality do not even occurred until 1968, so the Haw had to wait for the federal Clean Water Act of 1972 for the dramatic turn around to begin.[155]

J.F.D. Smyth, who slept with the Bailey family as mentioned earlier, was here gleaning material for his book on America. Unlike William Byrd, John Lawson and others, he was not impressed with the lush Hawfields that stretch between the Eno and the Haw, but in 1784, English audiences would read his description of an impressive Haw, "about twice as broad as the Thames at Putney," and learn it carried its name all the way to Cross Creek (Fayetteville).[156] Here is his view of the Haw Valley from the west:

> The Deep River below and directly before me winding in beautiful serpentine meanders, as if issuing from underneath the mountain where on I stood, and spreading for a long way under the eye, as if drawn upon a carpet; with the dark chasms containing Reedy River and the numerous other branches of the Haw; but the whole, as is constantly the appearance in this country, an immense unbounded forest, so perfectly overspread and totally covered with thick and seeming impenetrable woods, that the thinly scattered human settlements can scarcely be ascertained by the eye, lost and confounded in the magnitude, grandeur, and immensity of the general objects which compose the surrounding, pleasing, awful scene.[157]

We are not thinly scattered today as we paddle on by the Cedar Forest Golf Course to the final rock garden before Swepsonville where some kayaks perched. Sam, Ruth, and I make it through to a jammed take-out. Yet, no one is in a hurry. Brian Baker, in his Yee Haw T-shirt, is still standing, being congratulated. He is able to smile when Sam jokes, "So will we be doing this next Saturday?"

Up in Smoke: Back Creek and Haw Watershed

Rushing by Alamance Community College on I-85, you might have caught sight of Gail Galbraith's or Daniel Sigmon's students monitoring runoff's damage to the Haw. Less visible is harm from the air. Mercury is a main concern, and most of it comes from Duke Energy's and Progress Energy's coal burning power plants and others far upwind.[158] One source of mercury, however, is the plant nestled in the hollow between Back Creek and the Haw. Here, trucks from 20 states bring 21–26 million pounds a year of medical waste to Stericycle for incineration.[159]

Today two plumes of smoke catch the late sun's light as they drift over woods and water past Stericycle. There should be only a trace of vapor, but mercury, dioxin, and other toxins the plumes may contain are the issue.[160]

Mercury, housed in muscle tissue, is a potent neurotoxin that damages the brain, kidneys and lungs and the immune and nervous systems.[161] It's especially harmful to brain development in the very young. When one reads that ⅛ teaspoon of mercury is enough to contaminate a whole 20 acre lake[162] and cause a fish advisory, it's hardly reassuring to know that Stericycle's allowable mercury emissions are 3.82 pounds a day even though their actual emission may be far less.[163]

What's worse is that mercury is transformed by bacteria in rivers into more lethal methyl mercury. The result is a fish advisory, long in place for streams south of I-85, that now covers the whole state.[164] Older fish that have eaten smaller fish contain the most toxin.

Dioxin, notorious for its use in Agent Orange, is stored in fat. It can cause cancer, damage the immune and nervous system and disrupt hormonal signals and sexual development. Because of its toxicity at very low doses, it is one of 12 chemicals to be eliminated or strictly controlled by the UN Treaty on Persistent Organic Pollutants.[165] One variety is produced when polyvinyl chloride (PVC) tubing and medical supplies are burned. Stericycle is permitted to put 14 ounces of hexachlorodibenzo-p-dioxin in the air each year — though that amount is deadly.[166]

Those who live by the smoky stacks first raised concerns. After Sam Kiser moved here in 1993, he found that his voice, his livelihood in big bands and quartets, was troubled by bronchitis and losing range. When the Division of Air Quality (DAQ), two counties away in Winston-Salem, did not respond to his reports of black clouds foaming out, Kiser brought in TV news cameras and held meetings, inviting company officials. Kiser's concerns were for others' health as well. Downstream, the Haw becomes drinking water, and Kiser remembers roaming barefoot with friends in the 1940s and scooping up water from the Haw's branches.

A neighbor, Martha Hamblin, came to lead GASP, the Group Against Stericycle Pollution.[167] When her husband had bladder cancer, his doctor suggested its cause was environmental, as he was not a smoker.[168] They worked with David Mickey of Blue Ridge Environmental Defense League and the national movement for autoclaving of wastes.

In 1999, Stericycle purchased the plant from Browning Ferris Industries (BFI) and added $2 million in scrubbers to the stacks. Yet, six violations followed from 2000 to 2004 with only minimal fines assigned. Then in June 2004, a mercury emission read 12.9 times the federal level. Stericycle pleaded ignorance because waste arrives boxed. This is hardly reassuring as a *Times-News* editorial pointed out.[169] Nor does the fact that they were caught once at this level tell how often similar emissions occurred. This time DAQ assigned the small, though maximum, fine of $10,353 and denied them the right to burn any dental waste, the likely culprit.[170]

Since then, DAQ compliance engineer Ray Stewart says Stericycle's mercury emissions have dropped under permitted limits by a factor of ten. However, because Stericycle has passed recent emissions tests, it is monitored for mercury only annually, at an announced time.[171] Dioxin and other heavy metal emissions are monitored every other year.

Many want more reassurance and better ways to deal with the waste. The best is to eliminate the source — for hospitals to use alternatives to mercury and PVC plastics that release dioxin. G.M, Gerber, Mattel, Brio and Helene Curtis have switched to other plastics. Consumers can help by avoiding the V or 3 symbol[172] denoting PVCs and by using rechargeable batteries to avoid mercury.[173]

There are alternative treatments for medical waste. Stericycle itself has pioneered a radiowave technology, ETD (electrothermal deactivation) which disinfects with low frequency radio waves and uses it in Argentina, Australia, Brazil, Japan, Mexico and South Africa and in Concord, NC. Why not in Haw River?

All attempts to bring the voices of Stericycle into this chapter were refused, but their Web site suggests they know what should happen: "We have championed the use of non-incineration treatment technologies for a long time...."[174]

Ray Stewart of DAQ agrees. "Sure, I would much rather have someone autoclave it than burn it." He is, however, quick to add that if GASP is concerned about air quality, they should go after the open burning of plastics that goes on routinely in backyards across the Haw River Basin. Though he hears GASP's concerns, he adds: "In terms of regulations, we can only enforce to the point of what the permit requires.... Stericycle is operating much better than their predecessor."

When I ask what mercury emissions should be permitted, Stewart reminds me: "DAQ, as a regulatory agency, can only enforce regulations that apply to a facility as they are written. As private citizens of the state, everyone in NC has a civic duty to express themselves to their elected officials and push to have such regulation toughened if necessary."[175]

In April 2005, the Burlington *Times-News* reported that the American Lung Association had dropped the Triad and the Triangle out of the top 25 worst places for air pollution.[176] However, their ranking only dropped to 27th and 30th, and what goes into the air goes into us and the Haw.[177]

Headwater Communities: Stagg and Back Creeks

Back Creek cuts low in its channel as it joins with the Haw just below Rt. 54; its flow is blocked to make Quaker Lake. A trip up its waters reveals county history and threats to health.

A century before Alamance split from Orange County, Carole Troxler tells us in *Shuttle and Plow*, the first courthouse, a jail, and stocks were built on Back Creek in 1752 by Marmaduke Kimborough, but his profits came from the tavern and lodgings he also built. John Saunders, an English merchant, visited in 1753 and found the entertainment to his liking but not the drafty conditions of the log cabin: "Our lodging room was verry Airy and verry light notwithstanding we had never a window." After 1754, Kimborough's profits evaporated when the courthouse moved to a place later called Hillsborough.[178]

Back Creek is one of seven places where the gold fever struck in Alamance County. When gold was discovered on the Pleas Dixon Homeplace (Trollingwood Road), John and Thomas Dixon sold stock in not one but two mining companies. Very little Haw River gold resulted, and this and other Alamance mining companies had shut down by 1884.[179] The hopes weren't entirely foolish; due to its volcanic history, North Carolina was a leader in gold mining until the California rush.

High up in Back Creek's headwaters on a branch, Stagg Creek, are the Occaneechi Tribal Grounds. Below the dividing ridge of the Dan and Haw River basins, the Occaneechi Band of the Saponi Nation settled after their migrations from the West and then European encroachment.

Occaneechi had held a strategic island on Roanoke River and a grip on trade routes until Nathaniel Bacon's attack in 1676. This led some to migrate back to these lands and Stagg Creek in the next century. Forest Hazel, tribal historian, explains what followed: "Due to increasing intolerance of Free Indians after the American Revolution, between the year 1790 and 1820 the tribe migrated to their ancient homeland in Caswell and Orange (later Alamance) Counties where they established their farming community of Texas. As a result of racism, the tribe remained distinct."[180]

On a January day as freezing and grim as the Piedmont gets, Forest Hazel and Gaither Jones greet me with no complaints about the cold. Their spirit is in this Stagg Creek project of re-assembling a tribal village of the 1650s, a log cabin from the 1880s, and a 1940s farmhouse lived in by elders. This village and the June pow-wow re-capture a culture in more vital contact with the natural world; simple authentic structures show another way to live.

In 2007, the log cabin and farmhouse are only posted photographs, but the village is framed out and surrounded by a cedar stockade. Forest Hazel is one of the Occaneechi making this happen: "This is a learning experience," he says with a caretaker's scrutiny of the fence. "We now know that when you weave in wet branches they dry and drop some. We'll want to thicken this in the spring. We think stockades like this kept out deer and gave some protection from raiding parties ... but not if they had guns. Then villagers would have been trapped and shot at, like fish in a bucket."[181]

The enclosed space on this wide field is inviting. In the center is a fire pit; a round hut, two long houses, and two cedar covered work areas rim the circle. A sweat lodge and barbeque pit are yet to come. Forest Hazel eyes the stones around the fire pit skeptically. "With 30 or so people here, there would have been no need for a fire guard, no grasses would have survived, the area would have been dirt and mud." The circle of this village is maybe 40 yards in diameter, similar to the Jenrette site on the Eno; they have worked with archaeologists at UNC–Chapel Hill to base this project on their evidence.

My thoughts move to the Haw and Forest Hazel anticipates: "Very, very similar, the Occaneechi and the Sissapahaw or the Saxapahaw were different villages of the same people. Linguistically and culturally they were the same. Those names came from European explorers or settlers who would ask 'What is this place?' and then ask at the next town and get another name, but they were all culturally linked. There was no difference between groups. Names were much more important to settlers than Native people."

I enter the long house. It's small for the eight or ten people who would have slept here, but comparable, as Forest Hazel points out, to the pioneers' log cabins. More of life was outdoors. Shelters were for sleeping and perhaps eating, with small fire pits inside.

Across the way, tobacco grown for rituals is drying. In the adjoining round house, the temperature rises comfortably; a typical second tightly woven layer would have made it warmer still, Hazel says.

Forest Hazel grew up here in the '60s, an Occaneechi when people didn't celebrate their Native heritage. "You kept quiet about it, but then there were people who would make sure you didn't forget it." It was 1984 when Forest Hazel met with Withmore Jefferies and

Forest Hazel stands in the Occaneechi stockaded village re-constructed on Stagg Creek. In the background is a dwelling which would have provided considerable warmth with two tightly woven layers (photograph by the author).

others to gather their community and envision this project. Formal recognition as the eighth tribe of the state came in 2002, then the purchase of these 25 acres for this village, cultural center, orchard for heirloom apples — all for their heritage, "for the children," and for economic development.

We look out over the old tobacco fields and some magnificent oaks; a heron descends to the pond on heavy wings. Stagg Creek cuts a ravine between two fields so they rise as hills. Forest is convinced that if you bulldozed either hill, you would find Native traces. They might have fetched water in clay pots, bark buckets daubed with pine pitch, or "animal stomachs hold water well." Deer that left prints in the dirt near us would have traveled along the creek bottom then and turkey too, but Forest doubts there were buffalo. "There's no archaeological remains, none in Hillsborough, there would have been some sign. Yes, they were in the Cherokee area, and maybe a few small herds wandered down this way." I ask about the naming of Buffalo Creek. "Probably there were buffalo there, but it was so remarkable."

"Ahkontshuck" or "amanishuke" are the words that the Yesah, the people, used to describe their Piedmont, "the high or hill land."[182] Even on this day of steely winds, it is good to be here, and certainly it is at the June pow-wow when on the hill across, Occaneechi in full dress circle as they beat the rhythm of the drums into the earth.

Septic Racism

"Work Set to Begin on Sewer Project" may be a dull headline, but what took place in the headwaters of Back Creek, west of Mebane, is a dramatic version of a repeated story with river and human health implications. Here it involves the West End Revitalization Association (WERA) and Omega Wilson, its president, and the documentation of e-coli and fecal coliform in Back Creek and public waters. Their fight for safe water is one that needs to be waged on other Haw tributaries, and they were successful here against what the Department of Justice calls "patterns of historic racial discrimination."

I met with Omega Wilson in WERA's offices in a historic African American Community. The cornerstone of the Mebane First Presbyterian Church across the street reads 1868; adjacent graves hold former slaves. This is West End set up outside of Mebane city limits during Reconstruction to keep freedmen apart, yet within working distance of white homes and factories.

When Omega Wilson graduated in 1968 from a still segregated Central High School in Graham, he did not expect to be back one day leading a "safe sewer lines" fight that would gain him a reputation in national public health circles. It began with the struggle over a road. The NC Department of Transportation's plans to drive the 119 Bypass right through two historic African American communities, West End and White Level, came without warning to the residents despite over 16 years of official planning. The news hit hard and provoked action, as Omega Wilson describes:

> The water issue became a part of the strategy because when we filed a complaint to the U.S. Department of Justice in 1999 ... we were focused on the physical impact of the highway. But the Department of Justice said, "What is the most urgent thing?" and I said, "Stopping the highway," and they said, "No ... what is the most urgent thing based on public health statutes? Clean water and clean air, safe drinking water, avoidance of toxic substance, and control and management of solid waste disposal." That's not all of them. Those are five.[183]
>
> And I said, 'OK, I got you." ... So you have to look at it based on how the government looks at it ... and when we did, we determined that we are only four blocks from downtown, but without sewer. Meanwhile, the city was running miles of pipe out to Mill Creek [a new high-income development], right next to White Level [a low-income African American community], but not including it.... The Department of Justice described that as a "Denial of basic amenities under a pattern of historic racial discrimination." West End residents were closest to the Mebane Sewage Treatment Plant, built in the 1920s, and had never had access to it. We filed a complaint under the Civil Rights Act, Title VI.

Omega Wilson is quick to emphasize Title VI is a tool for everybody and goes on to explain how a favorable ruling on their appeal put a moratorium on the 119 Bypass until progress was made connecting those long redlined out of the sewer system. This moratorium was extended several times. Hooking to a sewer system is vital for keeping septic waste out of yards, drinking water, and the Haw's streams; older homes all over the watershed may not be on land that perks or has room for a septic system. Wilson explains:

> You have regulations in Alamance County now that you have to have a lot of about an acre to have a septic tank and the land must perk. But that was not always the case.
> The Haw River Assembly became one of our partners. They [HRA] were looking at the bigger river, and so what we were finding is that the pollution in these failing septic tanks was not just staying in people's yards. The rain would wash it into ditches that were perennially wet from [leaking] sewers and, of course, where do these streams go? They wind up in tributaries coming out of Mebane and into the Haw River. And so you have what the newspapers refer to as nutrient-rich waters: that's [rich with] animal waste or human waste.... And that became a health crisis.

"We don't have a health crisis." Of course, this is the city and county responding, so with the help of the EPA and UNC–Chapel Hill's labs, we tested this drinking water and well water and stream water.... And we found human waste in the streams over 300 times what the EPA allows under the Clean Water Act. Someone said "Well, it's farms; it's cows." Well, cows have been gone for many, many years....

Then we actually found residents who had kidney failure or kidney disease as a result of fecal coliform, based on medical evaluations at UNC–Chapel Hill's hospital.

So the information was submitted to EPA, and EPA says "OK, you got a public health violation under the Clean Water Act."

What we [also] found out was that human waste, based on our tests, was in the municipal water. And that was not revealed intentionally; we just wanted to show the residents ... how nice and clean your water is going to be compared to the mess we find in your well water. And it blew us back because we thought we must have got the samples mixed up. We did them four times and UNC–Chapel Hill's School of Public Health lab said, "no, no, no."

"But this water, God, it looks so pristine." We finally had to tell [the lab], "No, number b74-2 is city water."

And what we found out was that the city had this problem for several years and had not reported it, so we drove all this stuff out of the closet; we found out it was not only a problem for this community ... it was also a problem for the larger community.... There is something about that that is so elementary, but we seem to have difficulty getting beyond it.

And people were so placid and pleasant about this, even white residents did not realize. "We are supposed to get a notice [on contamination] in the mail?" And there were residents in this community just watching [West End] community like watching TV and didn't realize that the process of denying information to this community was also impacting them.... When the rare notices came out, they would be written so you couldn't understand what it was — in some 19 letter words.

The breakthrough point was when EPA actually started coming here in 2004 ... and basically they said [to Mebane officials] that if you don't start addressing these local issues ... tapping local sewers on, then you don't have to worry about the local community suing you, we will sue you [the City of Mebane] in federal court for violation of constitutional law. And they said that face to face to them and that created a lot of tight throats and very visceral attitudes, but things started to change.... And the 119 bypass got pushed farther and farther back.

As houses were installed on sewer lines and others lined up to be installed, county officials pressed for the go-ahead on the bypass. Wilson continues, "It's kind of like, 'Mom, I didn't clean my room, but I took out the garbage and washed the dishes, and I haven't combed the cat, but can I go play now?' So it is kind of like that now, but they are saying 'Omega, are you all right? Is it cool?' And, of course, we are saying, 'No, no, no, not yet.' But they persist, and we persist."

And the Haw persists, with septic systems leaking into its many streams headed to Jordan Lake and on to coastal wetlands.

The 119 Bypass has not yet been built. Plans in 2010 move it slightly to the west still near West End Community and Back Creek waters. However, through projects, grants and research led by WERA, 100 of 500 homes have been able to tap on to Mebane's sewer service. This rare success earned WERA one of the first U.S. Environmental Protection Agency's National Environmental Justice Achievement Awards in 2008. Omega Wilson now serves on the EPA's Environmental Justice Advisory Council.

Culture Clash: Swepsonville and Mitchum Site

> But their name is on your waters
> Ye may not wash it out.
> — Lydia H. Sigourney[184]

Looking at the Haw's calm waters by Swepsonville, it is hard to imagine that period so innocently called the Contact Era when Native peoples met Europeans. Trading paths conveyed deerskins exchanged for Europeans' beads, iron tools and weapons, and trade became a way of life for some. Like "free" trade later that silenced mills, with advantages came problems — smallpox, addiction to alcohol, and war. Iroquois raids added misery.

Contact likely came to the Haw in distant vibrations building to major impacts. There are no reliable records of face-to-face meetings though early explorers likely crossed near Swepsonville, but from explorers and archaeologists we can glean something of the people who lived by the Haw 400 years ago in the Contact Era.

These people of the Haw are known by names others gave them — Sissipahaw, most commonly. James Mooney suggests they were a strong group because they gave their name to the river and had control of the Haw Fields.[185]

Trawick Ward and Stephen (Steve) Davis, authors of more recent research, *Time Before History: The Archaeology of North Carolina*, see the Sissipahaw as a backwoods people, for little evidence of trade is here. The Mitchum site, below Chicken Bridge, is the only one where a few trinkets have been found — not much to compare with 325,716 trade artifacts at Upper Saratown on the Dan River or 12,911 at the Fredricks site on the Eno.[186] They also remark how little notice Lawson took in 1701 of the "Sissipahau," only mentioning them as being down river, while he eagerly visited other Indian groups. This may mean the Sissipahaw were no longer a "viable social entity."[187]

Much earlier, Juan Pardo on Juan de la Vandera's 1567–68 expedition mentions the Sauxpa, and this is sometimes cited as the first contact, but DePratter's 1983 study puts Pardo no closer than the Yadkin.[188]

The first two European explorers are thought to have forded the Haw near Swepsonville following the Trading Path's braided route from Hillsborough to Salisbury (just south of the path of I-85 today). First came John Lederer from Germany in 1670 and then John Lawson from England in 1701; both were 26.

Their routes are uncertain because the two men were mapping a new world as they went. When Lederer traveled, some still hoped a short route to the Pacific lay just beyond the Appalachians. John Farrer's 1650 map showed California "in ten dayes march with 50. foote and 30. horsmen from the head of the Jeames River [Virginia]."[189]

Even Lawson, viewed as a more reliable guide, had little idea that the Native peoples he visited were in new territories, having moved to avoid raids or filling in lands of a weakened group.[190]

John Lederer came on assignment from Governor Berkeley to explore into the Piedmont and is often given as the first European to cross the Haw (around Swepsonville) and name the Shakori, whom he said lived on a rich soil, perhaps Hawfields. However, there is debate about where he was or if he was here at all; Steve Davis at the UNC–Chapel Hill Research Laboratories of Archaeology suggests that Lederer was on the Eno, not on the

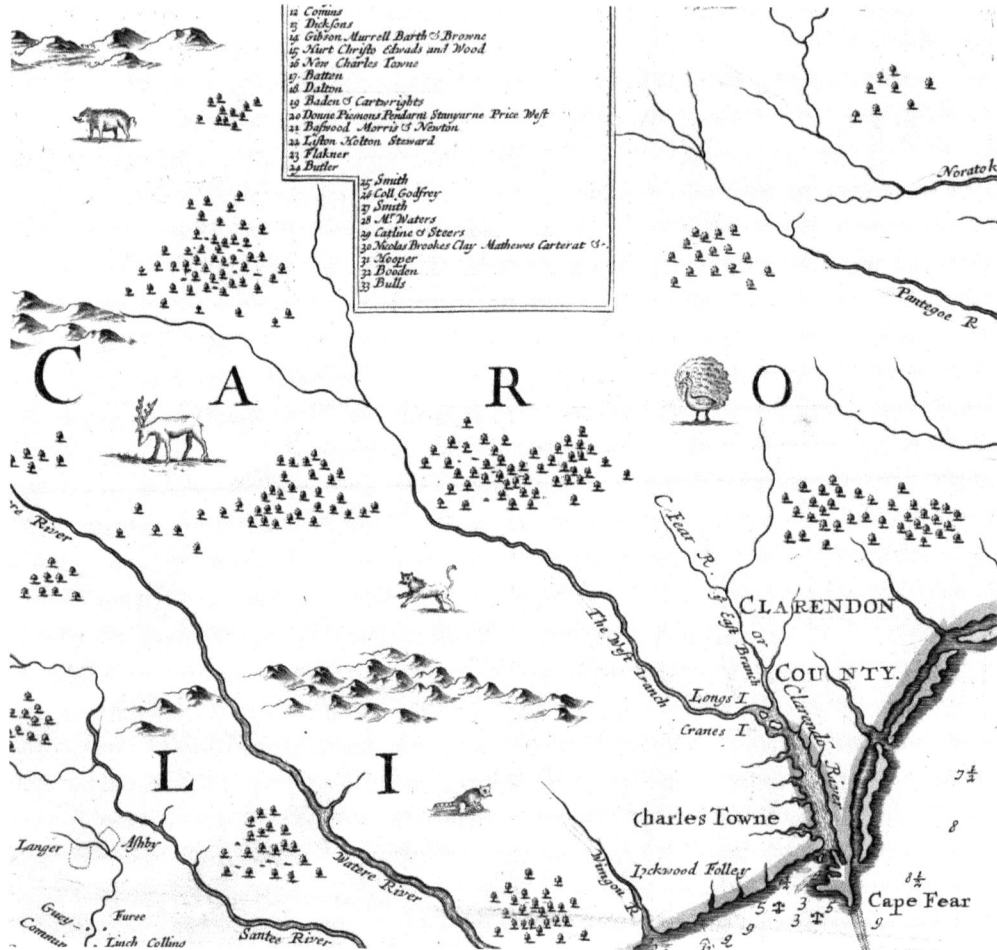

A section of a "New Map of Carolina" c. 1685 by John Thornton reveals how much was yet to be discovered by Europeans. The names Cape Fear *and* Clarendon *vie for the river (courtesy Outer Banks History Center, Manteo, North Carolina).*

Haw, when he spoke of the Shakori and that "Shakor might be the Jenrette site at Hillsborough, beside the Fredricks site." He does note the Shakori were "culturally related" to the Sissipahaw.[191]

Since early impressions of the people of the Haw are scarce, let us put doubts aside and hear Lederer's few, unkind words, on the Oenuchs,[192] whom he says the Shakori were like:

> They are of mean stature and courage, covetous and thievish, industrious to earn a peny; and therefore hire themselves to their neighbours, who employ them as Carryers or Porters. They plant abundance of Grain, reap three Crops in a Summer, and out of their Granary supply all the adjacent parts.... Some houses they have of Reed and Bark; they build them generally round: to each house belongs a little hovel made like an oven, where they lay up their Corn and Mast, and keep it dry. They parch their Nuts and Acorns over the fire, to take away their rank Oyliness; which afterward pressed, yield a milky liquor, and the Acorns an Amber-colour'd Oyl. In these, mingled together they dip their Cakes at great Entertainments,

and so serve them up to their guests as an extraordinary dainty. Their Government is Democratick; and the Sentences of their old men are received as Laws, or rather Oracles, by them.

Fourteen miles West-Southwest of the Oenocks, dwell the Shackory-Indians, upon a rich Soyl, and yet abounding in Antimony, of which they shewed me considerable quantities. Finding them agree with the Oenocks in Customs and Manners, I made no stay here.[193]

We have, regrettably, no record of the Native peoples' impression of Lederer, whose words suggest the culture clash. For Native people, the earth is alive, and plants, animals, even rocks, are spirits; for the Europeans the new land was to be colonized, its raw goods and people turned to trade and wealth. Sharing and giving were enmeshed in communal Native ways; explorers and traders were pursuers of wealth in a complex economy built on individual saving and acquiring.[194] John Lawson would try to bridge this gap.

John Lawson's journey took him from Charles Town, S.C., through Swepsonville, to what became Lawson's Creek near New Bern; from these explorations, he wrote "the only book to come out of proprietary North Carolina."[195] Lawson's entrepreneur's heart thrilled to what he saw of the Haw and its lands, though getting across the river near Swepsonville was rough:

At last, determining to rest on the other side of a Hill, which we saw before us; when we were on the Top thereof, there appear'd to us such another delicious, rapid Stream, [the Haw] as that of Sapona [Yadkin], having large Stones, about the bigness of an ordinary House, lying up and down the River. As the Wind blew very cold at N.W. and we were very weary, and hungry, the Swiftness of the Current gave us some cause to fear; but, at last, we concluded to venture over that Night. Accordingly, we stripp'd, and with great Difficulty, (by God's Assistance) got safe to the North-side of the famous *Hau*-River, by some called the *Reatkin*; the *Indians* differing in the Names of Places, according to their several Nations. It is call'd *Hau*-river from the *Sissipahau Indians*, who dwell upon this Stream, which is one of the main Branches of *Cape-Fair*, there being rich Land enough to contain some Thousands of Families; for which Reason, I hope, in a short time, it will be planted. This River is much such another as *Sapona*; both seeming to run a vast way up the Country. Here is plenty of good Timber, and especially, of a Scaly-barked Oak; And as there is stone enough in both Rivers, and the Land is extraordinary Rich, no Man that will be content within the Bounds of Reason, can have any grounds to dislike it. And they that are otherwise, are the best Neighbours, when farthest of.[196]

Lawson continues his praise for the land on his way to the Eno, passing "great Gangs of Turkeys" and traders, with 30 loaded horses, who agreed that they had never seen

20 Miles of such extraordinary rich Land, lying all together, like that betwixt *Hau*-River and *Achonechy* Town [on the Eno] ... [Natives'] Cabins were hung with a good sort of Tapestry, as fat Bear, and barbakued or dried Venison; no *Indians* having greater Plenty of Provisions than these. The Savages do, indeed, still possess the Flower of *Carolina*, the *English* enjoying only the Fag-end [the coast] of that fine Country.[197]

Though he came to know the coastal people best by the time he wrote his book, Lawson's observations likely reveal much of the world along the Haw.[198] He writes of a world where wolves were kept as pets, bear fat made hair beautiful, and passenger pigeons, now extinct, flew in flocks that could "obstruct the light of day."[199]

Lawson tells how central rivers were: "When these Savages live near Water, they frequent the Rivers in Summer-time very much, where both Men and Women very often in a day go in naked to wash themselves, though not both Sexes together...."[200] As soon as the child is born, they wash it in cold Water at the next Stream."[201]

And rivers were food:

> Their Herrings in *March* and *April* run a great ways up the Rivers and fresh Streams to spawn.... Their taking of Craw-fish is so pleasant that I cannot pass it by without mention; When they have a mind to get these Shell-fish, they take a Piece of Venison, and half-barbakue or roast it; then they cut it into thin Slices, which Slices they stick through with Reeds about six Inches asunder, betwixt Piece and Piece; then the Reeds are made sharp at one end; and so they stick a great many of them down in the bottom of the Water (thus baited) in the small Brooks and Runs, which the Crawfish frequent. Thus the *Indians* sit by, and tend those baited Sticks, every now and then taking them up, to see how many are at the Bait.... By this Method, they will, in a little time, catch several Bushels, which are as good, as any I ever eat.[202]

However, neither abundant, healthy waters nor the Native cures that impressed him stopped the horror of European smallpox from "sweeping away whole Towns."[203] Rivers, like the Haw, became death traps: "Their running into the Water, in the Extremity of this Disease, strikes it in, and kills all that use it."[204]

Lawson, often a sympathetic observer, is amazed by silence in meetings and respect for elders as much as their unpossessive sex lives. "Never love-mad" and "Married Women unconstant" are two of his section titles as he sees beyond easy labels of promiscuity or sinfulness.[205] He sat at their celebrations, awestruck as they danced "for several Nights together, with the greatest Briskness imaginable," to the cadence of drums and rattles and human voices.[206]

Lawson observed with a foot in two conflicting worlds. He never stopped looking with a capitalist's eye at all around him, yet he did take in the new culture, "so contrary to ours, that neither do we or they fathom one anothers Designs and Methods"[207]:

> They never walk backward and forward as we do, nor contemplate on the Affairs of Loss and Gain; the things which daily perplex us. They are dexterous and steady both as to their Hands and Feet, to Admiration. They will walk over deep Brooks and Creeks, on the smallest Poles, and that without Fear or Concern ... and as for letting any thing fall out of their Hands, I never yet knew one Example. They are no inventor of any Arts or Trades worthy mention; the reason of which I take to be that they are not possess'd with that Care and Thoughtfulness, how to provide for the Necessities of Life, as the *Europeans* are.[208]
>
> They say, the *Europeans* are always rangling and uneasy, and wonder they do not go out of this World, since they are so uneasy and discontented in it. All their Misfortunes and Losses end in Laughter ... unless some of their Kinfolk and Friends have lost their Lives.[209]

This attitude is partly explained by Lawson's description of a mutual aid community. Lost canoes or burnt houses would be rebuilt together. "They say it is our Duty thus to do."[210]

Lawson pleads for the English to be better than the Spanish who "set up a Christian Banner in a Field of Blood."[211] Native peoples "are really better than us, than we are to them; they always give us Victuals at their Quarters ... but [we] let them walk by our Doors Hungry.... For all our Religion and Education, we possess more Moral Deformities, and Evils than these Savages do, or are acquainted withal."[212]

Yet, his book is a guide for those seeking to manage Native peoples and exploit the land. He was, after all, employed by the Lords Proprietor to survey *their* lands and so signed his book as their "most obliged, most humble and most devoted servant,"[213] and as surveyor general, he worked to develop Bath and New Bern on Native lands. He was killed for this encroachment. Tuscaroras are thought to have jammed white oak or pine splinters into his skin and set them on fire.²¹⁴ It was a death he had described in a section titled "Indians Cruelty to Prisoners of War."²¹⁵

Meanwhile, the people of the Haw were enduring their own tortures. Stephen Davis suggests that "within 20 years of Lawson's journey, Piedmont North Carolina lay largely abandoned, its native inhabitants having become victims of disease, rum, warfare, and slaving raids."²¹⁶ Warfare was with Europeans and with Iroquois who "traveled south each winter to make war on the Piedmont Indians."²¹⁷

We are left to imagine how the people of the Haw fought to endure. Bewildered scattering and regroupings followed, the wretching displacements of a people bonded with land, rivers and each other. Where to turn? To any kin nearby, failing that, to any Native peoples.²¹⁸ The Sissipahaw fled, Stephen Davis concludes, mainly "to the towns of the Catawba nation in upper South Carolina."²¹⁹

Few Native peoples remained in the Haw Valley to contest when the first settlers lurched down the Great Wagon Road and travelled on to the Haw in the mid–1700s, but that doesn't mean there were no practicing cultures. Some held on, and do today, blended into the Occaneechi of the Saponi Nation.²²⁰ Ward and Davis point out that North Carolina has the most Native Americans of any state east of the Mississippi.²²¹

Signs of Native peoples during the Contact Era were found at sites where the Haw now enters Jordan Lake and at the Mitchum site in northern Chatham County, "the only Contact Period village found in the Haw River drainage"²²² thus far.

Trawick Ward and Steve Davis think that the Mitchum site, not Saxapahaw, was probably where the Sissipahaw whom Lawson refers to were located.²²³ When Steve Davis and others from UNC–Chapel Hill dug at Mitchum in 1983 and 1986, they found peach pits, showing a Spanish influence, a few gunflints²²⁴ and, in two graves within the stockaded village, 1,351 glass beads, brass bells, and some copper and brass beads.²²⁵ Without the iron knives or hatchets found elsewhere, these trade items are viewed as a harmless collection, not necessarily signifying direct dealing with Europeans.²²⁶

Steve Davis conducted a study of 33,033 pottery sherds, from the Wall and Fredricks sites on the Eno and from Mitchum. The potsherds at Mitchum point to two periods of occupation: net-impressed fragments of the pre–Contact, late Woodland era and simple, stamped and brushed sherds of the late 1600s Contact era were found.²²⁷

Post holes at Mitchum marked a small stockaded village and one large 18 by 24 foot oval building.²²⁸ Pits and depressions hold further clues. Several deeper pits for food storage or hiding valuables suggests a small community under strife and on the move — not a permanent long-term settlement, but one where a group would stash valuables to return to later. Cob-filled pits at Mitchum may have been used, as Steve Davis suggests, "to smudge clay pots to water proof them," or they could suggest an increased interest in scraping deer hides for the European trade.²²⁹

Steve Davis has been on every excavation on the Haw since he explored the Mitchum site in 1983. As associate director of the research laboratories of archaeology at UNC–Chapel Hill, he now directs the excavations that first enthralled him on a fifth grade field trip to Town Creek. In an interview, he details this work.

> We've worked on sites in southern Virginia and South Carolina, looking at the descendants of the people who lived in North Carolina in the 1700s, like the Sissipahaw, the Saura, and

the Shakori. Remnants of these groups wound up joining the Catawba. [We're] currently examining 18th and early 19th century Catawba sites in South Carolina.

Who was on the Haw in the Contact Era? I suspect that at that period, it's the Sissipahaw. Collectively all we know is that, say between 1650 and the early 1700s, the tribes in this area are the Sissipahaw, Shakori, Eno, Occaneechi and the Adhusheer. And the Occaneechi weren't here in 1650, so strike them.

The earliest record we have of anyone coming into this area and naming names is John Lederer. He doesn't mention anybody on the Haw where he crosses. Now the archaeological evidence we have for the Contact Era is nowhere near Saxapahaw. Those sites are near Chicken Bridge, where the Mitchum site is, and further downstream near Jordan Dam.

How were the Sissipahaw impacted during the Contact Era? My suspicion is that, sometime between 1650 and the early 1700s, smallpox epidemics hit here as Virginia traders began to traverse the Carolina Piedmont on a regular basis. Prior to that there really isn't a whole lot of trading contact. In fact, in 1650, the English were effectively prohibited by the surrounding Native population from trading in the interior. An account from the 1660s suggests that much of the trade centered upon the Occaneechi village located on the Roanoke near present-day Clarksville. That situation probably held until Bacon's Rebellion in 1676. After the rebellion, the Piedmont was opened up for direct face-to-face trade, and one consequence was that you have sick or potentially sick traders coming into the villages. And that led to Lawson's comment in 1701 that there is not "the sixth savage living within two hundred Miles of our Settlements as there were fifty Years ago." So it was just a devastating period for the Native population.

The trade artifacts from the Mitchum site suggest that it was occupied sometime between the 1660s and the 1680s. What we didn't observe in our excavation on the Haw are large numbers of burials. We did, however, discover several looted graves. After epidemics hit, village populations would have declined dramatically.

The [Sissipahaws'] movement was probably like their neighbors on the Piedmont. During and after the Indian Wars of the 1710s, there was such turmoil. Not so much with the Tuscarora War, but in the later wars, those who fought against the British were severely punished, and they were increasingly exposed to warfare from the Iroquois.

Were the Iroquois down on the Haw? The Iroquois raided throughout the Piedmont. When Lawson crossed the Haw River, he encountered a trader named Massey who warned him to watch out for the Iroquois and to avoid going to Virginia. And this raiding continued throughout the 18th century with the Catawbas being their primary target. It's kind of ironic because many native peoples in the Piedmont sought protection from the Iroquois by joining the Catawbas, only to be harassed there.

Did the Sissipahaw go to the Catawbas? What exactly happened between 1710 and 1720 is uncertain, but before 1710 tribes or remnants of tribes were scattered throughout the Piedmont, and by 1720 the Piedmont was largely vacant. The Catawba deerskin map of 1721 provides a socio-political snapshot of Catawba society at that time; the Sissipahaw are depicted as a small group attached to the Succas. James Adair was a South Carolina trader who says he heard twenty languages there [at the Catawba settlement] between 1736 and 1743, and he lists the Sissipahaw among them.

It's a difficult question when they become Catawbas; at this period, the Indians themselves probably recognized different tribes, but to the English, they were increasingly regarded as Catawba nation. Probably the greatest social leveling event for the Catawbas occurred in late 1759 when a smallpox epidemic struck and killed about half the population.

It's hard to trace the Sissipahaws' path at this time, as their archaeological record is pretty scanty. Much of the archaeological evidence informs us about their material culture, in terms of the way they constructed their houses, the stone and bone tools they used, and the kind of pottery they made.

What can it tell us about the people of the Haw? Archaeologically, the collection of artifacts from the Mitchum site looks very similar to the Jenrette site, which we think is a Shakori village and which would indicate that those two groups were very closely related. They may have developed out of a common culture, and linguistically they might not be that different. Their languages, probably Siouan, might have been very close.

Small villages [on the Haw] were probably just groups of extended families. They didn't live in large nucleated towns as did some other groups at this time period. These are the hillbillies. [Davis laughs.] Yes, it was rural.

What sites on the Haw have you worked on? I worked on all of those that were investigated by UNC's Siouan Project during the 1980s. I first went to the Mitchum site in the spring of 1983; we were told about it by a couple of brothers from Chatham County named Royce and Jimmy Reeves who were artifact collectors. When we got out there, it looked like a mine field, so my first experience out there was to map and document the damage. Later that summer, we discovered the Fredricks site and started working at the Wall Site [both on the Eno near Hillsborough]....

I took a small group of students and went over to the Mitchum site, and we were very fortunate. We selected an area and dug a small excavation, and one of the first squares we dug contained evidence of an oval house [an oval pattern of soil stains representing a wall constructed of vertical posts]. That was the first oval house we had seen in the Piedmont. All the other sites we had worked on contained circular houses, and there was a lot of debate over whether it was actually a house pattern. We later found similar houses at both the Fredricks and Jenrette sites, so Mitchum was no longer that unusual.

Is there research on the Haw now? Nothing in the last several years.

If you had a grant? I'd probably go back to the Mitchum site. We mostly worked in an area that was heavily disturbed, but toward the end of our second field season, we got into a less disturbed area. It would be nice to learn more about the community pattern; that is, how the town was laid out.

Do you find yourself drawn to that time as Lawson was? Yes, I'm fascinated by that time when Indians and Europeans first met. But there are a lot of things that Lawson found fascinating, such as religious beliefs, conjurers' practices, and Indian sexual habits, that don't lend themselves well to archaeological study. Lawson wanted to go out on a grand adventure; it was a difficult time, but he had a ball. I just like traveling in time. In essence, that's what archaeologists do. We're time-travelers of a sort. Our perception of the past is very different from Lawson's, very much based on what has survived. I get excited about seeing how artifacts are distributed across a site and looking at patterns of refuse disposal, looking at the little things that give you insights into the lives of people who once lived there. Often the exciting things to archaeologists are the nuts and bolts of daily life, identifying patterns, putting the puzzle together.[230]

A Visionary and a Scoundrel: Swepsonville

At Swepsonville, the Haw's confluence with Alamance Creek drew Native encampments long before the Contact Era; then new settlers came, and the first mill was built by John Armstrong c. 1764.[231] The Hermitage followed, a plantation and social center of the early 1800s overlooking the river from Alamance Creek's first bend. Archibald Murphey, one of the Haw's most notable residents, made it famous. In 1800, before Murphey married Jane Scott, he purchased the Hermitage for 300 pounds from her father, who had built it. This son of Caswell County, born likely in 1777, had done so brilliantly at the new university at Chapel Hill, he became professor of languages a few years after his graduation.[232]

When he moved here with his wife in 1801, he began twelve years of building and developing the Hermitage that added 2000 acres to the original 95. He had the mill put back in operation and added a sawmill; both ran on the same waterwheel, river right, and were managed by an enslaved African American known as Jerry the miller. A distillery was

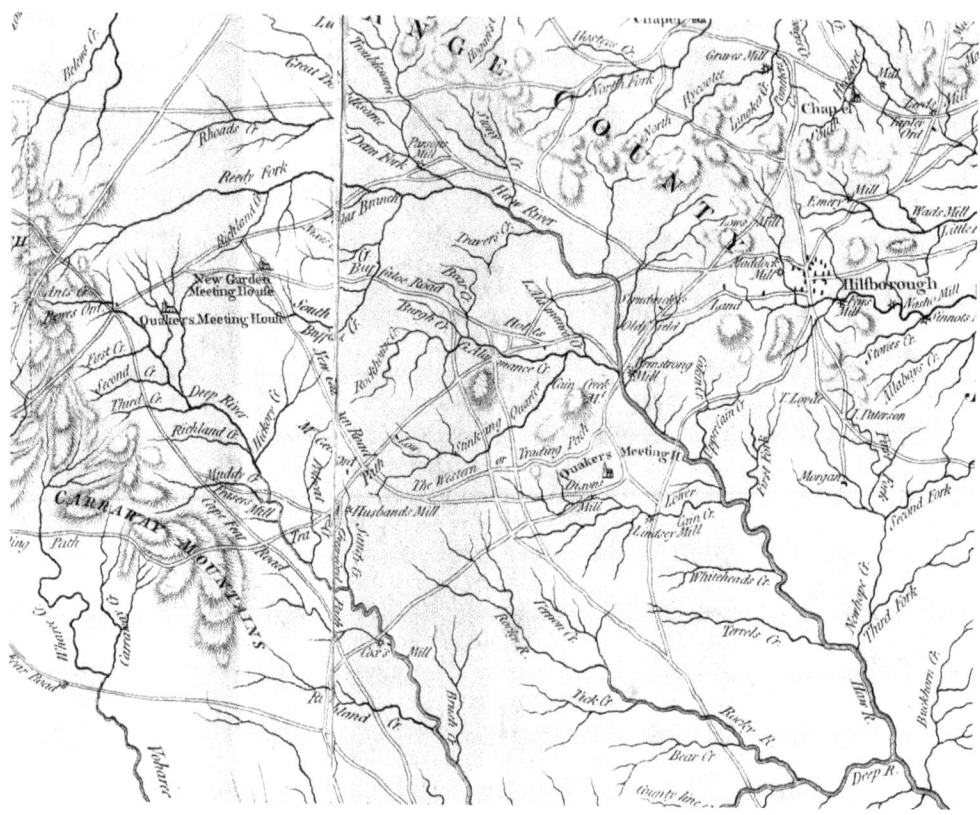

"An Accurate Map of North and South Carolina" (1775) by Henry Mouzon reveals the Haw Valley at the time of the American Revolution with key mills and trading path routes noted. Armstrong Mill was at present day Swepsonville. Above it, Strudwick's Old Fields marks the northern section of Hawfields (courtesy North Carolina Office of Archives and History, Raleigh).

next, which sent whiskey as far as Petersburg, Virginia. (The still's remains may yet be on Alamance Creek.)[233] In 1811, a store was built which Jesse Turner remembered fondly: "When a boy, I often went to Murphey's Mill mounted on a horse with corn or wheat to get ground, and while my grist was being ground I sauntered up to the store of Scott and Murphey where the genial Thomas Scott was always to be found. He nearly always gave me a newspaper to take home and read."[234]

The store's records show considerable building materials charged to Murphey's account. Indeed, he added two wings to the original building, all while studying law, then working as a lawyer, traveling the circuit. Over 100 enslaved people did the work,[235] directed by his wife, whom he instructed: "The Waggons must be kept employed. Otherwise I shall not only loose the expense of maintaining the Teams, but shall not be able to get my flour etc, sent in good time."[236] For some years, the Hermitage was a thriving plantation and highly integrated project; offal from the mills fed the hogs who also fattened on acorns in the Haw's bottom woods.[237]

A man of vision and yet often ungrounded, Murphey came to the Hermitage in 1801 when Thomas Jefferson became president and had similar beliefs in progress[238] and willingness to go into debt.

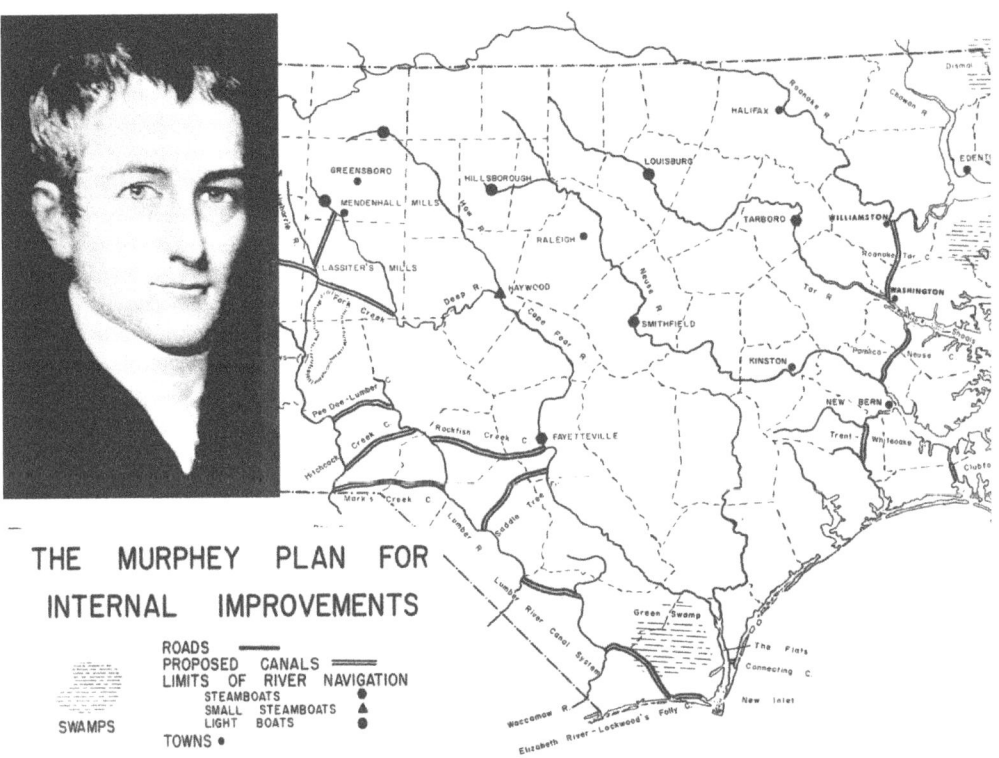

Archibald Murphey's portrait (insert) looks over a map illustrating his "Plan for Internal Improvements" connecting rivers with canals. Note that the Haw would be passable up to High Rock, where Murphey owned a resort. It is first reliably channeled there (courtesy North Carolina Office of Archives and History, Raleigh).

The Haw had its own powerful effect on those visions. In the summer of 1819, seven years before his plan for a bridge over the Haw at Swepsonville was carried out, he wrote "Memoir on the Internal Improvement" for the legislature, laying out how the state's eastern rivers could be arteries of transportation joined by canals, one of which would join the Deep and the Yadkin. The Haw would be opened to High Rock, where Murphey owned the nearby healing springs and health resort of Lenox Castle — jointly with banks and creditors. Murphey also speculated in land at Haywood, farther down on the Haw, where he had great hopes for river transport and a development boom. Let me not create too harsh an impression, Murphey was an entrepreneur in the most public-spirited sense of the word, a "true-believer" capitalist, whose scheme for transportation would help all to prosper, including himself. Biographers credit him with a vision to bring North Carolina out of its "Rip Van Winkle" days and recharge commerce. His ideas were needed; in 1820, North Carolina had the lowest per capita income of all the states,[239] and the opening of the Erie Canal in 1825 showed that amazing feats could be done.

Murphey served as a state senator from 1812 to 1818, twice ran for U.S. Senate, wrote an important paper urging public education (with public funding for the poor), and fundraised for a written history of North Carolina. While none of these plans were realized in his life, he sparked later change and brought people together for the public good. The

Hermitage on the Haw was a meeting place for law students, lawyers, politicians, Chapel Hill faculty, friends and family.[240]

Yet, like Jefferson, he treated inconsistently the enslaved people who created his plantation. On the one hand, he was capable of writing to Thomas Ruffin protesting his new overseer's harshness[241] or of drawing up a plan for freed slaves to have land in California. On the other hand, he gave Ruffin permission to beat Bridget, his beloved helper whom he begged for in his last days:

> Your flogging Bridget has given me no offence I assure you. I had ordered her to go from Wm Foust's to the Haw-fields Meeting House and then on Home and not to call at your quarters. I gave her a pass.... I read it to her, and told her to go on directly to the River — I knew you disapproved of her being at your quarters.... I am sorry that you should have thought it possible that I should take offence at your chastising her.[242]

Though entrapped in his culture's blindness, Murphey also had a spirit that was admirable; he once assured Ruffin: "You cannot expect to live in a World of Care and trouble, and be free from Anxiety and Inquietude ... every such Affliction has a moral tendency ... to teach us the Virtues upon which our principal Happiness is bottomed, Fortitude in Adversity, and Resignation in Affliction."[243]

Archibald Murphey had the spirit of the river — he kept rolling on through hard times that were never far away. He and Jane lost three children; she was a semi-invalid; he suffered from arthritis, and there were the debts, the Panic of 1819, more undertakings, and more debts. In 1821, he was forced to sell the Hermitage for $24,000 to his former law student and friend, Thomas Ruffin, who allowed him to live there until 1829. By then, his plans to find gold on Alamance Creek, using a new, expensive extraction method, had yielded nothing.[244] His creditors finally closed in, the year his wife Jane died.

Imprisoned for debt, Murphey bore his experience with dignity (it may have helped repeal the debtor's law). The sheriff left his cell door open for more light for reading and ventilation; friends visited. He emerged homeless but not friendless and found lodgings in Hillsborough, where he died in February 1832. Due to the state of winter roads, his body could not be buried beside Jane's in the tulip and jonquil bed at the Hermitage.[245]

After his death, the mill and the ford at Swepsonville came to Thomas Ruffin (1787–1870) though for a while it remained "the Murphey Place where Judge Ruffin lived."[246] Thomas Ruffin, from Tidewater, Virginia, was as prominent as Archibald Murphey. He was influential in developing the state bank, the railroad, the University of North Carolina at Chapel Hill, and the North Carolina Agricultural Society. Three years after moving to the Hermitage, he became chief justice of the North Carolina Supreme Court.[247] It was also Judge Ruffin, E.M. Holt's neighbor down the creek, who gave young Holt the courage and funds to overcome his father's objections and start Alamance Cotton Factory.[248]

Though his law office was in Hillsborough, after 1852 Ruffin's base was on his plantation by the Haw; politically, he worked by discouraging discussion of policy,[249] and slave policy certainly. Thomas Ruffin topped the list of slave holders in Alamance County: he had 99.[250]

It was Ruffin who led the call for the Confederacy. Carole Troxler has retrieved late 19th century historian Sallie Stockard's record of her father's memory of this scene by the Haw, which captures the roaring spirit of a misbegotten vision: "Picnic Tables had been set up on the lawn. Barbecued meat had been cooked in outdoor pits over green hickory and oaken embers.... A piano, banjo, violin, harp and bugle, all together or separately, made music that swelled on the summer breeze, adding to the enjoyment. Large bouquets of flowers filled jars and crocks of every kind...."

Then a hush as Thomas Ruffin spoke:

> My friends and fellow citizens, ladies and gentlemen. What a great pleasure it is to see so many of my neighbors and friends here today. I hope you all have enjoyed yourselves. Let everyone who has, say Aye.... What a roar! If those Yankees could hear you, they would turn and run. Do not hesitate to send them packing if they come this way. This is your home. The Yankees want to take it away from you.
>
> See this handkerchief: Looks as white as snow does it not? Well, with this handkerchief I will be able to wipe all the blood that is to be spilled in this War. There will not be enough to stain it.... Come on boys, who will volunteer to whip the Yankees if they ever come to the Old North State? ... Come on boys, Enroll! Volunteer! We want as long a list as possible here today.[251]

Young men signed on to join the Alamance Regulators, Company E, 13th NC Regiment, while organizers for the Union made plans to rally at Alamance Battleground.[252] The state was divided.

Troops counted on their communities for support, and Ruffin did not disappoint. When his son pleaded for clothing supplies from Chickahominy River in May 1862, six days later the request was met with cloth purchased from E.M. Holt.[253]

Other neighbors, like the enslaved Rodgers family at the confluence of Little Alamance Creek, saw a different side of Ruffin, as Carole Troxler relates. Isaac Rodgers, a boy at the time of the Civil War, reports that Ruffin's workers "were glad to be hired out in the neighborhood, for they were not fed enough at home" and that after the Civil War, "Some of the slaves stayed on the plantation, but father and me made us a little shack, and I raised melons and sold them. It was the first money I ever had. It was green shin plasters for 10 cents, 25 cents, and a dollar, and let me tell you, I was sure proud of that money."[254] Isaac Rodgers married and "went to railroadin' and run all over the country"[255]; Judge Ruffin went into bankruptcy and sold the Hermitage.[256]

So where can we find the Hermitage today? In Swepsonville, people look puzzled when I asked about it. By knocking on doors along Alamance Creek, I learned it only remains as the basement beams in the house the present family calls "the Ruffin Homestead." Thomas Ruffin lived there through the Civil War until the spring of 1865, when a company of Union soldiers camped out and trashed the place. With enslaved labor gone and much to be rebuilt, the place decayed. Even the rock wall and most headstones of the family cemetery were carried off to be ground down for the county's first macadam road, from Graham to Burlington.[257] *Sic transit gloria mundi* the Romans might remind us.

When George W. Swepson purchased Ruffin's Mill with George Rosenthal in 1867,[258] another segment of the Haw's history at Swepsonville began. Swepson named the town after himself and the mill Falls Neuse, peculiarly after a paper mill he owned on that river.

George Swepson had a knack for causing trouble and traveling lightly from it. At the age of 23, he likely eloped with a 16 year old from a prominent Caswell family, Virginia Yancey.[259] They settled in Raleigh, where Swepson spent $241,713.39 from 1868 to 1870 to lobby the state legislature, leading with General Littlefield what was known as "the Ring."[260] Treating money for the expansion of the Western Carolina Railroad as his own, Swepson speculated badly, the state was defrauded of $4 million, and the railroad delayed; however, the Bragg Commission's investigation into the scandal was delayed as well. Legend goes that Swepson "disappeared from Raleigh in the dead of night with the $4 million in bonds in the cab of a Raleigh and Gaston Railroad engine, tipping the young engineer two dollars when he arrived at his Haw River destination."[261]

While *swepsonize* was a new verb in Raleigh for playing foul, he and George Rosenthal, a drygoods merchant who knew the popularity of Alamance Plaids, teamed up to produce plaids right in Holt territory below Haw River.[262] Mill construction began in 1868. Most of the workers came from the eastern part of the state and labored a twelve hour day for $1.80 a week or less. The wage was standard, but the hours were the longest around. Swepson thought this was fair as housing was free,[263] but the workers would strike here in 1887.

By then Swepson had evaded a murder charge. Not surprisingly, he fought with the Holts about infringing on their copyright of Alamance Plaids. Swepson allegedly circulated slanderous letters about Adolphus Moore, a Holt brother-in-law at Granite Mill, linking him with Mary Bivans.[264] Then, on January 27, 1876, Dolph Moore was found outside Swepson's Haw River home, barely alive. Two days later, he died without identifying his murderer. Swepson somehow escaped conviction,[265] but found it safer to live in Raleigh. In 1880 when fire destroyed the wooden mill which Rosenthal had heavily insured, rumors spread that the suspected murderer was also an arsonist. Swepson's scandals ended with his death in 1883.

Ashby Baker, an associate, took over and rebuilt the mill, renaming it Virginia Mills for his first wife. When fire consumed the mill again in 1892, Baker rebuilt in brick and enlarged it. When business prospered during World War I, wages and bonuses reflected that success and one of the earliest profit sharing plans in the Southeast began.[266]

In the Depression, the mill closed down until Baker's second wife, Minnie Baker, became president, the first woman in North Carolina to head a million-dollar enterprise.[267] Virginia Mills produced fancy dress goods, upholstery, and drapery materials until the early 1970s. On October 4, 1989, it was destroyed by a fire so huge it was reported visible in Danville.[268]

Raymond Herring, Swepsonville's Mayor

"He was a rascal," admits Mayor Raymond Herring about the man who named and once ran Swepsonville.[269] Raymond Herring became the first mayor in 1997 after he led the drive for incorporation to preserve Swepsonville's independence. After a hot summer day of outside work and with other mayoral duties waiting, this mayor still finds time for his river town's history.

> The reason [George] Swepson built in this location was that they already had textile mills around. That's why we are here.
>
> He would bring the cotton by train to Haw River, off-load it onto a barge and float it down to this location. Now he had a large dam across here, a wood structure, and he built canals by the river near the bridge and floated the raw cotton down to the mill, off-loaded the cotton, and then took whatever goods he produced and put it back on a barge and with poles and mules took it back to the town of Haw River.
>
> Then later, Virginia Mills actually owned this whole community. Houses, store, the mill, they owned the people basically. They would not pay them in money, they were paid in scrip. They tell me Mr. Baker oversaw it all; he would be in church taking a headcount. He was a true dictator. He selected who got what house and when.

I hear stories from ladies who still live here that they would come with their parents to the mill and work alongside of them. A lady who was 87 years old could remember going when she was six years old with her parents.

There was a community school, but she never went to school. She went to work.

Who's going to check and see? I know my dad — his schooling was through the second grade; his dad ran a sawmill, so he got into construction. My grandfather would take him to school in the morning and then pick him up at lunch. He never got to stay the whole day.

That's just the way it was. That's why you had children; it was your labor force. My mother's mother worked here at the mill. She lived in a company house. She got paid in company scrip.... She said they lived hard, but everyone did at that time.

It was a great job. It's still a good job. Cause most of the people came off farm land, mere existence. I worked at Saxapahaw when I was a kid. I thought it was great I had a job. There were ten of us. I was self-sufficient from the time I was 17. I grew up on a farm, and there was no income and no money on a farm. Steady employment, structured lifestyle, probably decent housing for that period of time — the stability, it was better.

Sure, there were sharp divides like you say. Right here is a prime example, the house next door, where the Bakers lived. You can look out the window and see the servants' quarters, right there in this partitioned house, one side was for the men and one side was for the women. You could see the difference between the Bakers' house and the mill houses. The haves and the have-nots. It was quite obvious.

It was a hard life, but I don't think you questioned what was going on if you wanted to stay in employment. You kept it to yourself as you would now. You may have talked to your friends about it, but you didn't take it to the management. It was obvious how these people lived vs. the others. Right behind here is the carriage house, the horses and the carriages. They had the great life. They traveled to New York. Their servants would take them to Haw River, put them on the train, and they would be gone.... That's where the business was taking place, in New York.

The mill closed in 1971 or -2. At one time, they had 1200 employees; that was in the late sixties. New management came in; the Bakers were aging out, and they hired outside management. They filed for receivership, froze the assets. This was the start of the decline of the textile industry in Alamance County....

Workers were limited in the new jobs they could find because their background was textiles. It was devastating. People had to commute, some as far as Winston-Salem, to get a job. I don't think any left. These houses, a number of them, stayed in the family. That's in what I call the old town, Main Street. There is not a large turnover. And that's why we want to keep our identity; that same atmosphere of a small town.

We did not actually become a town until 1997. This came about ... because Virginia Mills owned the sewer system. And they put the sewer lines and water lines in at their expense and when Virginia Mills went bankrupt, there was no management and no money. So the state allowed us to run it until about 1983, and, of course, the environment became a big issue and rightly so. The Haw River was just a sewer line basically.

This talk of the river moves us to the mayor's Carolina blue pickup truck and a river tour. First stop is 1075 Main Street which Herring tells me is "the old stage coach house, mail drop, where they changed the horses out at and stayed the night at. They forded down there near the power plant on the old trading path. If the river was up, they might stay there for a week."

We swing by the baseball fields where home runs could land in the Haw. "Every time the Haw floods, it floods my ball fields," says the man who maintains them. Baseball was important to mill towns and rivalries were heated.[270]

We pass the supply pond built by Virginia Mills, then on to the powerhouse, barreling past a No Trespassing sign. "I know those people," the mayor says, referring to the folks in the house once reserved for the man on 24-hour call to keep the water flow clear of debris. Inside the powerhouse, huge windows give light. Looking down into the funnels that held turbines driven by the Haw, I sense a whirlpool of power. The Haw drove the mill at least partly until the last decade before it closed.

Barges like this, called bateaus, moved raw cotton from Haw River to Swepsonville and then transported cotton fabric back. Gasoline tugs made the work easier around 1900, and improved roads and trucks changed the whole mode of transport after 1910. Blacks were not employed in the mills until after the 1964 Civil Rights Act, but worked outside the mill, as here (Burlington Times-News *photograph, courtesy Haw River Historical Association).*

We travel back north to the old mill, where it is the power of fire, not water, that dominates this vast, empty space that once held the longest mill south of the Mason Dixon line.

Across the bridge somewhere in the steamy tick-filled woods, river right, are the markings of Swepson's old canal. It doesn't take Raymond Herring long to find them, maybe 35 wide and 6 feet deep at their clearest. Without this canal, barges of raw cotton and cotton goods would never have made it by the rock-filled river section just before Swepsonville. In the twenties, Corbett trucks with solid rubber tires took over the job.[271]

We move on to the new Swepsonville River Park where the mayor stands by the river, an old dam spiking up behind him:

> We purchased that 18 acres with the full intent of leaving it as natural as possible. We're not giving up on the idea of someday getting rid of that dam. The consensus in the community is that the dam should go. It's backing water up, and it's useless. You got the City of Graham and then you got south Burlington dumping their waste in Alamance Creek, and it's all damming up here in our back door. That's some 30 million gallons a day between the three facilities discharging, and I would like for it to move on down, just pass on through.

There's a wonderful breeze off the river on this day of wet heat. Has the mayor ever been down the river? "No, but I would like to do that."

Has he ever swum in it? "No, no, no."

With a healthy river, this spot is paradise. "Yes, it's an untapped treasure," says the mayor, who is working full time to make it so. The county, Haw River Assembly, and Elon University students have joined in, and the park now has campsites, a 2½ mile path, and canoe access.

At Mayor Herring's urging I call on Duard Farrell, who "knows as much about Swepsonville's history as anyone." Farrell grew up, played baseball here, started as a service boy at Virginia Mills at 17, and then worked there as a manager of quality control from age 23

Remnants of two dams at Swepsonville are just above where Mark Chilton believes one of the prime Indian trading path fords was. In the background is the new bridge at Swepsonville. On the left is the public park (Sönke Johnsen).

until his retirement. His father hauled cotton on the barges; his grandfather first came to Swepsonville in the 1880s "along with five or six brothers, and they all worked in the mill too. In 1906, mother and father moved to a two-room Virginia Mills house when they were married." It grew as the family expanded to six. "They would apply to the mill management for additional rooms, and they would add them."[272]

Duard Farrell wants to set the record straight about the old stage coach route that forded here, so he drives me by Haw Creek, "the site of a lot of arrowheads," to establish how the road from Hillsborough crossed on the east side of the creek, cut behind the first house on Ball Park Road, digging groves 3–4 inches deep, then "passed the stage coach house on Main Street. Right behind was the barn that they kept the horses in, went down the north side of the ball field and then turned left until it crossed the Haw below the second dam. That was the route."

Duard Farrell maintains that Boywood Road was not part of the Old Stage Coach Road, so we head to Southwick Golf Course, where he commandeers a golf cart and drives us up to the 13th teeing off spot and a high and beautiful view of the Haw, winding by its wooded banks. Directly above the dam is the trough of the old road I climb up and down, always offered the arm of this Irish gentleman.[273]

Duard Farrell is old enough to have swum and fished in the Haw as a boy: "We used to go down below the dam there Monday mornings; they'd start the wheels up to run the power plant on Sunday night, and the water would stop running over the dam and go through the race instead. And the water under the dam would drain out of those holes, and we'd get in there and scratch around under the rocks and catch catfish with our bare hands and have catfish for supper on Monday night. That was late twenties, early thirties."

He remembers the coming of electricity and telephones. "From early to the middle of the twentieth century, that was a time of real progress, not gimmicks but real improvements." To understand what he means, we have to picture a time when ice was stored in sawdust in a brick building by a pond near the bridge. "My understanding is that Mr. Baker carried the key to the ice house. It was only to be used for sickness."

As we drive back to Main Street, we pass the Baker house from which the mill owners surveyed the village. A young couple had just bought it. Duard Farrell wishes he had known it was for sale: "To them, it is just a house."

Massacres and Dynasty: Alamance Creek

A major tributary, Alamance Creek pours into the Haw below the Swepsonville bridge, river right, past places of Revolutionary struggle and cotton mill history.

Up Alamance Creek past Rt. 49, just before Gun Creek, is one of the oldest fords, a strand of the Trading Path, now a gulley trailing up a hill.[274]

Overlooking the creek, in 1771, Governor Tryon and his forces camped before the Battle of Alamance, where he defeated all but the spirit of the Regulators. Their movement, a call for just laws and fees, is still recorded in history texts and strikes an empathetic chord today. In their own words, they were "obliged to pay Fees regulated only by the Avarice of the officer; obliged to pay a Tax which they believed went to inrich and aggrandise a few ... for the Men in Power, and Legislation, were the Men whose interest it was to oppress, and make gain of the Labourer."[275]

Edmund Fanning symbolized the abuses and alien coastal culture the Regulators loathed. His new house in Hillsborough stood for the despised Tryon's Palace that farmers were taxed for. Fanning was convicted of extortion — tripling the registration charge for a deed, but his punishment was another injustice — a penny for each abuse.[276]

The eloquent Quaker Herman Husband headed the opposition; he was kicked out of the General Assembly after announcing, with a bag of money in hand, "Here are the taxes which were refused to your sheriff.... We pay to honest men, not to swindlers."[277] Regulators could be unruly. When the sheriff took a mare for one man's unpaid taxes, the Regulators retrieved it, leaving some bullet holes in Fanning's new roof.[278]

In 1770, Regulators erupted all over the state, but the biggest riot was at a trial in Hillsborough. Fanning "climbed high on the judge's chair for safety," but was dragged down the steps and spat on in the street.[279] Governor Tryon threatened force and advanced to Hillsborough. He had trouble raising a militia of 1,000.

Regulators had easily twice that number, but not his firepower.[280] And they failed to block his crossing of the Haw downriver, a strategic moment: "Accounts were ... received that the regulators were advancing ... and fears were entertained, they would reach Haw river soon enough to obstruct his passage, the ford of that stream being so easily defended, that, on that contingency, the crossing of it must have cost a great deal of blood."[281]

Tryon crossed the Haw without battle on May 14, 1771, and camped at the high bluff on Alamance Creek before the famous battle in the curve of Stinking Quarter Creek that Alamance Battleground commemorates today. Only half the Regulators had guns; they planned to show force, then negotiate. Tryon gave ultimatums. Confusion and gunpowder ruled.

Fanning, a villain to the end, fled the battle and headed to New York, where Tryon was installed as the last colonial governor in a position his sister is alleged to have purchased for him.[282]

Tom Magnuson sees the Regulators' defeat as ending frontier culture: "It was an anarchist society; three cultures were living together, largely tolerating each other.... The Regulators were the last hurrah of that other way. Then the government came in and shot them into submission. They finally got a monopoly on violence."[283]

Ten years later, February 24, 1781, the Revolution was fought at its ugliest by Alamance Creek. Pyle's Defeat which took place north of the creek, was more aptly known as "Pyle's Massacre" and "Pyle's Hacking Match."[284]

Sometimes acting like rival gangs, Crown-loyal Tories and Revolution-supporting Whigs were eager for new recruits before the showdown between Greene and Cornwallis. Carole Troxler uncovers the details in *Pyle's Defeat: Deception at the Racepath*. Pyle, a local doctor, showed his Loyalist colors and rallied a militia when Cornwallis came to Hillsborough. He crossed the Haw at Swepsonville and forded Alamance Creek en route to join Tarleton at the beginnings of Little Alamance Creek (near City Park today). After fording, Pyle's men paused near an old racetrack.[285]

Behind them, crossing the Haw at Butler's Ford near Alamance Creek, came Lee and Pickens, Revolutionaries affecting a British look and wearing green coats. When Pyle's men mistook Lee for Tarleton, a Tory, Lee headed toward Pyle asking the Tory soldiers to make way while his men lined up parallel to them. As he shook Pyle's hand, his troops in the back started firing.[286]

Ten minutes later, the slaughter was over. Lee's forces lost one horse; an estimated 200 of Pyle's men were dead. Some died still confused, "crying out, 'Your own men, your own men, as good subjects of His Majesty as in America.'"[287]

Today, on Anthony Road, periwinkles mark one mass grave.[288]

Alamance Creek was the site of one more battle where Lee harassed Tarleton at Clapp's Mill by Beaver Creek, near where the marina on Huffman Mill Road is today. There 16 Loyalists and 8 Revolutionaries died[289] — low numbers, but huge losses in small communities. Then the men, neighbors with guns and different causes, went on to the Guilford Courthouse, using the Haw's streams for direction and relief from thirst, and hunger, as the Haw's tributaries led to mills where food could be pillaged.

Alamance Creek also drove the Holt textile dynasty. In 1837, E.M. Holt, fascinated by machines and influenced by mill owner friends in other counties, built Alamance Cotton Mill where the present bridge on Rt. 62 is. Holt started in partnership with William Carrigan but was soon on his own path, putting textiles in the Haw Valley largely under Holt control. By 1880, the Holts owned 4 of the 6 Alamance County textile mills, and by 1919, 23 out of its 27 mills.[290]

Back in the 1830s, the South was shipping raw cotton north and thread and fabric back. The market was ready for local production. Holt built 15 log buildings along Rt. 62 — then a planked road, passable in winter.[291] Weaving machines were added in 1848, and frame houses, still there today, replaced the log ones. In 1853, a French dye expert was engaged by Holt's son Thomas so colors would hold for plaids. Alamance Plaids were a major success, replacing the sturdy material that had sold at 7 cents a yard.[292] Thomas Holt was able to buy Granite Mills in Haw River, develop an estate there on 400 acres, and acquire the influence to become governor.

The mill on Alamance Creek churned out plaids until ginghams came into style around 1900. Spinning was only interrupted twice: once when looms were hidden during the Civil War[293] and then by fire in 1871.

A tour of the Holt residence, now headquarters for the Alamance Historical Association, reveals mill owner and mill worker life. Scrip from the company store and the payroll show pitiful wages that were standard. So in 1884, when E.M. Holt asked as he lay dying, "Do you remember any instance in my life in which I ever took unfair advantage of any man or woman or child? If so, tell me, for I want to make it right," those around reassured him before his final sleep.[294] In retrospect, the mill village resembles an upgraded plantation system,[295] but let us not feel superior, the issues of fair pay and workers' rights have yet to be resolved in our times.

This early 1900s Alamance Mill postcard's caption reads: "The first Colored Cotton Fabric manufactured in the South was woven in this Mill, built 1837 on Alamance River. Burned and re-built 1871." Note E.M. Holt in a tophat surveying the women workers below him (North Carolina Collection, University of North Carolina at Chapel Hill).

In 1926, the mill was bought by John Shoffner, a Holt factory boy who became chief of Standard Hosiery.[296] The new company name adorned the school buses, playground, bowling alley and men's community bathhouse they provided. After Shoffner's death, Kayser Roth took over. Sweaters USA ran there until 2005. Then the weaving industry was gone, along with remnants of Holt's factory, demolished in widening the Rt. 62 bridge.[297]

Downstream at the Rt. 49 bridge in Bellemont, Lawrence Holt and Lynn Banks constructed a factory (still visible) and mill village, before they built Glencoe in 1880. Their stream of enterprise reveals Holt family agility and power. After receiving one fifth interest in E.M. Holt's Sons Cotton Factory (Alamance Mill) in 1873 and helping to manage that mill and the new mill at Carolina, they built Bellemont Cotton Mills. When a profit showed, Lawrence Holt sold his interest, financed E.M. Plaid Mills in 1883, and became part owner of Altamahaw Cotton Mills in 1884.[298]

Today, the force of Alamance Creek no longer drives mills and fortunes, but dammed, forms Lake MacIntosh and fills pipes to Burlington's homes and businesses.

Flatwater World: Swepsonville to Saxapahaw

The dam at Saxapahaw backs up the river to Swepsonville and creates a wide, smooth Haw with islands and some of the river's safest paddling. Access at Boywood Road bridge and a portage around the second dam ease what used to be a formidable challenge en route to Saxapahaw. (The Puryear Portage, named for the man whose grist mill once ground there, is river left, but beware—both the dam and powerhouse intake are hazards in high water.) One winter day, following advice from a dauntless, and as it turned out, outlaw paddler, I struggled at the bottom of an eight-foot ditch by the second dam, dragging and pitching my Kevlar canoe along. It took a gorgeous paddle to Saxapahaw to smooth my ruffled feathers.

The hydro-electric plant and dam where I labored are now owned by Mark Chilton, mayor of Carrboro, whose witty and much needed *Historical Atlas of the Haw River* evolved out of his days as a kayaker and real estate lawyer. He gave the easement for the upcoming Puryear Portage and is always interesting on the subjects of rapids and dams:

> I have attempted to inventory the whole Haw River basin, and I find on all its tributary streams about 160 mills, and I believe there may have been as many as twice that number, 300 or so mills on the Haw at one time.[299]
>
> I want to comment a bit on where you find mills. The way the mills worked is that they captured the power of falling water, the force of gravity pulling the water downstream and that meant that it was important to be where the water was crashing down—where there were rapids. In fact, in the parlance of the 18th and 19th century, and perchance before, they did not use the term rapids; they would use the term power and say, 'Well, there's a power there' because they saw every rapid as a potential site for a water mill. In fact, if you were to travel down the Haw River starting from say up in Rockingham County where it begins to become somewhat rapid and follow it all the way down to its mouth, you would come across 25 or 30 different sets of significant appreciable whitewater rapids in that distance. And in every one of those sites at one time or another, there was some type of grist mill or cotton mill or sawmill.
>
> Good mills sites don't move; they are where the rocks are exposed. If the rocks are there, there are chances of a mill there. In fact, based on my research it seems likely that in one time period, say about 1850, that there was almost no free flowing section of the Haw River, that every rapid had been made into some sort of dam and that the river itself had been made into a stair step of pond, dam, pond, dam, pond, dam going the whole way down.
>
> Now there are definitely some negative things to be said about that level or that extent of hydro power because it certainly interferes with the natural ecological aspects of the river. And so in some ways one piece of good news is that a lot of those dams have been destroyed by floods and that much of the river today is free flowing although there remain along the Haw River a number of these dams.
>
> [Part of] the downfall of the water and grist mill was the railroad; coal became much more accessible and people became much less dependent on local goods. The technology of coal-fired steam power ... was here in North Carolina in 1890 and was much more practical. A hydro-powered mill was like a generator, but a generator that can't be moved around; you had to move your goods to it.... The shift in the 1880s drove grist mills out of business altogether, and hydropower was abandoned by the 1950s except at Ossippee on the Reedy Fork and Glencoe.

Mayor Herring and others may want the dam out, but Chilton hopes to restore power at Swepsonville to join dams at Altamahaw, Glencoe (pending), Saxapahaw, Bynum, and Jordan (pending) in harnessing the Haw for electricity.

Just below Chilton's powerhouse is Spirit Island, privately owned by Joe Jacob of Haw River Canoe and Kayak Co., so paddlers with permits can picnic or camp among the trees of the Haw. On this day at winter's end, just the trees' barks make interesting company: the smooth muscular shafts of ironwood and beech, the warty cobble of dogwood and hackberry, the curving wooden feathers of white oak, and sycamore's shadowy patches.

The banks I pass are full of plants that hold the land together and can tolerate an underwater life, even for weeks: tag alder, buttonwood bush with its round balls, and smooth, yellow-barked black willow. (Chewing that bark gives the same effect as aspirin, *salix* being the common ingredient.)[300] There are even some river oats, like their relative sea oats, another great gripper of soils.

Soon Haw Creek, river left, feeds into its namesake; just above it was Hunter's ferry in the early 19th century and just below was a ford that had its perils.[301] On February 20, 1818, The *Raleigh Register* reported: "Joseph Steele and Joshua Wody, on the 3d instant, were drowned in Haw River, from crossing John Thompson's Ford, one on horse, in a state of intoxication. Both have left wives and families to lament their premature end."[302] Fording was a learned skill; the act could be fatal.

Many button bushes later, Varnal's Creek wraps around Cane Creek Mountain and flows in under fans of holly and ironwood branches, river right.

A long thin island, filled with bamboo, cuts the waters like a narrow ship moving upstream; heavy logs, smashed on to its prow, hold and collect more — the Haw's creative force at work.

After a power line crosses the river, below Meadow Creek, river left, was Clendenin Ford and Cedar Cliff Mill. A bridge was also here from 1880 to 1890 bringing business to James Newlin's grist and saw mills. It is no easy task to keep a structure standing against the power of water; modern river bridges are due for replacement every 50 years. The island served as part of the mill race until its five-foot dam was flooded in 1938, when a new 30-foot dam in Saxapahaw changed the river level.[303]

Cedar Cliff Mill

One fall Sunday, members of the Trading Path Association gathered by the Haw, river right, to find traces of the road and crossing that served Cedar Cliff Mill. More than forty were on that mission, led by Tom Magnuson.

We study maps poised on car hoods before trailing up a ridge. Tom shouts briefings along the way: "Creeks are good for pack horses.[304] They are no good for a wagon. They are a perfectly good ramp to get down into the Haw if you are backpacking or on a pack horse. One foot step down horses like, more is tough. Pack horses were very important before wagons, before 1720."

We climb almost to the crest. "Native peoples, and the military, travel along the crest, about six to ten feet down from it. 'Don't be seen.' They disciplined themselves, made a foot wide path, walking heel to toe."

And just below the ridge is the road we seek, a shallow dip with a three-foot-wide bottom. It's thickly leaved below and above by tulip, beech and maple; a cattle fence runs the middle.

"These get filled in, but you can see the berm, the shoulders. When cattle get in on a road bed, it disappears fast. Thought I saw some cobbles.

"It's interesting to see how they persist and how they erode. Roads are flat, maybe ten feet wide. What you look for now is a V notch with a set of shoulders." So we take in this ditch's full meaning. Tom adds: "This is a channel of the Great Road to Salisbury." Natives trading Yaupon berries from the coast, Occaneechi wanting the iron and glass of the English, Quakers trudging to Cane Creek walked this path. "It was never a single road. Like the Ho

Chi Minh Trail.... They needed a number of ways; some were great in dry weather; they needed alternatives when the water was high."

Below by the Haw is another ditch, but this one "is a natural dike; when it floods, it overflows here and then loses energy."

Looking across the Haw, we imagine the continuing road. "If you trace Cedar Cliff Church and Mt. Willen Road on the far side onto Highway 57, it is a straight as a die. This is the ford. I like straight lines. I believe in straight. When we find fords, we can surmise the roads." Though this ford disappeared under the 1938 dam's backwater, we got closer to the path today. On a future day, a different crowd will be back prodding out the past.

Farther down the Haw, I canoe by a fringe of trees — a coy, futile disguise of a clear cut. Dirt will soon be flowing into the Haw here. Janet MacFall of Elon University's Environmental Center is interested in what that dirt, i.e., sediment or silt, reveals.

"I study dirt," she likes to tell people, but it is the carbon levels in the Haw's hyporheic zone, where water flows very slowly through soil, under and along the river, which fascinate her now. "Nobody's looking at the hyporheic," but evidence of high carbon content in that zone could be another important gauge of river health. "Carbon is the basis for the food chain. It's basic."[305]

Today, dirt is getting in the way of the Haw's health. It keeps the phosphorus count high because phosphorus binds to dirt. It also shelters UV sensitive bacteria by blocking sunlight.

"In the Haw, from Burlington to Bynum, there is an increase of sediment, a decrease in fecal material and the same grams per second of nitrogen. No one is thinking about this."

The dirt carried by stormwater is high in phosphorus and nitrogen, both contained in fossil fuels. One of MacFall's students found they could ignite soil that was right by the road; others tested and found traces of octane 15 feet away, all on its way to the river.

MacFall would like to see biological filters: patches of absorbent wetlands near roads and a 300-foot riparian buffer along the Haw and its creeks. That's why she gets grants, $1.5 million to date, and works to develop the Haw River Trail. Her hope is that the trail will increase awareness that will lead to buffers along all the Haw's streams.

"We could turn the Haw around in five or ten years if we have the political will to do it." My hearty colleague never runs out of energy or hope. "Even if we just leave it alone, there's a phenomenon that's really remarkable — the banks are collapsing back into the river. No one knows why. The drought may have stressed the trees. Some are sliding into the river and rooting there. The river is connecting back to its floodplain." That's healthy for a river because water doesn't race through, scouring banks, releasing dirt. And the floodplain absorbs nitrogen and phosphorus.

A healthy Haw's banks would slope gently to it, MacFall says, be "integral with the river," dropped only a foot or so below the bank, not deep cut and bare. This logged slope above me is headed in the wrong direction.

Farther on, Whitehead Creek sends its current into the Haw from off the eastern flank of Cane Creek Mountains. Miles on, Cane Creek comes off the south flank, once turning mill wheels all the way.

An osprey rushes feet first into the river then passes overhead clutching a small fish. The bent angle of its wings, like a wide W, identify this bird found round the world. I lean back to watch. The fish rides aerodynamically head first.

Far less picturesque are the turkey vultures one sees along the Haw. Leaving the asphalt at a road-kill site, turkey vultures aren't beauty in motion or in face, but they are grace in motion spiraling high on air currents, one or more on different flight patterns. Their meals look gross, but the birds clean up carcasses. An experiment with a hidden carcass by local naturalist John Terres, who wrote *From Laurel Hill to Siler's Bog* about the natural world by Morgan Creek (below Chapel Hill), confirmed that they hunt primarily by smell.[306]

It's rare to paddle on the Haw and not see a heron; they always shift downstream as if pursued and without a reverse gear. Lifting off into the air, heron may lurch like pterodactyls, but they cruise beautifully, long necks folded back. Tall and elegant on land, their neck sends us messages; a straight neck — or worse, a straight neck with the bill down — means clear out of my territory, and it is wise to do so for the striking power of heron is fatal to more than fish. If a heron comes toward you wings lifted, the message is even stronger. But heron appear romantic if you catch a pair courting with the tips of their beaks clasped together, heads swaying back and forth.[307] Perhaps I will see that one day.

We flatwater paddlers never impress anyone, but time can allow us these pleasures. A tree house with a terrace roof and a sycamore over the Haw with slats nailed up its trunk show others relish river time here.

Just before the Haw widens at Saxapahaw Lake are the high fields of the Paynes, who plowed corn and tobacco here for generations. Their sense of homeplace — family, land, and river together — repeats up and down the Haw.

Some settlers' claim on their land by the Haw was threatened in earlier times. Mark Chilton writes about one Samuel Strudwick who arrived from London in 1764 with papers in hand. He had inherited 30,000 acres of the Hawfields, river right from roughly Rt. 54 to Cane Creek, from his father who had acquired them from a debtor whose own possession was questionable. This increased tensions.

Samuel Strudwick worked from 1764 until his death in 1795, to take hold of *his* land. A number of settlers, who were mostly Quaker and Scots-Presbyterian, did come to terms with him, but others resisted, spurred on perhaps by revolu-

A nest-building osprey grips a twig. Their angled wings make it easy to identify them from below (Elaine Chiosso/Haw River Assembly).

tionary spirit of the times—first Regulators' activity and then the Revolution itself. They broke surveyors' chains and scared them off. One settler threatened to bring a posse, if need be, to haul off peaches "to distill" from an orchard by Strudwick's house, and another took out a patent on Strudwick's house leaving Strudwick fearing a "posse [would] drive us into the woods." One wonders if the land gains were worth what he gave up in anxiety and time. In any case, his son carried on lawsuits after his death, and Chilton reports they did better than most in court and many settlers were forced to pay.[308]

On this day, all seems calm. Below the Payne's fields beavers have settled in, blocking a creek and creating a pond at river's edge. In spring, when the Haw banks hum a faint olive and Judas trees flash lavender, an adjacent water slough will turn green with lizard's tail, violet, chickweed, and gill-on-the-ground. Then cormorant city will be back. The dark birds cruise up to their necks in the Haw, hold forth on recumbent tree skeletons, then reconvene in high limbs until, restless, their wingtips flick overhead.

The horizon flattens to the long straight line of a dam, and my paddle ends on once popular Saxapahaw Lake. Older residents remember its glory days: "Motor boats on it Sunday morning, you'd hear them like crazy."[309] "See forty boats out there and people skiing. It was 20 feet deep then, now its filled in."[310] And another one remembers the lure of factory lights across the lake where they didn't have to chase cows.[311]

Today an absolutely essential take-out is the dock, river left, at the Ben Bulla Scout camp, named for the man who argued with B. Everett Jordan to get it built in 1962. To assuage Jordan's concerns about vandalism, cement blocks and steel frame windows ensure Boy Scouts will have a place by the Haw for years to come.

Saxapahaw

The water is wide at Saxapahaw and hills slope easily away, so the shallow Haw provides a ford. Tom Magnuson can point to where wagon wheels cut tracks on the island Buddy Collins bridge crosses over.

Explorers' attempts to name the Haw's Native peoples here led to the town's name. It was changed from Sissipahaw to simplify spelling, which Ben Bulla, town historian, notes, judging by the incoming mail, was not of great help.[312] Bulla also wrote *Textiles and Politics* about the life of U.S. Senator B. Everett Jordan, who ran the mill here.

Bending over maps with me one afternoon, Bulla fills in more local history. The river spawned the two industries of Saxapahaw—mills and ice which was stored in dug-out caves along the banks, probably on both sides of the river near the bridge. A grist mill, river left, was operated by the Thompson family in 1782.[313] Near where the community center came to be was the blacksmith shop. A few houses up the road to Swepsonville was the first gin mill.

If you buy gas or pick up a snack, you are at the old dyeing rooms of the 20th century cotton yarn mill Sellers Manufacturing Company.[314]

John Newlin started a cotton mill with an overshot waterwheel here in 1844. The water power was worth the commute—four miles plus fording which meant rigging a boat to a cable in high water.[315] It wasn't until 1875 that the first bridge was built.[316] As Ben Bulla explains, the mill needed to be on the east bank, as the west was too stony to cut a race through. Mote's Creek supplied red clay, and sand came from the western bank.

The mill itself was "a brick one-story affair that had fireplaces for heating and was lighted with candles and oil lamps. There was an arched opening underneath the floor where the water from the race ran over the overshot water wheel that turned all of the moving machinery by means of a line shaft. With such a shaft arrangement, it was impossible to stop one machine without stopping the entire plant."[317] The river wafted in humidity to control static electricity and keep the cotton from breaking.[318]

The mill's race was built by African Americans which is remarkable only because Newlin, a Quaker, opposed slavery. This created a puzzle for historians. All agree Sarah Freeman had bequeathed to John Newlin 42 slaves to carry to freedom. A cynical version had Newlin making them build his exceptionally long mill race first. A villainous version has a family member selling the slaves in Fayetteville. Carole Troxler found the record clear: the registered deed of emancipation for 42 people, many surnamed Newlin, in Bellefontaine, Logan County, Ohio, on December 10, 1850. This affirms they got to freedom, and Newlin avoided a $42,000 bond for their emancipation. The delay that led to their building the mill race came about because Sarah Freeman's will was challenged in court from 1839 to 1850. Ironically, Judge Ruffin, who disagreed sharply with John Newlin on slavery, ruled in his favor because private property rights trumped all else for him.[319]

When Newlin died two years after the Civil War, his sons struggled in difficult times

Saxapahaw Mill Workers assemble for a company photograph around the turn of the century (Gary Mock, www.textilehistory.org).

then sold both cotton and grist mill to E.M. Holt in 1873. This Holt, White and Williamson enterprise joined together two Holt sons-in-law who named the cotton mill White, Williamson after the patriarch's death and produced "ginghams, tubing, and some flannel and outing goods." These were sold up North and traded as far as South America until depressed cotton prices of the twenties forced them to sell.[320]

Charles V. Sellers was the buyer, and his nephew, B. Everett Jordan, became manager of Sellers Manufacturing. As Ben Bulla tells it, B. Everett Jordan arrived in 1927 via dirt roads and a single lane bridge, passing corn growing on the island in the Haw, to find a dilapidated mill, "mostly cats and dogs and dirt and dust," and have his entry blocked by a conscientious worker who had no idea who Jordan was.[321]

Jordan plunged into the work wholeheartedly, removed outdated weaving looms because he did not know anything about weaving and focused on "yarns — combed, carded, nylon blends, various combinations, dyed and natural, mercerized, stretch, and other specialties" for international sales.[322] In 1938, he got 2000 horse power from the Haw by building the 30-foot cement dam.

Ben Bulla, "a longtime colleague," captures how B. Everett Jordan ran his mill.[323] One worker, Greef Smith, overwhelmed with medical bills and about to lose her house, remembers his generosity: "He said, 'Do you want it?' I said, 'Sure, I do!' He said, 'Well, let me worry about that.' 'That's a second daddy what is a second daddy.'"[324] Staley Gordon has positive memories too: "One time the mercerizing machine broke down, and he put on some old clothes and came down there, and it took 24 hours to get it going; that's how long he stayed down there. I did too. I got so tired and sleepy that if I leaned on a post I would go to sleep, but he was still going strong."[325]

B.E. Jordan was hard driving in other ways: he did not tolerate unions. Someone canoeing under the bridge in 1934 would see National Guardsmen with machine guns and live ammo on the bridge and on the roof of the mill. This was Jordan's response to a possible visit from the union-run "Flying Squadrons" who were sometimes able to close mills down.[326]

Leon Madden, an employee, says the guns were on the hill, and in Swepsonville too. He adds that White and Williamson, the former owners, blacklisted anyone involved in unions and didn't even let them have ice.[327] The Squadrons never appeared, but workers certainly got the message.

The man whose name is on Jordan Lake and Dam was blunt about strikes, saying: "Y'all can put the union in, and we'll shut down!"[328] John Jordan, his son, emphasizes what was here for the workers — doctor, gym, community center, baseball field — plus, "This mill ran more than any other mill. People had to work."[329] Leon Madden adds, "It was Jordan who pushed Sellers to keep the mill running when he wanted to shut it down [when] we made a game out of counting hobos, so many came through looking for work."[330]

From whatever combination of loyalty, guns or threats, the union never carried the vote here. And B. Everett Jordan went on to Washington to serve as a senator from 1958 to 1973 and be listed in *Parade* magazine as one of the ten wealthiest senators.[331]

Hoover Dixon, who rose to the rank of mill manager, describes life around Sellers Manufacturing. His family moved in 1943 to the Haw's banks, where he lives today in the shade of tall trees. "We were tobacco farmers. Of course, in the winter time I worked in the mill. That was cash for shoes and clothes. When my three brothers went off to World

War II, I became number one plow-jockey and had to finish high school later. This was a farming community; back then, everything was cotton, hogs, tobacco."[332]

When his brothers returned, Hoover Dixon had a stream of mill jobs: yarn packer, winding room, quiller (separating and putting the ends on bobbins). "That was a high paying job; everyone wanted to be a quiller. [But] there was no air conditioning then, it was hot, dirty, dusty, unsafe. We had one shaft running through the floor ... and a belt slapping from the ceiling with no safeguards."

Dixon returned from the army in 1950 to work in the dye house as a gofer: "They really didn't need me, but I'd just come back, and they made me a job. I was a handyman, I reckon. I was a local guy. I'd load yarn, run dye machines, weigh dye, work in the lab, match shades. Then for ten years I was plant manager. I took courses, and Sellers supported me. They didn't pay, but I got my education."

Years later, Dixie Yarns bought Sellers out and Hoover Dixon went to work for Charles Craft, where he found "the more they made, the more they would give you back. My first year there, Charlie Bouie handed me a bonus check for $10,000." Hoover Dixon still thinks

> it was a good job here too. Don't get me wrong. Everybody believed in Senator Jordan. He made jobs when there were no jobs out there. High volume, low profit. He would keep the mill running, jobs going, but times change; that was way back then. He'd give a bonus, maybe a hundred dollars or something. He had a retirement system; they were good people; now I'm not criticizing.[333] We never voted a union in over here. We had several, two or three votes. I think it was probably loyalty to the Jordans who went the extra mile to give people work. Then unions were no doubt what put the economy moving and the best thing that ever happened to people. One person can't do much by themselves. But they were soon taken over by people who wanted something for nothing....
>
> Everybody knew Everett Jordan. I went to the same church as him, was in the same Sunday school class. Fine gentleman. If you ever needed anything, you'd go to him and you'd get it. He was kind of a daddy figure to everybody. He liked it that way I think; well, I know he did. Lotta places paid better salaries, but now, where did you get the personal attention that you did working for that one man. I was Hoover Dixon. I wasn't number 1835.

Ben and John Jordan

The old company offices and store look out over the Haw from the knoll at the end of the bridge. Long windows flood ample rooms and wood floors with light. Now the headquarters of Jordan Properties, where I meet John Jordan, its owner, this is still a hardworking place.

We join Ben Jordan, who sits behind U.S. Senator B. Everett Jordan's old desk. Portraits of the senator and their grandfather, a minister, overlook the room as Ben and John Jordan speak of C.V. Sellers, "Uncle Charlie," whom they affectionately referred to as "the man with the money," who sent their father to Saxapahaw in 1927.

The history of Saxapahaw is their mutual passion, and we start with this building—the old company store.

> JOHN JORDAN [JJ]: We never did own the "company store." We rented the space to a family that came here from Gastonia; they always called it "The Company Store"; I don't know if you had to shop here, but you didn't have much choice.[334]
>
> BEN JORDAN [BJ]: As John said, it was very convenient. They sold everything from overalls to plow points, ice cream, groceries, meats, and workers could charge it, and it came out of

their pay. But if you bought too much and you didn't have enough wages, you had no income, but you had your groceries and your shoes. That probably was abused in many cases, and it is also unfortunate that if the mill was on short time, your wages would not cover your bill. Mostly they just worked it out, but sometimes if they owed too much, they would move away.

I ask about growing up by the river.

BJ: I, and probably John too, we just played in the river and walked across in the summertime and fished in it and swam in it. And I don't know a soul who died of typhoid, and I am not at all sure why none of us did.

JJ: The same as Ben's. I went to school across the river, and whenever it got warm enough, you would always walk across the rocks when the water was down, and you could fish there. My parents had a rule that anything we caught, we had to bring it home and clean it and eat it, and I have eaten a lot of fish out of the Haw River.

BJ: I used to go swimming above the dam, but that was before this one. You went across the gates to the race ... and all us little boys would just dive in there and swim. My folks didn't think too highly of that. After the [1938] dam, you couldn't get to it, but I was a big kid then. It wasn't the thing to do. But before, it was just boys playing in a river, and we were all buck naked, too.

How did being the sons of the head of the mill make life different? John remembers getting whipped for bragging at church.

JJ: I was telling about the two mules my father had bought for $500. That was a lot of money then. See, I wanted people to know about the $500. He did not like that. He was a quiet man. He kept things to himself.

BJ: We knew that he ran it, but that's just about all. Other than that, we did everything that everyone else did. I could not wait for the first of May to go barefoot, and I almost never put my shoes on again till going back to school. No, not even going back to school. I never knew there was any difference, and there was no difference. Everyone called Daddy Mr. Jordan. And I always called the people in the mill who were my parents' age, Mr. and Mrs. That's almost an unheard of thing now; everyone is Molly or Tom or Ben. You don't even know the last name. Wasn't that your experience, John?

JJ: Yes, [my father] had a philosophy. He wanted employees and employers to be very close. We all lived here, went to church here, went to school here.

The philosophy is good and bad. We certainly grew up together ... but then when you got to be management, it is hard to have to criticize or correct a friend. I remember one time I got after Rachel about something ... and she said. 'John you can't correct me; I used to change your diaper.'

Spencer Love [head of Burlington Industries] was just the opposite. He did not want management and employees to live together. It was imperative ... for that one reason ... not to have the difficulty of correcting a friend. I only had to fire someone twice, and I would rather have been whipped.

I ask about river, power and the mill.

BJ: A new dam was built in 1938. I heard so many times that the wooden dam burst. You would lose electric power, so it had to be repaired.

Getting the diesel engines into Saxapahaw was not easy, they had to be transferred in along planks laid across the road. They would have crushed any road between here and Swepsonville. They couldn't cross the wooden bridge in Swepsonville, so they came on NC 54 where there was a concrete bridge to where they could get on dirt roads.

I don't know why they built a mill around here. Well, I do know why they did, but everything had to be trucked in whether it was coal or diesel fuel or cotton or yarn going out, and there were no paved roads.

JJ: Duke Power in their great wisdom made it cheaper to buy all of the electricity from them than run on hydro-electric. We had switched over to Duke at this time, and the diesel engines broke, not infrequently, and we were selling them for junk, and my father wanted to know when it was built, and I said "I don't have a clue."

And he said, "It's on the boiler plate."

And I said, "Well, I haven't looked at the boiler plate."

And he said, "Well, why haven't you? Be curious, be curious, be curious. If you are not curious, you will not learn anything," so that stuck with me forever — be curious.

BJ: Glad he didn't ask me. I knew they had one, but not what's on it.

Was B. Everett Jordan in charge of the mill until he went to the Senate?

BJ: He was it. He was in charge until he died.

JJ: Until he drew his last breath.

BJ: And that was in '74, now that was somewhat facetious but not much. He had several people in charge, theoretically it was me (laughs), which was only in theory.

JJ: He'd come back from Washington once a month and have a meeting with Ben and the rest of them and tell us what to do, and he'd say, "I'll be back in a month, I'll be back," and, of course, you don't run anything "in a month" like that. One of the most frustrating things was we would be sitting right here and...

BJ: Yes, he sat right here and leaned back just like I did.

JJ: ...discussing whether we'd buy new spinning frames or whatever, and the phone would ring, and there would be some little old lady with a social security problem and an hour later....

BJ: He would never not take a phone call, never not take a phone call.

JJ: That's right, which is, you know, commendable; he was an extremely constituent-oriented public servant, but the meeting was called to buy spinning frames. So it was extremely frustrating to those of us who were trying to keep the business going.

When was the mill sold?

BJ: March 1978. He died March 15, 1974, but we sold it in 1978. In March, too. To Dixie Yarns in Chattanooga. They shut it down in '94.

Lastly, we talk about Jordan Lake.

JJ: After the dam was complete, the environmentalists got the court injunctions to keep the gates open ... and Daddy said, "I'll be the only person in the whole wide world who has a mudhole named after them."

BJ: At the dedication, I was surprised how much it had filled up, how big it was. That was a few years after Dad died.

JJ: We were there, but he was not alive when they did the dedication.

On the way out, John Jordan jokes with me as he nods toward the Haw: "Now you do know this was once the Jordan River, the whole of it." I assure him I have seen those 16th century maps.

An old Saxapahaw Mill workers joke when the Haw was low:
"Call Swepsonville and tell them to flush their toilets."

Health of the Haw

How's the health of the Haw at Saxapahaw now? The pollution in the river that even made it hard to dye cotton thread has turned around since the Clean Water Act of 1972.

Cynthia Crossen, singer and head of River Watch for the Haw River Assembly, has gathered a crowd to check on the river's health. It's July and hot, but all that lifts away as cool flood waters recede past the island at Saxapahaw. Her voice like the river draws you as she explains monitoring the river by counting bugs. It's the Benthic system which scores macro-invertebrates, bugs without a backbone and large enough for the eye to see, by their tolerance to pollution. Top-scoring stoneflies, caddisflies, ruffle beetles, water pennies,

At a river monitoring session in Saxapahaw, all ages hunt for microinvertibrates off the island in the Haw. Background right are the Saxapahaw River Mill Apartments built in the old factory (photograph by the author).

mayflies, Dobson flies and gilled snails get three points for being pollution sensitive. Pollution tolerant ones, aquatic worms, blackfly larvae, pouch snails that open left and leeches earn one point per bug found. Fifty teams all over the Haw's basin check out their site each season looking for 24 macroinvertebrates, and send the scores to Cynthia.

Cynthia mentions other tests for temperature, transparency, pH (acidity or alkalinity), and then she releases us. Nets scrape rock puddles and swish river depths, the murk is dumped into plastic basins, and heads go down to see what moves. A midge larvae (1 point) gathers a crowd. A father uncovers a crayfish (2 points) and dumps it

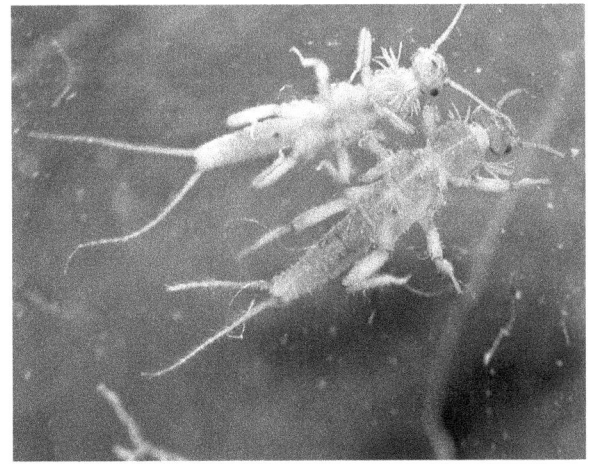

Two stonefly larvae wander together. This variety, Nemouridae, are identified by their neck gills and are ¾ to 1½ inches long. The Haw River Watch Macroinvertebrate Manual *explains that they live in well-oxygenated water until they transform, in anywhere from three months to three years, into flies. Highly pollution-intolerant, they indicate good water quality (Cynthia Crossen, copyright Haw River Assembly).*

into the basin of my granddaughter Sierra, who swells with pride; at 6 inches, it's twice the usual size. A leech elicits screams and scores a one because diversity alone is a sign of health. A mayfly (3 points) and clams the size of a thumb print (2 points) appear along with some unidentified scurrying spots.

Cynthia adds intriguing bits of insect news. Adult mayflies have no mouth parts, so in their winged form, they must court, mate and lay eggs in just one day. Caddisflies spin underwater webs to filter in nutrients. Under the magnifying lens, she shows us how a mayfly's fluttering gills breathe. All this, while kids and adults jump from rock to rock, fall in, yelp, turn silent in wonder, lose a net to the river and tour the basins.

Our final score, the Haw's that is, is low; Cynthia usually finds only fair water quality here, and the Haw is high so searching under stones is hard. The Haw's turbidity is only 13; that's the number of inches of river water in the transparency tube before the Secchi disk pattern at its bottom is obscured. Runoff after our drought-smashing rains doesn't make for clear waters.

Our pH test results were near ideal — a bluish 7.2 — close to neutral and the 7.0 humans and river creatures need. Nitrogen is 0.25 and phosphorus is 0.1 — not alarming, but you want zero for reasons to be explored downriver.

Everyone grabs nets, basins, charts and tubes and heads over to Saturdays in Saxapahaw where the music, games and a farmers' market are gearing up. The man who found the crayfish now helps his wife sell jams, parents settle into lawn chairs and enjoy the band, and Mike Holland rules over the kids' water slide, where black plastic coats a dip in the land, hoses create a stream, and an overhead hand trolley turns the whole thing into shrieking bliss.

It goes on. Tours of apartments in the converted old mill can be arranged; boats can be rented from Haw River Canoe and Kayak Company. Once again, the river is central.

If Mike Holland has his way, it will always be. This pioneer of the Haw River Trail has an office by the old mill, full of charts and maps where he plots ways to reconnect people with the river.

Dr. Mike Holland is the spirit of the Haw River — at flood stage. A UNC post-doc in environmental oncology and a guerilla activist and Tom Paine orator, Mike sees the Haw like no one else. At the Jordan Lake Rules Hearings, Mike is at the door in full hazard-gear festooned with orange caution tape and a snouted gas mask handing out a flyer against sludge. I just say, "Hello Mike"; who else could it be?

His path to the Haw was by way of a farm in Alabama, an eighth grade NASA camp scholarship, a year in a body cast from "nineteen-year-old motorcycle silliness," and his post-doc at UNC–Chapel Hill. What better place than the Haw to study the environment's human health effects?

"You could tell how an entire region was doing by the water quality; the land, air, water all came through this one pipe [the Haw] down to you."[335] So he has been paddling, and fighting for, the river ever since, working on the Haw River Trail, public put-ins, clean-up, Sierra Club 4-C's program, CERES (the Hollands' networking organization for environmental education and activism), landfill fight, Mountains-to-Sea Trail, Heritage Trail, the list goes on. Mike is the gadfly and initiator, a motor boat swerving down the muddy waters of the Haw, leaving a clear wake and rocking other boats.

Not that every idea happens — this one from a March 2003 e-mail hasn't yet.

Just a wild thought, but try to envision an area of the Haw river, in Swepsonville as a corporate sponsored whitewater park with year-round Olympic training facilities and search-and-rescue training camp, maybe run in conjunction with Alamance Community College and the Graham/Haw River/Burlington/County Park and Rec....

Instead of removing the upper dam in Swepsonville, we could simply move the rubble around a bit and manufacture rapids, which the bugs and fish could use as well as boaters.

The whole idea of rapids and oxidation — the breath and healing force of the river — can get Dr. Mike jumping out of his chair, scattering maps, to display the drop and spray of the Haw with a wide sweep of his arm.

Dr. Mike loves getting people on the river to connect the dots: "The more they use it, the more they would fall in love with it and understand it. Then they'll protect it when someone tries to take it away."[336]

Mike got people out for the Haw River Paddle Trail which he co-opened along with William Nealy and others:

In 2000, the City of Graham opened up one access point and the Haw River Trail opened up five. Now I was going in my own kayak that day, and someone said, "You need to take this woman," and I looked down, and she had gold heels on, and I said, "O God, can I just please take my boat," and they said, "No, you must take her; that is the mayor's wife." Turns out she was the best paddler there. She was like a motor boat, so we beat everybody, and we beat William Nealy and the mayor. Fun trip. We had corporate leaders taking canoe rides in their ties and getting dirty down by the river.

Mike has a vision for the landfill too, just below Saxapahaw and a factor in the Haw's health.

The whole area is threatened with a mega dump and the kinetics of that are that it's ten miles to the interstate.... There's about a hundred and twenty million dollars worth of fillable space.... It's a big cash cow.

It needs to be turned into a green landfill and become the first truly sustainable landfill, and we can do that with commitments from industry. We would have jobs for people who are out of work. In Chatham, you don't have to throw your old TV set away. You can put it in a shed, and if somebody likes it, they can just take it. It's not brain surgery, yet we don't do it. That's not re-cycling that's re-using. We won't take green wine bottles. We have plenty of brown beer bottles in this county, but not enough green wine bottles. But that's not recycling; that's just thinking about recycling. We only take a few kinds of plastic.

In a better recycling program, the bags get busted open and every piece of that garbage is gone over.... And the little trickle that could not be sorted would go into the actual landfill which could then last for three or four hundred years. And Saxapahaw would have an energetic thriving company, a place we'd use as public gardens, where other cities could send their people to learn....

Mike also worked to make the Haw part of the plan for the Mountains-to-Sea Trail and helped create an additional route from Cane Creek to Jordan Lake:

The Mountains-to-Sea Trail will spark the public's imaginations ... and be a master trail for history tourism, cultural tourism. People will drive down, and one parent goes into the antique store, and the other takes a kid on the river, and at the end of the day, they go to a bed and breakfast.... We can develop an alternate form of economy and jobs for our kids. And the Haw's future?

Let me start with the bad news first: The sprawl community. There's a sprawl union — the Chambers, the realtors, the Home Builders' Associations. There's nothing evil about them;

they are just trying to maximize their profits.... If the whole world were one big laid-out golf club that would not bother them in the least, but that's not a sustainable existence, and it doesn't look smelly or smoky or bad, so to them, it isn't. And so I picture this region being an expanded version of Cary.

This is my evil future vision: between Greensboro and Durham, straight line along 40/85, mainly one double interstate, probably double decked; you will drive in the dark or in the light, and we will have cityscapes that extend over this interstate, with a Wal-Mart probably every mile. The river will be encased. Greensboro's going to connect to Burlington. The Haw will be in a pipe around through Graham to Swepsonville in one big mega city, and that's the evil version.

The good version is, there is no good version because ... right now human nature can change for an individual, but it can't change for a population. And the proof is there are 6 billion people. We have not yet realized the fallacy of trying to feed six billion with an imperfect program; that's a recipe for disaster.

But you are working on it, so what's your hope?

It's like that old story about starfish. This old man comes to the beach and there's this little girl pitching starfish into the ocean. And the old man asks, "Little girl, don't you realize you can't save all the star fish on this beach?" and she says, "That's ok; I can save this one," and she pitches it into the ocean.

We will probably in the end save a corridor of managed riparian buffer, and I hope we save some creeks, like Alamance Creek and Cane Creek, Reedy Fork Creek. Two years ago, I opened up a trail from Northeast Park in Guilford County in conjunction with their Economic Development Office and Parks Department, and that comes down two or three miles on Reedy Fork and on down the Haw. You can be in Glencoe and have dinner and camp, so it's a great example of how these creeks are going to become the next focus for the Haw River Trail. Those are pathways of hope through this community. And this community will need that because it is going to be like a concrete steam-roller in the next twenty years.

Spirit Streams: Cane Creek Mountains

Many are standing knee deep in a river and are dying of thirst. — Bruce Holt[337]

One way to know about the Native peoples along the Haw is through Bruce Holt, a healer who was born by the Haw's streams on Cane Creek Mountain. I have watched students encircle this man with long grey hair in a purple shirt and a necklace of eagle and bear claws to absorb his calm and deep attention.

"You know I have been called back to the light twice?"[338] He was saved by the intercessions of Native rituals and healings; this shaman's "individual path to power" is now to help others learn of Native ways and healing.

Without knowing who I was or asking a fee, Bruce Holt agreed to meet Janet MacFall's and my students by the Haw to help us in our study of the river's people. Picture this tall man with a swarm of students around him seated at a table. Professors are putting map after map in front of him, each with a question about where something is. As the maps pile on, Bruce raises his hands to stop the confusion. "Where am I? Where am I?" he laughs with mock concern.

"Oh, I am here." He settles his hands on his heart. Everyone gets his point. His roots go deep and ... by implication, so do ours.

"They were a river people," he says of those who lived before in a spirit world, healing through herbs and rituals, as Bruce Holt does now, offering the smoke of sage in four directions on the banks of the Haw.

On a later visit to his house, I ask questions that he enlarges by his answers. Is there a Native group that he identifies with? "I identify with everybody in the world. I have 6.2 billion sisters and brothers."

He points to the statues in his and his wife Nancy's garden: St. Francis and the Virgin Mary. "We're not Catholic, but they are good spirits."

Were there particular river spirits? "Spirits are everywhere. Nothing ever leaves, everything is energy. Everything has spirit. River spirits are everywhere."

How did he come by his Native spiritual ways? "Well, my mother of course," but then he makes clear that the spirits themselves were his teachers.

Bruce shows me an ax he found along the Haw. Dr. Coe, professor of anthropology at UNC–Chapel Hill, confirmed that it was a Guilford ax (6500 B.P.). The peoples were connected enough that the styles were the same. "You see this snake with a tail in its mouth. This dates back 1800 years. And Indians used it as a symbol of eternal life because a snake sheds his skin and lives on. We do the same thing."

Another snake is a talking stick. "People sitting in council would use this. No one spoke until he had this in his hand."

Bruce Holt keeps his own creations separate; one, an ingenious rattle and drum, sounds like rain as Bruce shakes the beans in the gourd. He beats on the deerskin stretched across its wide top. Thunder. The effect is strong. The idea came to him in a vision. He wraps string stripes on two spears — his symbol not to use them as weapons.

Bruce Holt grips a talking stick; the snake, a Native symbol of eternal life and wisdom, gave one the right to speak (photograph by the author).

Finally, Bruce Holt shows me a rusty hoe he reattached to a wooden shaft. "I had a vision that I should search there, and my boys put their money together and bought me a metal detector, and I found this. I show it with my other artifacts and honor the ancestor who used it."

A week later, Bruce Holt takes me to Cane Creek Mountains[339] where he was born near the crossing of two Native paths which, like river confluences, are sacred places. I am stunned by the high view. Even with a thunderstorm waiting to explode the haze, thin ribbons of hills go on and on past the plain below. This high place, on a worn Precambrian nub of the Uwharrie Mountains,[340] was certainly sacred.

We stop by an old quarry littered with chiseled stone tools. "This spot used to be 100,000 feet into the ground [referring to the vast erosion of Piedmont surfaces]; soon it will be covered with grass." Bruce has been at work collecting ancient tools and displaying them on an altar of stones. "Let the spirits know I respect their efforts." He demonstrates how they would shatter a small boulder by lashing another one to a sapling and springing it loose.

Bruce thinks he has found the ceremonial mound mentioned by Juan Pardo in 1567, here by a two room chamber between two quarry sites and Cabin and Wells Branches. "I scaled it off, mound to the river, and it's about five miles. Now you know that's nothing to people who want to go out and worship."

Bruce Holt shows me where anthropology student Stan Powell and he had a vision. They asked permission to be there, placed tobacco on a cement altar, and smoked a pipe with sage while Bruce drummed: "And we heard this sound ... a woman humming to a chant. Then a male voice started to chant with her and then there was drumming with it; a deep bass drum joined the singers." Bruce starts to drum.

"And then a horse goes neighing, then it gets quiet, and the horse started right where Michael was building his house and Bryan [his sons]. And in between the two sites was about a 100 foot ravine. And that voice, the horse spirit, went right across that ravine. I can show you the tree where the horse spirit passed. It told me so much about life. Not only does man have eternal life, but so do all the other creatures. We do live on."

Bruce shows me the spot where hail fell all around them in a circle but not on them. Another time 26 people were here and they all heard spirit children laughing. "Exactly like when a boy is rolling on the ground and being licked by a puppy. That kind of mirth.

"It's not about me. It's about possibilities. If I can get in touch with these spirits, then anyone can do the same thing.... So everything has its function in the universe. We can all be murderers or healers. We have to find our individual paths to power."

Certainly this spot was visited by Native peoples. Living in a wooded land, how strongly they must have felt the release to space that comes in this high place. Bruce has built an altar here of beautiful proportions, reflecting his spirituality and humor. A plow's metal disk will hold water, and the rods from an old Plymouth hold torches. He tells me that below, on Wells Branch, he found the site of Grandpa Eiseley's old still. "I told the boys they ought to clean that up and make a monument."

I ask about the rituals of the peoples here 450 years ago, but he waves the question away. "No, it's not like that. It is the spirits that are here now that matter, and the healing available to us now; not the minor variations time played on them."

As we go, I am introduced to healing spirits. "Oh, boneset." He greets the low fuzzy white brackets. "It's good for a wound that is slow to heal, and boneset, well, the name tells you." A patch of nearby daisies are good for ague, an old-fashioned word for flu. Rabbit tobacco was used to treat asthma; you make these into a tea. Bruce Holt did Ph.D. work in ethno-botany at NC State; as a boy, his cough syrup was mullein, pinetops, and wild cherry bark, sweetened with honey.

A plant is more than just a plant. They are spirits. It's all energy, we're all connected. If you're aware of it, it's aware of you. Uncle Bird used to say you could make a tea out of the first seven plants you saw. It doesn't matter. It's that you're entreating their help.

The Native peoples didn't have a name for God; they'd say Great Spirit or Holy Mystery. It's all much larger than anything we can understand.

1930–2009

Runoff: Haw Watershed

Driving home from Cane Creek Mountains, the thunderstorm erupts. Lightning flashes five times in the same pattern. I pull off in the near-empty parking lot of BMOC shopping center where the wind and rain sweep and crash about, bouncing high off the asphalt before running to drain pipes that rush water to a stream — so directly that the sudden flood will scour banks.

On Cane Creek Mountains and all along the wooded Haw, the pelting rain is taking a different course. Its hard drive onto the tops of oaks, tulips and maples is broken by one splay of leaves after another. There is no high bounce off the soft ground of leaves and twigs. The rain will sink into the soil or run in rills down to streams. Certainly the Haw will swell and even flood as all the overflowing streams drain into it, but the curve of that swell will be more gradual and extended than the sharp, high peaks of river's rising caused by asphalt's fast runoff.

In the woods, over half of this rain will transpire back into the air. Tree roots are waiting to suck it up. One full-grown oak can pull in and breathe off through its leaves 170 quarts of water a day.[341] One hectare of evergreen forest can exhale 10 tons of water in the first hour after a heavy storm.[342] This vapor is the woods' breath and an important flood control. The few lovely Bradford pears planted in thin, concrete-rimmed islands at the mall cannot compete with woods for evapotranspiration.

Parking lots may steam and gather a few puddles, but the rest will drain quickly, causing bank erosion and possibly flash floods. After one inch of rain, one acre of asphalt can pour 27,000 gallons of water into the Haw if none evaporates.[343]

Something else important is going on when water rushes off the asphalt and lawns of Burlington. Rain runoff picks up and carries heavy metals (e.g., from brake pads), paints, solvents, fertilizers and other debris of urban life. Gasoline rainbows in gutters are one sign of this toxic runoff. In the woods, rills will carry organic matter in limited quantities good for stream life. Any gravel washing in provides places where fish spawn.[344]

A large collection of roofs and roads is a problem for the river — just developing 10 percent of land with hard, impervious surface brings a decline in water quality.[345] Most people think of pollution coming out of a pipe, but this non-point source pollution — runoff — is worse and comes from everywhere.

Two under-road pipes speed up water flow and cause the blasting of stream banks shown here, sending more sediment toward the Haw. The lack of a stream buffer means that manure from the cows will eventually run off into the Haw, raising nitrogen and phosphorus levels (photograph by the author).

But there are solutions. For Patrick Beggs of Ecosystem Enhancement Program it is captured in a word: "disconnectivity: not channeling rain from the roof down a pipe across an asphalt driveway into a gutter pipe to a creek."[346]

The best idea, Beggs suggests, is rain gardens with plants that soak up water at each urban home so it can pause and settle some of its load before moving on to the river. Also, runoff, rather than rushing directly to a stream, can be channeled through wastewater treatment to reduce some pollutants and sediment. Stream buffers absorb three times the pollutants of grass.[347] A pond near me below a fertilized field covered over with algae in a few months after a narrow tree buffer was removed.

In cities, retrofits help. "You don't have to go in and tear out city blocks and add ponds," Elaine Chiosso of Haw River Assembly explains. "What we are talking about are adding dry wells, ten feet deep and filled in with rocks, which create instant water storage. You can take a parking lot and dig up six spaces at the end where all the water is running to and put in a rain garden. You know those islands of plants in parking lots ... make them lower so water runs into them. There are so many ways."[348]

Collecting water at each house in 600 gallon cisterns would stop runoff's rush. Perhaps we should start with labeling storm drains with their creek and river names to help people make the vital rain-river connection. Then more people would want gravel parking lots, porous pavement and grass roofs, like Duke Medical Center has, or rain water re-cycled to mop floors or flush toilets.

With the hardened surfaces of city and suburban sprawl spreading over four times the

A healthy stream is cooled by its woodland buffer, and does not flow low in deep-cut, bare banks. Surrounding woods absorb rain and lessen its force in streams. Granddaughters Sierra Herold and Jamie Lee Potts look on (photograph by the author).

area covered 60 years ago, it is time we make changes to revive rivers.[349] The good news is these innovations also make urban life more pleasant.

Vestiges: Alston Quarter

Rivers hold mysteries and Alston Quarter below Saxapahaw holds two of the Haw's — a plantation and a mill village that both have vanished. This tract passed to John Porter in 1728 and to Roger Moore and his son George in 1751.[350] But the records are still buried for the Alston plantation which stretched from roughly two miles below Saxapahaw to Cane Creek, river left. Joseph John Alston, "Chatham Jack," was reputed to own 40,000 acres in Chatham County and, by an 1810 count, 168 enslaved workers. The family spread out in smaller plantations, one of them likely here. Today the Alston Chapel Holiness Church carries the name, but it has morphed into Austin and Alyston elsewhere.[351]

In the 1920s, a whole mill village was here except for the mill; falling cotton prices intervened before J.W. Menefee's River Falls Cotton Mill was hammered on the Haw's banks.[352] Only four chimneys are left of the big house where W.J. Brewer, or O Boy, as he was called in his Alston Quarter days, grew up in the 1920s and 1930s. He remembers those days by the Haw:

> The only thing I can really tell you about is I was born there in 1923 and raised there, but I left there when I was 16 years old.[353] Anybody be down there, after they stay long enough, they will run away. What I meant by that is, you didn't know what a tractor was back then, you used mules to plow. You followed them for years and years, and you'd say when I leave here, I ain't never coming back.
>
> No mailman, no paper, no electric lights, no radio, and the only way we could tell it was 12 o'clock was to stop, and if you could stick your foot's shadow out like that on your head, it was 12 o'clock.
>
> On the farm, there is something to do every day in the year. I carried 13 dozen eggs and walked 3 miles and 6 tenths with one of them 6 quart milk buckets and carried them up there to Crutchfield's for 65 cents, and he sells to the man that comes around and sells butter and eggs and goes on and makes more. I traded them out for 5 pounds of sugar and pinto beans and something else. You'd only get what you don't raise.
>
> I had to walk 3 miles and six tenths to get on the school bus, and I had to walk 3 miles and six tenths back to the house, and then I had to do my chores; we didn't have nothing luxury; we had to cut wood, I mean stove wood and fire wood.
>
> The house was built for electricity, had bathtubs, but there never was no electricity turned on, never was no water running. Everything broke loose and Mr. Minifee went bankrupt.
>
> I lived in the big house…. As far as I know nobody ever lived in it [except] my granddaddy, Fletcher Brewer. It had big columns in front. Downstairs it had six rooms and seven rooms upstairs and the third floor wasn't finished. The ceilings were so high you would have to put a ladder on a table to change a light bulb. It was going to be a boarding house. They built about 20 or 30 houses; they were going to start a mill down there, but they never did get the stuff from the river up and in place.
>
> All my granddaddy did was look after the place. He could grow what he wanted to and didn't have to pay no rent…. They had at one time 2000 acres, and they growed tobacco, corn. We had watermelons on the river bottom. It makes huge watermelons. We sell 'em for 25 or 40 cents, big old watermelons, they'd weigh 50 or 60 pounds, they would.
>
> I would say [the mill village] was a half a mile from the river … not like Saxapahaw Mill. We lived back up cause that old river then could get wicked. Since they built Saxapahaw

Dam the water hasn't been high down at that end like it used to be. They held it all back. No, I did not play in no river.

When that river is low, it has a lotta rocks, gives you about fifty ways to jump from one step rock to another. Then come a lotta rain, the water come out and look like an ocean. It can get weird that river.

Granddaddy went and got him two goats, and it wasn't two months that word got to Saxapahaw that there are so many deer at Alston's Quarter that you had to run them out of the way. People started coming in there in droves. That's what started all the fires. And they knew the old man was old ... and they would fool with him. He stayed down there until about 1950.

Then they sold the land I believe to Jefferson Standard Life Insurance Company in Greensboro, and I think they just used that for bird hunting for the crowd to come down, we called them city slickers, to hunt, you know, in their time off. And after that it was sold to West Durham Lumber company in West Durham, North Carolina, and after that, I can't tell you who got it. The lumber company cut all the timber off. Duke was in there; they had like a bunch of trailers over like in a bird sanctuary, maybe chase a rabbit and find out how he goes and what he does. Then they let this old landfill come in and just about take it over.

[The big house] did burn and [left] nothing standing but the chimneys and the silos. Must have been in the late nineties. The roof is still there on top of the grass, ain't nothing but weeds around it. That tin didn't burn, just roll up and ... there are four or five chimneys, they had a fireplace in each room.

I know I seen markers on a tombstone, 1800s and I thought, "O Lord, that's way back." Up from the spring house a pretty good way ... there's a graveyard there where all the slaves are buried. The kind of markers it had on it was just like a big rock, cut your initials in there, stuff like that.

I went to show my nephew where those graves were, and we got up there right close to the big house and I said, "See this here big place. I use to plow and chop cotton in here." My nephew just looked up at the trees, then at my son, and said, "Watch out! He is getting ready to have a stroke." My son didn't believe me either.... There ain't a field down there nowhere ... nothing but woods and just four chimneys and a silo made out of this dark maroon brick.

I left at 16 and come to public works here in Graham, hardest work I ever did in my life, and it wasn't nothing but a city slicker job. That was the tiredest you'd ever be — the tension you'd take or the pressure.

I loved it, I loved it. I'd rather took and lived down there than in Elvis Presley's mansion. Whippoorwills and katydids. Shoot you could go down there and get joy; you don't never have no headache; nobody don't never bother you or nothing; eat three meals a day, fresh food. But at 16 you wanted out.

Yeah, I think everybody ought to go to Alston's Quarter. You could get out and holler and nobody would hear you. And that's in the past now; it's all grown in now.

Landfill

My afternoons of searching never bring me to the four chimneys, but they do take me right past the landfill on Austin(!) Quarter Road. Vehicles, 1800 of them, come in here each week with garbage from Alamance County's homes and industry. Roughly 3.5 million gallons of leacheate — the inevitable drippings of a landfill — leave each year headed for the East Burlington Wastewater Treatment Plant.[354]

Siting the landfill here caused an uproar in the years before 1994. That simple statement collapses months, even years, of people's lives into one quick sentence. Many fought to protect the Haw from the inevitable time when toxins escape the protective barriers. Groundwater has only 1000 feet to go before it is in the river; accidental fires can lift toxins into the air. However, the battle was waged and lost.

Those defending the landfill argued that it was state of the art, a far cry from the junk or burn piles of early 20th century dumps or even the dirt covered landfills of the fifties. Alamance County's 534-acre landfill is excavated, has a liner of "2 feet of compacted clay, a betonite mat, 60 mil HDPE liner, geotextile fiber materials, leachate drainage material (geonet/#78 stone)" then pipes catch the drippings which are sent off for wastewater treatment; compacted waste is covered with almost a foot of topsoil; add to that eight wells that monitor methane gas, twelve that test ground water, and seven that sample surface water.[355]

Despite all this, one protector of the Haw surveyed it and said: "You're telling me that a pile of rock and this orange fence is the technological edge?" After all, any mercury, cadmium, or lead needs to be contained for thousands of years, and even best management practices cannot guarantee that. (Landfills of the Roman Empire are still leaching.)[356] Heavy metals in the leachate sent off are not broken down by wastewater treatment and may be piped back into the Haw or seep back from bio-solids.[357] They are monitored at the wells to see if they register above EPA standards, leaving us to rely once again on the EPA's questionable standards.

We are spewing forth great bundles of trash each week, 552,000 pounds a day.[358] Trash must go somewhere and be managed, but right next to the Haw was an unfortunate choice — the river soon becomes drinking water.

Just when the wound of the landfill was crusting over, it was reopened by plans to privatize it and to take in trash from a 60 mile radius. While these plans could yield a fine income and help neighboring counties, the idea of tracking more toxins to the Haw met with sharp resistance and was curtailed.

Greg Thomas, director of Solid Waste at Austin Quarter Landfill, has a different role to play in protecting the Haw — managing the landfill safely. He draws the focus to what it does do: meets all requirements and acts as a major recycling center, e.g., 25 tons of plastic, 27 tons of magazines and 304 tons of newspaper in one year.[359] Add to that a paint and pesticide day to sequester those wastes elsewhere. And he makes sure I know that the county-run landfill is self supporting — running not on tax dollars, but fees.

Perhaps we should be paying more for sending on our garbage. Some researchers suggest that the real costs would mean 15 to 25 cents more each day per person.[360] This financial pressure could make us less tolerant of over-packaging, less ready to buy and toss.

Reduce, Re-Use, Recycle adorns the Alamance County landfill's web page, but the local paper reports that only 57 percent of Burlington residents currently recycle weekly.[361] Greg Thomas is aware of the problem and offers a fact sheet that promotes recycling: one can of motor oil down a city sewer contaminates a million times its volume in the Haw; Americans pour away 120 million gallons of recyclable motor oil each year, forcing the extra purchase of 1.3 million barrels of oil per day. Yet, five gallons of used car oil, car batteries, or five tires off the rims can be recycled at no charge at the landfill.

Solutions require thinking, narrow and wide, about what we buy and what is produced. Will we watch our flow of trash, six pounds a day per person on average,[362] and change our habits? Thinking wide, we need to lobby the Environmental Protection Agency (EPA) to require proof the new chemicals added to the waste stream — seven each day for the past 25 years[363] — are not toxic.

Native and Techno-Haw: Saxapahaw to Chicken Bridge

Forecasts mention thunderstorms, but it's a fine May day for a paddle from Saxapahaw to Chicken Bridge, so I shed any worries onto my paddling companion, Bob Brueckner, past president of the Carolina Canoe Club. He's lean, full of energy, only his grey swirling afro tells you he's not a teenager. Today's decisions are his, and, for insurance, so is the kayak I am using.

The river is very low, but as I learn on the shuttle, Bob is evangelical about nature and paddling: "Previously I took hikes; half the reason I got into paddling was to get out in the woods. Then I did get hooked on whitewater," Bob says, revealing his dark side. "Well, whitewater is what gets me out on a 25 degree day in February. The scenery is nice and still is, but...." The "but" means the part of the Haw we paddle today is off his charts. Kayakers' Upper Haw doesn't even start until Chicken Bridge, where, I am told, Bob Brenner says "the best whitewater east of the Mountains begins."[364]

"Probably is, it's got the most volume and rapids and about 12 miles of whitewater. If it has enough water in it, it's great to run. And it's close. It's lots of paddlers' first whitewater experience."

That gets Bob reminiscing about the early days of Carolina Canoe Club (CCC) and canoeing on the Haw:

> CCC began, I think it was 1969, and there were maybe a dozen people in it. Today there are about 1000. But 1969 was before my time. It was Bob Brenner; he was the first president; William Nealy was still in Alabama, but he and Howard DuBois and Paul Ferguson were part of it too later. They were essentially the first explorers, the guinea pigs, doing the first recorded descents. They had no guidebooks, no designated access areas. You just put in at bridges and hoped the landowners weren't going to shoot you. Bob Brenner wrote the first guide book from these trips. People went out without helmets, then they finally had them, but they were construction helmets.
>
> When we first started [on safety], we had this elaborate rope system for extracting people from the river, so you would set up these 27 point Z drags. You were supposed to construct Tyrolean slings across the river [like mountaineers]; then people realized in real life you don't have that much time. You have someone stuck on a rock in the middle of the river, and it's 40 degrees, and they are going hypothermic, so strategies changed to get rescuers to the victim as soon as possible. But having trips and teaching people how to paddle and teaching safety skills — that's a lot of what the club still offers. The club now has a range of trips for beginners, novices, intermediates and advanced boaters ... at least one trip per week.
>
> Before, you went and looked at bridge pilings [to see how high the Haw was]. Now you can look on the Web site[365] that updates every 15 minutes, so we are really spoiled these days.

Bob even has his own gauges. When he pulls out his GPS monitor, I know I am in for my most technological paddle. He has Paul Ferguson's guide memorized, plus topos. We are running under the water level Ferguson says is absolutely essential — 200 c3/s; that's the number of imaginary boxes filled with a cubic foot of water passing a point every second. Two hundred sounds like a lot, but it's a wide river.

From the shade of the bridge, the Haw is a radiant blue mirror. A Y-shaped branch suspends from a telephone wire, a reminder of how high and wild this river can be. The mill with its 28-paned windows stands above us, its bricks warm in the sun.

Accustomed to running groups, Bob hits the safety basics, mostly for a river that's up.

Don't tie it down if you don't want it. If you swim, it's all garage sale.

Stay away from strainers. They let water through, but not solid objects, like you. If you are caught in one, try to climb on top of it.

Have maps with you to show you roads if you need to exit before you intended.

Wear a hat. Use sunscreen, drink plenty of fluids. Dehydration can lead to fatigue, heat exhaustion, heat stroke.

Look at where you want to go — not at the rock, but the bead next to the rock. It's uncanny, but that makes a big difference.

And with that final advice, Bob swivels his yellow kayak, and I hustle after.

At .6 mile: Within sight of the mill, a funnel of rocks has one outstanding, the only obstacle; I hit it.

Bob has seen it before. "I know where you were looking." I deny it, but later admit looking where I want to go helps. The funneling rock formation is likely a fishing weir, endlessly rebuilt, directing fish into a net or basket.

Mote's Creek comes in river left; next the landfill goes by, its only marker a gravel road, then pasture land where a few cows have cooled off in the Haw. The mother is the color of the red clay she stands on, her calves are two black shapes against the ruddy bank. Despite their ecological damage, their beauty holds me.

"No cows ever attacked me," Bob says forgivingly.

Rocks are everywhere; even Bob comes to full stops on sleeper rocks. Last week-end, he went to run the Nantahala, but "the river was broken" — the dam release kayakers run on met with turbine problems. Bob is philosophical and patient on the pre–Upper Haw as well. He spins in his boat easily, catches eddies snugly, and whistles along. Technical readings point to what he yearns for more of. "Our running speed is 2.6 mph, average speed 1.6, maximum velocity 79.3 mph." Even under his helmet, ball cap, and strap-on glasses, I can see an eyebrow go up. "I don't entirely trust this GPS. Well, maybe the clock."

I enjoy the lazy Haw as do guests at the River Landing Inn and Benjamin Vineyards hidden in the right bank's woods. The inn and its swaying paths are far above the ordinaries of the past. Both inn and winery benefit from the beauty of the river here: isolation and abundant woods with elderberry blooming in profusion. Ferguson rates the "scenery" an *A*, Bob tells me. The only human signs are those of river love: stairs down the bank, hammocks, odd chairs, and a hut on stilts.

River right is the trail of rocks that mark the bridge replacing John Woody's Ferry which ran here before 1759.[366] Governor Tryon was relieved to cross here without resistance on his way to battle the Regulators, May 12, 1771.[367]

We pass Mary's Creek, river right, the site of the Union School, for girls and younger boys, one mile south of Woody's Ferry; in 1818, advertised as an institution "in which will be admitted 10 or 12 who may be taught ... the following branches of Literature: Reading, Writing, Arithmetic, English Grammar, with the Art of Scanning Poetry, Geography, Drawing, Painting, embroidery, and other kinds of needle work. Terms, $16.50 for board and tuition a quarter, which must be paid in advance.[368]

Soon a long island, river left, signals Alston Quarter plantation and Minifee's mill village and the big house, with its four chimneys hidden in the woods. We take the left channel by the island, an emerald world of vines and leaves ... and tires.

Bob is familiar with tires; "I saw 7 or 8 in there." Hard to explain when both Guilford and Alamance allow free drop-off of five tires.

"Looks like we need to do a clean up here," he says. It all comes down to the river, as Bob knows who has been part of HRA clean up days with hundreds of others on foot or in boats. One year, 161 tires and 552 bags of trash were pulled from the Haw full of TV's, balls, fridges, lawn-chairs, toys, pipes, couches, boogie boards, signs, needles, condoms, paint cans, patio umbrellas and endless cans and bottles.[369] Plastic bottles are my main gripe. The 30 billion that became waste in 2006 cost us 17 million barrels of oil and 110 billion gallons of water to make. Only 20 percent are recycled, so here they bob in the Haw.[370] A "deposit" bill would change this.

Bob's pet peeve lies elsewhere: "Fishermen who put all their garbage into a bag and then leave the bag. Why can't they do the next step?"

At 3.54 miles: We hit another ledge, and Bob recognizes an old dam in the straight wall of rocks.

"I like this section; it would be good for training." The kayakers' unacknowledged Haw is winning him over. "The Haw is a lot clearer here than when it leaves Burlington. Two feet below, the bottom is sharply visible."

At 3.97 miles: We lunch at Little Saxapahaw Falls where the white splash on rocks and dismembered crayfish parts show a heron preceded us. Falls and ledges like this make demands in their own technical way, but do not spell excitement. Bob likes his rapids in the Class II and III range. "Fear," he explains modestly, though he does like moving water. "I'm lazy." Yet through this 90 degree day Bob paddles effortlessly in a stubby kayak not made for tracking.

Below the island and before Cane Creek, we travel through the Box Elder Bottomlands where water-tolerant green ash, elms, sycamore, river birch, and sugarberry abide along with ... box elder.[371] It's the only maple with compound leaves, but the seed pods will look familiar.[372]

At 4.67 miles: Cane Creek is river left (call it "upper" because there is another Cane Creek below, river right). Up this one, Morrow Mill churned in years past.[373] Now, north of Rt. 54, Orange Water and Sewer Authority's 72 foot high dam holds back Cane, Toms, Caterpillar, Turkey Hill and Watery Fork Creeks to form Cane Creek Reservoir, a four-mile lake in a beautiful park. When the 2007–2008 drought sparked contingency plans to pump water in from the Haw, a better plan to re-use water was already in place. Reclaimed water, not drinking standard, will take care of UNC–Chapel Hill's cooling system — 6 percent of their use.[374]

The wide and shallow Haw continues on, bordering Orange County for three miles while the land on the right turns to pasture. Above it, wet-slate clouds bank the sky. I hear thunder; Bob tells me it's a jet.

At 5.68 miles[375]: We sight Old Greensboro–Chapel Hill Road bridge and descend remnants of a man-made dam. Bob points out the take-out, a miserable incline of raw clay. The Haw River Trail and Alamance County Parks and Recreation are working on better access.

At 8 miles: Lower Cane Creek streams in river right through quirky loops descending from its namesake mountains. Before Quaker settlers, Native peoples lived here in the Woodland era, 3000–400 B.P., on two now excavated sites. In Woodland times, the People of the Haw became increasingly foragers and cultivators of its floodplains.[376] By the end of the period, they were primarily an agrarian people, who had added beans and corn to their

The Haw is seen here from Greensboro–Chapel Hill Road (Old Greensboro Highway) bridge; well-buffered banks protect the river as rapids add oxygen to the water (Sönke Johnsen).

diet, crafted pottery, hunted with bow and arrow, and settled in hamlets and small villages along the Haw and other rivers, grouping and identifying by rivers.[377]

There are other Woodland settlements, a village-size one is on Back Creek, but its dimensions were nothing like Town Creek on the Pee Dee, where ceremonial buildings and mounds suggest hierarchy of life and labor and possibly a need for protection. The people of the Haw are thought to have been "egalitarian shifting settlements" with relaxed tribal boundaries.[378] Their thin dispersement may have been due to this security[379] as well as the Haw's narrow floodplains.[380]

The signs of late and brief Woodland habitation at the Guthrie Site on Cane Creek, near Route 87, come in its pit features and pottery that are primarily net-impressed.[381]

Seven centuries ago, corn, sunflower, squash, and beans might be growing here at water's edge. Foraged food like hickory and acorn nuts were likely soaked in water and wood ash, and rinsed in a creek before being dried and pounded into meal.[382]

A few artifacts are all we have to rough sketch a life with grave pits to suggest their spirituality. On the Pee Dee River at Town Creek, the People of the Fire celebrated the Green Corn Ceremony each year, fasting, bathing in the river, destroying old food and fire to start anew.[383] Did the People of the Haw as well?

How did the Haw's Woodland people celebrate the wonders of their life in abundant nature or under skies thick with stars? Had they heard the Cherokee story that the dense slice of our galaxy, the Milky Way, was cornmeal scattered by a dog caught stealing and yelping home?[384] Without the false security of our endless night lights, they were open to a miraculous universe, floating in its majesty, night after starry night.

We have more record of the Quakers who built the first meeting house in 1751 (still there at Snow Camp Outdoor Theater). Six years later, there were five meeting houses.[385] Mills spread everywhere too: Stafford/Henle, Guthrie, Lindley/Sutphin, Allen, Holman's, Ward Mill and Dixon's Mill where Cornwallis filched grain — all ran on Cane Creek waters.

Take the branches of Cane Creek, and there are even more mills. One is the exceptionally high McBane Mill which Ruffin Hobbs bought and transformed — blacksmith shop into sculpture space; saw mill into furniture factory.[386]

At the Cane Creek Cotton Manufacturing Company, John Newlin first experimented with cotton weaving in a mill built by Peter Stout around 1770 for grinding grain.[387]

Lindley Mill which Thomas Lindley and Hugh Laughlin built in 1755 at a "c" curve and sharp drop in Cane Creek is still run by Lindleys, nine generations later, grinding organic grains under electric power. Joe and Teresa Hadley Lindley are co-owners, both descendants of original settlers, as they discovered while dating.

The Battle of Lindley Mill took place on September 14, 1781, when Brig. General John Butler ambushed Loyalist Col. David Fanning on his way to Wilmington with his prisoner, Governor Burke, newly-elected by a Whig Assembly. The four-hour battle led to a gruesome score: 60 patriots and 40 Tories dead, by one account.[388] Burke went on to Wilmington, a temporary captive; five weeks later the war ended.

Cane Creek marked the Chatham County line until 1897 when, in a less river-centered time, the General Assembly changed the boundary to a straight line and dropped it 2¾ miles south.[389]

Below Cane Creek, Bob and I pause on rock formations, called tombstones, to cool off in the Haw. Hundreds of tadpoles feed in rock puddles. Below, there are fish wider than a hand-span, and we have passed half a dozen men fishing from the shore or boats — all omens of river health.

At 9.5 miles: River left, rocks are strewn so thickly you could walk across on them. Above, audible power lines cross the Haw. "I've only heard them once before — Quebec hydro lines — never down here. They're like a June bug, a really big June bug," Bob says. The sound of crackling arcs sends me under them at maximum speed.

At 10.67 miles: Jeanne's Falls. Bob flows deftly through, but I walk, thinking of the bottom of his boat. White streaks on rocks indicate it is too late.

At 11.06 miles: At a gushing rapid, river right, I study the rock and the strainer behind it instead of the bead through ... and hang. Fortunately, I know to lean toward the rock. Unfortunately, as Bob tells me, that advice no longer applies if the gunnel is under water. However, nothing lost to the garage sale.

At 11:39 miles: We head right of a long island to explore Terrell's Creek. Timothy Terrell, here in 1754, put his name to another Terrell's Creek, below river left.[390] With spelling that slid from Terrel's to Tyrrett's, finding river addresses can't have been easy.

At this Terrell's Creek, rocks covering opposite banks mark the old road to the ford. The sharply rising cliffs not only suggest the mountains but support mountain plants, galax, mountain laurel, and wintergreen (chew the leaf and you will recognize the taste). Close to the Haw are herbs as lush as their names: bloodroot, mayapple, blue star, black cohosh, and green dragon. Red-tailed hawks cruise above and river otters below; neither live where they are crowded by development.[391]

Terrell's Creek leads on to Baldwin's mill, first operating in 1790 when Washington was president. It began the bond between mills and Hobbs family that has held for four generations, as Grimsley Hobbs, Jr., tells it: "Grandfather [Richard Juniors Mendenhall] Hobbs was a professor at UNC and something possessed him to buy the mill.[392] He just fell in love with it, still standing, but in the process of falling down. He did buy it in 1941 and started work on it with his three sons. The dam was repaired in 1949, the mill put back in operation in 1958."

One of those sons was Grimsley Hobbs, Sr., future author of *Exploring the Old Mills of North Carolina* and president of Guilford College. He helped strip it down to its hewn timbers, 45 to 50 feet, "the limits of what timber they could get."

The mill anchored a small village of shops, grist mill, sawmill, blacksmith, and dry goods store. There is even an opening through one of the stone foundations that is reported to have been for a still. Dances and social gatherings were held there.

The Hobbs homeplace is preserved "for historic value and just the love of it." It is a high maintenance relationship. "Mills like ours make no money, but take a lot," Grimsley Hobbs notes stoically. The diagonal crack in one of the stone foundations requires repairs. The mill will not be grinding until that is done.

"When mills run, they vibrate all over," Hobbs explains matter-of-factly. He knows first hand. "I've lived in a mill or on a mill site much of my life; that could be one reason I became an architect." In one mill in Indiana, the belts ran through his bedroom. When the mill was grinding, "Everything was moving." Hobbs knows the smell of millstones and grain, too: "You get rock dust in the flour; I know I can smell it."

Rain splatters penetrate the trees, and with black clouds, we move on. Bob has devised shelter for whole groups when skies over the Haw let loose. "The one thing you don't want to do is be the highest object, and you don't want to go near a tree. It's best if you can find a little field, just get in that as far as you can. [Lacking that] get into a creek if you can get up on some rocks. It's somewhat counter intuitive, but the rocks aren't going to conduct."

Below Terrell's Creek, an island in the Haw ends at a stone ledge which helped form a 700-foot-long dam for Love's Mill, atypically river right. We slip and bump down through its rocks. Henry Armand London reports a bridge here in 1876.[393,394]

Chicken Bridge is our take-out. Not only are the pillars exposed, but six inches below. Ferguson's advice: do not attempt at this level. Our final reading is 200 cfs below runnable. (Bob only got out of his boat once to peel me off a rock.) Our stats:

4 hours and 34 minutes running time
2 hours and 26 minutes stopped time
1.9 mph average speed
2.7 mph moving speed
12.2 river miles
13.7 car miles (The river is quite straight here.)
79.3 mph still our maximum velocity

On the shuttle back, Bob pitches paddling.

> It's 15 times safer to paddle than to be driving in this car. [On the water], jet skiers have the most injuries and motor boats the most fatalities.
>
> If you paddle the Haw often enough, you'll build a relationship with this river. And I have

a lot of relationships with the people I paddle with on the Haw. One February, I was out with some friends on the Middle Haw. On a branch ahead of us was a bald eagle. He had a fish in his claw. We all paddled under him, and he didn't fly away. That's an experience. That's why you protect the river.

We've stopped exploiting it in some ways. Canoers used to see the river change colors all around them, dyes being released. But now there's more run-off. The data on the Haw show it rises higher and faster. That's wetlands loss: not much soaking in.

The next day Bob e-mails me; the Haw spiked to 1000 cfs that evening. The line of the waterdata.usgs.gov chart shoots up almost vertically. A mighty rain unleashed a very different river — Bob's Haw.

4

Water Power and Whitewater
Chatham County

River Dwellers: Rock Rest to Brooks Creek

The Haw's whitewater is marked by Chicken Bridge which is named for its collapse. In the early 1950s, Walter Hugh Campbell was on his way to town with a load of chickens when part of the bridge gave way under him. Chickens were liberated, but Campbell was pinned by a truss. Fortunately, his brother-in-law Harry Fox, behind him with his own chicken load, slammed on the brakes in time and waded across to rescue Campbell. That night flashlight beams crisscrossed the darkness as chickens were still being pursued.[1]

Chicken Bridge may offer the most literary inspiration of any place on the Haw. Doris Betts' novel *Souls Raised from the Dead* opens with a variation on the liberated chicken theme set nearer Chapel Hill. Doug Marlette's novel *The Bridge* describes Halloween on the bridge with mill workers carving pumpkins. (In the 1980s, crowds did set pumpkins' jagged faces to gleam on the Haw until police ended the fun, fearing a collision.)[2]

The old truss bridge that gave way caused a long detour until a one-lane wooden bridge replaced it in 1954. In Vietnam days, that bridge served as sighting practice for A-4 Skyhawks and F-4 Phantoms.[3] Their noise was real enough for community and canoers' protests to end that practice. A 1988 cement bridge is there now.

Terrell's Creek II, or Little Terrell's, joins the Haw, river left, at the end of Watermelon Island after the first big rapid, Sawtooth Ledge. In its bottomland shagbark hickory and willow oaks shadow dazzling red cardinal flower and white lizard's tail. The dragonflies found here indicate good stream health and sound made for fairy tale illustrations: great blue skimmer, fawn darner, black-shoulder spinyleg, rare Thorey's greyback, common sanddragon,[4] and widow skimmer.

Below Terrell's Creek, also river left, was the mill built by Henry Lutterloh in 1790 which came to be known as Dark's Mill before it was carried off in a flood sometime after the 1850s.[5] The remains of its dam upstream provide fun for kayakers today.

The mill was sold in 1811 to Edward Jones, NC solicitor general at one time, and his wife, Mary Curtis Mallett. They built a home, river right, c. 1801, and an estate that flourished with help from 11 enslaved workers.[6] Their name for it, Rock Rest, has held; the rock walls here are unusual. Captain Johnston Blakeley grew up here, foster son of the Joneses and captain of the *Wasp*, which had "many victories [in the] War of 1812," as the marker on Route 87 states. Blakeley's fate was cast with water. His brother and mother died on the

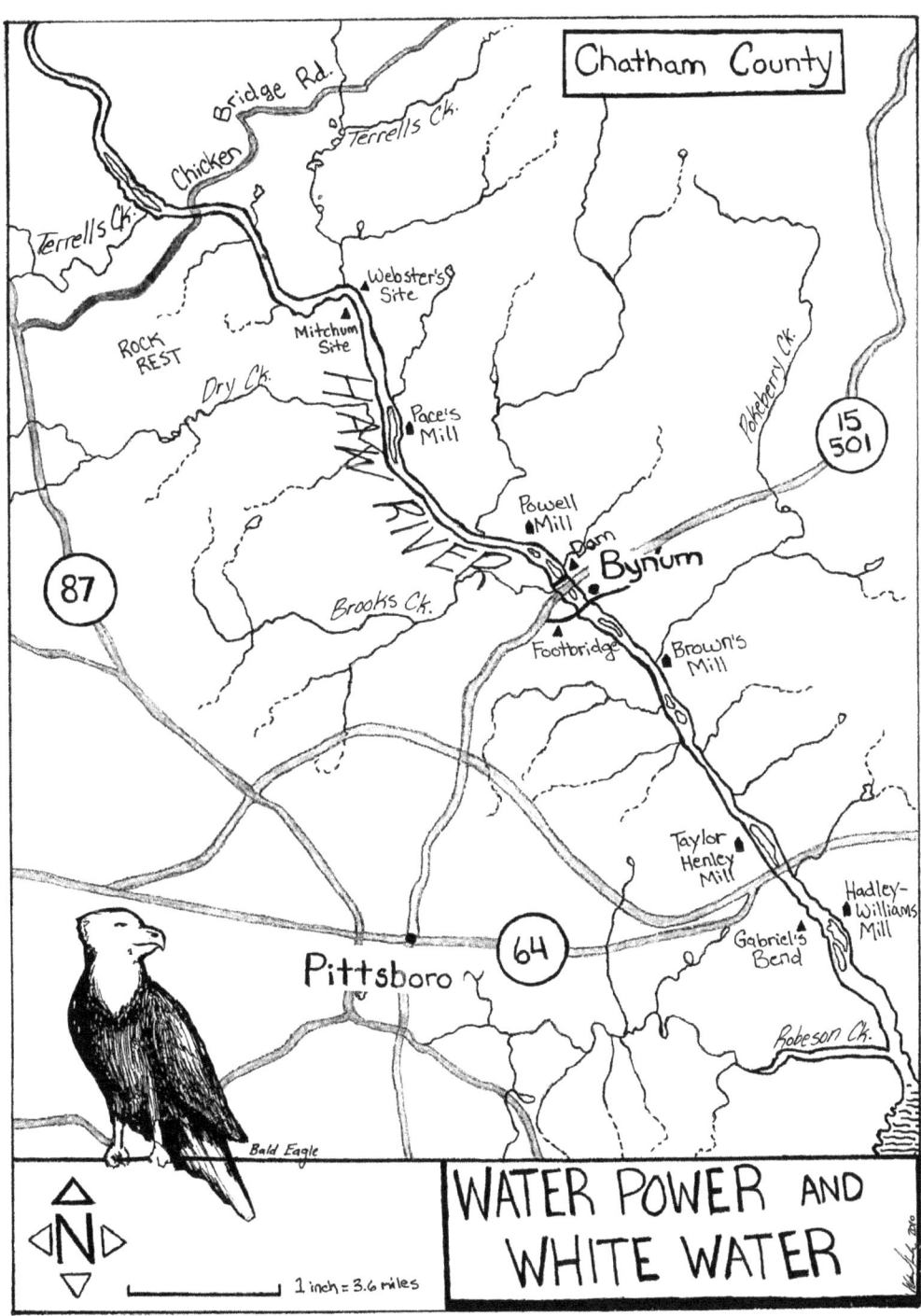

Water Power and Whitewater (William Rusch).

A male widow skimmer dragonfly lights on a rush near the Haw. Dragonfly larvae in streams indicate good to fair water quality; in mature form, they consume large quantities of mosquitoes. Ancient dragonfly, sometimes huge, appear in the fossil record 300 million years ago, even before dinosaurs (Sönke Johnsen).

sea voyage from Ireland in 1783; Edward Jones became his foster father, so he lived near the Haw. And Blakeley became a hero on the Atlantic,[7] burning or seizing British Trade ships, and then died in its waters when his sloop mysteriously vanished.

Two schools were once in the area. Mary Curtis Jones ran a school for girls at Rock Rest until 1831, when it moved to Pittsboro and became Kelvin School, in a less isolated location. In 1870, another nearby school, Rock Rest Academy, would feature its rural location as an "advantage to boys and girls. Is situated nine miles north of Pittsboro in a moral community and healthy country."[8]

Other families, Dark, Mann, Durham, Eubanks, West and Crawford, also settled by the Haw here. A grain mill, river left, most often known as Pace's Mill, went through name and owner changes that the road mirrored — West Mill, Cottons Mill,[9] Pace's Mill, now Rock Rest Road. This major road, now devolved into a ditch, led on to Graham and Chapel Hill from Pittsboro and Fayetteville. In the 1870s, Willis Durham started to gather lands along the Haw here including some of Rock Rest. They passed through his children to others, and a parcel was sold to a young couple, Jerry and Cathy Markatos, in the early 1970s.

Cathy Markatos says they were eager to "live on the land" and make clear "we were part of the solution and not a problem as some of the elders indicated."[10] The Markatoses sold parcels to like-minded activists and artists, "interested in cottage industry," who built their homes in their own original ways. A small amphitheater held plays, like King Lear with neighbor-actor Tom Marriott, and performances by the Carolina Chocolate Drops band and others. Rock Rest's informal community was strong; Cathy Markatos tells me, "At the 20th anniversary reunion, everyone came back."

Judith Peterson, who moved with her husband to the West homeplace adjacent to the Rock Rest community, is a gatherer of local history.[11] The artifacts she pulls out of a drawer attest to Native peoples living on the land. We walk by traces of a much later Pace's Mill community. Our first stop is the two-story schoolhouse. When the logs were core drilled, they dated back to 1819, and a deed from the West family grants "one acre for a school house." A doorsill, worn through, suggests rapidly moving school children. Judith Peterson figures that there were at least seven houses on this side of the river and more on the other around Pace's mill, plenty for a school.

Pace's mill and bridge were there by 1828, as Rev. Mann, one of the many Methodists, attests in his diary of January 5, 1828: "Old Christmas. They go on their way to James Kerby's. All this week is warm as summertime. O Lord save me from all evil. This evening I walked over Haw River to Isaac's [West] and stayed all night." Peterson's source, J. Lamont Norwood, notes: "Rev. Mann was devout in the extreme, but he did not walk on water. Although it is difficult for me to believe Pace's Mill Bridge was built by 1828, no other explanation seems to fit."[12]

We head on toward the Haw, stopping by Dry Creek where rock walls hold stream and soil apart. This was once farmed down to the creek, as Judith Peterson learned one day when a son brought his father under Hospice care for a last look at the earth he had tilled.

From here, Dry Creek takes an extravagant bend around a rise and enters the Haw. Along the road, Judith Peterson points out the ditch of the old mill road weaving from one side of the current road to the other. Down its final slope is the Haw, its waters full of mafic rock, rich in magnesium and iron, heavy metals weighted deep in the earth crust until a volcanic outpouring brought them up eons later.[13] Brush and young trees block anyone who might gallop out of the past headed for a bygone covered-bridge, one of many that Judith Peterson says were "washed out by floods fairly regularly." What stand now above the Haw are massive brick supports.

A tornado twice took Pace's Mill away from where it stood by the bridge across on river left. The *Chatham Record* of May 8, 1924, describes the second tornado with graphic detail:

> One of the worst storms that has passed over any part of Chatham took place at Pace's Mill, on Haw River ... when three people, a man, his wife and his son were hurled into eternity in the twinkling of an eye, besides seven others being wounded, some of them probably serious....
>
> Mrs. Blalock was blown through the debris straight up to the top or level of the hill, which is over 100 feet high. Mr. Blalock was blown 125 yards through timber and brush and his head and body were terribly lacerated. His head must have hit a tree or some hard substance, causing instant death. The baby was blown 200 yards from the house and was found in a gulley nearly covered with mud and water, its little face being just above the water thereby saving it from being drowned.
>
> The Pace old grist mill, standing nearby, a large three story frame building was completely wrecked, the big building being turned over and falling in the roadway.
>
> Across the river ... was a steel bridge, said to be one of the longest bridges in Chatham, and containing eight spans[.] [It] was partly blown down, only two spans being left standing in the middle of the river....
>
> A seven year old girl was almost literally scalped ... skin on the top of her head, almost as large as a person's hand, was torn loose and hung by one end.[14]

Just above Pace's Mill, mafic rocks redirect the river's flow. Their dark tones come from magnesium and iron sunk deep in the earth's crust until ejected as magma (photograph by the author).

The bridge was not rebuilt nor the flour and corn mill with its four sets of millstones; the old sawmill, wagon and blacksmith shop were no longer needed.[15] With no bridge, no mill and only the Haw left in the three part industrial equation, farmers had to put up with a detour on dirt roads over Bynum bridge.

The wide Haw is calm today as we turn to walk to the cemetery where the Wests and Durhams and Manns rest. The sun slides low through the young pines, successors to timber and cotton. On the thinning fieldstones, words for young women, simple and faint, evoke our longing to live.

> In Memory of Ira West
> Who Was Born June 16, 1821
> and died [May] 23 1852
> Aged 30 Years and 11 Months
> Blessed are the Dead which
> Die in the Lord from henceforth
> Yea Sayeth the Spirit that
> They may rest from their labours
> And their works do follow them. Rev XIV 13

Two sisters are side by side:

> Hear lies the body of Kezia Perry,   Hear lies the body of Jean Curl
> d. May 13th 1814.   d. June 19th, 1814

Judith Peterson wonders if an epidemic or childbirth accounts for this double loss, commemorated with two low mossy headstones and a double misspelling. The vinca on the ground was called "graveyard ivy" I learn as we head back, our thoughts staying on the morbid. The Petersons deliberately chose not to build on the Haw: "When the river is roaring, we can hear it from the house." They have seen a series of rescues, sometimes helicopters flying in from Fort Bragg to help out. "Four people went down the Haw, two to a canoe, and one of them had canoed once before, but they heard the river was up...."

Neighbors with a riverside house did install an outside phone for 911 calls. "They were afraid there would be an emergency, and they wouldn't be at home."

When I drive out along the river road, the wild turkey and pileated woodpecker noted in Hall and Boyer's inventory for Chatham County remain hidden, but the oaks, tulips, Southern sugar maple and shagbark hickory, redbud and buckeye are here. Canoeists who climb the levee may find pawpaw, that sweet mango of the South, and umbrella magnolia, unusual and "confined to igneous formations"; this is, after all, volcanic land. Underneath the Haw's waters, a mussel, the eastern elliptio, filters sediment as it breathes and feeds.[16] On this wintry day, I will trust the books it is there.

Downstream, Europeans settled lands, putting their names to streams; (Daniel) Drummond Creek and (John) Morgan Branch joins (Thomas) Brooks Creek which comes to the Haw, river right, within sight of highway 15-501. This and other history is eloquently documented in Wallace Kaufman's *Coming Out of the Woods: The Solitary Life of a Maverick*. Kaufman warns the reader that his time living in the woods led him to a conclusion, opposite of Thoreau's: "The preservation of wildness is in civilization."[17] While he thrills to much of the life he began in 1974 at SaraLyn, the land he divided and sold, he takes special pride in the water he drinks (certainly cleaner than the Haw in 1974):

> A few hills east of Morgan Branch, the common waters of the Haw River carry an endless load of both industry and city sewage. I can draw my water from my own rock or from Morgan Branch, but Pittsboro must take Haw River water and more than half of it has already passed through a factory processing line, a kitchen sink, a car wash, or at least one pair of human kidneys.[18]

Ties to the land go deeper still. Layered artifacts at the Webster site below Chicken Bridge disclose that Native peoples lived here on one of the few wide floodplains of the Haw in the mid– to late Archaic Period of 10,000 to 3,000 B.P.[19] Archaeologists have also found their traces in Altamahaw,[20] on a ridge above Alamance Creek,[21] below the island at Saxapahaw,[22] and at the Haw River sites now under Jordan Lake.[23] North Carolina has the most Archaic sites in the Southeast, and they are particularly dense for the mid–Archaic period, 8000–5000 B.P. when North Carolina, South Carolina, and Georgia stand out against the norm of a flat or declining population.[24]

As H. Trawick Ward comments: "In almost every plowed field some trace of the Archaic period can be found. The broad alluvial valley, the rolling upland hills, and the banks of small streams were all occupied, visited or utilized at some point during the 6,000–7,000 span of the Archaic period."[25]

This period is distinguished by a shift to a more settled way of life. At its start, nomadic people are thought to have moved in bands of 20 or many more in territories defined either by their river basin, migrating to the coast for in-season resources, or in territories crossing river basins defined by migrations to quarries for arrowheads and lithic tools.[26]

These hunter-gatherers adapted to a warming climate and smaller game and became more proficient gatherers in a Piedmont that Ward and Davis describe as "a cornucopia of plant and animal foods for hungry hunters and gatherers."[27] Forests of oaks, chestnuts, and hickories, provided nuts and cover for more food — bear, deer, opossum, turkey. Streams were full of "fish, turtle, mollusks" and drew ducks and other fowl in their seasons.[28]

Hunters developed atlatls, stone axes with weights for added power, to knock out prey at a distance. Arrowheads mutated in style from Palmer, Kirk, St. Albans, Stanly, Morrow Mountain, Halifax, Guilford, to the elongated Savannah River, which was perhaps a spearhead and a multi-purpose tool. Palm-sized river cobbles hammered tools or mashed nuts. Notched pebbles added weight to fish nets. Gourds and turtle shells likely served some functions of pottery, yet to come.[29]

Across the millennia of the Archaic Period as agriculture began along the Tigris and Euphrates, as Egyptian mummies were wrapped, and as Stonehenge rose over the plains of Salisbury, the lush resources of the Piedmont were drawing people to settle for longer times.[30] Yet, we have only a fraction of their artifacts; much remains unknown. Paddling the river or camping on its banks may be as close as we can come to our ancient selves — who once lived, bathed, drank, and ate by the Haw's fleeting streams.

Development: Dry and Pokeberry Creeks

> LEE: This country's real different.
> AUSTIN: Well, it's been built up.
> LEE: Built up? Wiped out is more like it. I don't hardly recognize it.
> — Sam Shepard, *True West*

It's taken almost a year to pull Elaine Chiosso out of her office for an interview. This busy schedule is part of the reason her co-workers at Haw River Assembly call her the "voice of the Haw." The "voice" also has a science education degree from UNC–Chapel Hill and enough musicality to have been the singer in local bands and a composer of Haw River songs. Add to this a passion for the natural world heightened, as Chiosso explains, "by watching the fields and forest I played in as a kid go under the bulldozer every day. Even the creek we loved, the state came in and channelized."[31] This was the 1960s in San Mateo County bordering San Francisco, a good place to prepare for being executive director of the Haw River Assembly (HRA).

Sitting in Pittsboro's General Store Café and talking about Chatham County housing developments might seem a long way from pursuing the Haw, but as Elaine tells you, "the Haw is really all the streams and creeks that flow into it," and development can deliver two of the Haw's worst enemies — sediment and excess nutrients — to it. "In our watershed, the water is either green or brown. Brown because of too much sediment or green because it has stopped raining and now the algae grow from too much nutrient." All of this polluted runoff flows via creeks to Jordan Lake and beyond.

If you played in a healthy creek as a child as Elaine did and saw it take the impact of careless development, you know what follows. Your stream full of fish, crayfish and other creatures turns to a lifeless orange of floating dirt and gill-clogged fish. On the bottom, fish eggs and mussels are coated and suffocated. One bad incident of mud flow is like heavy smoke inhalation for humans. And lest one have trouble picturing sediment as much of a problem, one study done during a storm in 1985 by Don Francisco, former professor of public health at UNC–Chapel Hill, measured 50,000 tons traveling down the Haw past Bynum in 36 hours.[32]

Elaine spreads a Concerned Citizens for Effective Communities map before us with new developments colored in like bright patches on a quilt. I ask her to talk about one or two developments to explain how mass-building and troubled waters connect. "Only *two*?" her look captures dark humor as her finger swings between upcoming Briar Chapel in zoned Chatham, river left above Bynum, to the expanding Chapel Ridge in the unzoned area west of the Haw on Dry Creek — two off a 13-page list on the Chatham Citizens website showing the growth from 2000 to 2004: 10,700 new houses with a likely 26,000 population increase.[33]

Even those in Chatham who haven't made the troubled waters–development connection resist the rapid transformation of country to "master-planned" villages designed by corporations run from Florida or California. Elaine explains the strange business of turning land to corporate profit. People fly over in planes, note open green patches; owners of large chunks are located, deals are made, and development rights sold. There's no guarantee of owners having contact with the land, like the Englishmen granted giant sections of North Carolina by the Crown.

We begin with Chapel Ridge (formerly Buck Mountain) because it is already built in 2007 — by the Bluegreen Corporation of Boca Raton, Florida, recently number 48 on Fortune's list of America's fastest growing companies.[34] Chapel Ridge's size, 790 acres, multiplies any negative impacts. Elaine and others fought for a more environmentally sound Chapel Ridge even when, as she admits, "with its no zoning status, there was so much less citizens could do. With other citizens, HRA did get them to do an Environmental Impact Assessment (EIA), but it was boiler plate, very poorly done." In it, Bluegreen Corp concluded there would be "no adverse significant impacts on water quality."[35]

Elaine Chiosso and others wrote a more detailed "Concerned Citizens Review." One of their top concerns was the "very high hazard of erosion" because 600 acres were going to be cleared and 300 of those had steep slopes. Bulldozed in large sections, the land would be a sediment disaster waiting to happen if rain transported tons of bare soil to the streams of Dry Creek.

After construction, a major concern of the review was the enduring runoff problems that come with 60 acres of roads, 210 acres of golf course, and 300 acres of building lots — where rain would pour off, picking up car oils, pesticides and fertilizers. The slope of the land also makes spraying treated wastewater on the golf course problematic; all run-off heads to Dry Creek, and down into the fractured rock of the Carolina Slate Belt that allows any contaminates, like fenamiphos planned for golf course use, to leach into creeks.[36]

Dry Creek is worth preserving. Pickerel hang out in the cooler waters of a swimming hole, and the creek feeds into some of the Haw's popular Class II rapids, which are also home to the endangered Cape Fear shiner.

How did the county commissioners react to the report? Elaine's eyebrows go up: "[They] could have cared less. They never turned down a development."

The North Carolina Wildlife Resource Commission also made recommendations for a 10 percent limit on impervious surface: Chapel Ridge will have 24 percent. Their request for 200-foot stream buffers was reduced to 100-foot stream buffers.[37]

When the project was underway, Chapel Ridge's plan for 700 houses swelled into almost 1601 houses on 1988 acres as other developments were attached. Elaine explains: "All of this was negotiated piecemeal *after* Chapel Ridge contracted to buy water from Pittsboro's water treatment plant on the Haw. They put in a pipeline to the intake just above Bynum Dam and then added in another development after the water deal was done." Chapel Ridge's claim is on 290,000 gallons of that water per day.[38] "It's not clear where all the water for these developments will come from, especially during droughts."

Is it possible to have good development? After a deep breath, Elaine takes a stab at it.

"Not if it is terribly dense development out in the countryside. It's better inside or close by a town or city, staying connected to the infrastructure, the sewer system or even a bus line. Somewhere you don't have to rebuild everything so there isn't such a cumulative impact — not all new shopping centers and parking lots in extremely rural areas as Briar Chapel is doing. At Chapel Ridge there are so far 2,000 planned homes; that's 5,000 people where there weren't any houses before. From 0 to 5000, you know?"

So is any development at Chapel Ridge bad?

"No, but it would have been nice if there were fewer houses and deep stream buffers and if they had really preserved the old growth forests and open land. Chapel Ridge will tell you the golf course is open space. It would have been nice to have someone come in and do one of the new nature developments, and this would have been a great place because there is spectacular scenery up there ... and if they had built fewer houses and really preserved nature and put walking trails through. There are some old woodlands as you go down toward the creek."

How about the on-site spray irrigation of treated wastewater?

"We're going to have to be open to more ways of thinking about wastewater. The only long term solution to the water shortage problems is "re-use"; there's no question about it. But treat that water to very high levels of purity so it can be used for everything we are not drinking. Do it the right way here, and spray it on things that can absorb wastewater, not turf grass."

And Briar Chapel, is it going to be any kinder to the river? According to Chiosso, it's also massive and includes an instant town: 470,000 square foot town center, 30,000 square foot village market, and another 12,000 square foot village center.[39] At a hearing, Chatham resident Sally Erickson likened Briar Chapel to an alien landing and asked those in power to consider how Pokeberry and Wilkinson Creeks and Chatham County could absorb the estimated 6164 people in 2,389 units on 1589 acres the Newland Corporation of San Diego is landing there.[40]

For balance, Elaine refers me to "It Takes a Village" in *Big Builder* magazine where the successful struggle for Briar Chapel's approval is described as "victory" and "resurrection." Profits aren't mentioned, but "galvanizing a silent majority" to elect three pro-growth county commissioners to help a depressed Siler City with an influx of 2,000 jobs and $6.4 million in annual taxes is.[41] My follow-up with the author yields no details on the nature or permanence of the 2,000 jobs, and a study by Mitch Renkow indicates that Chatham taxes covering the added costs of new residents are subsidized by farmers' and, even more, business's taxes.[42]

Elaine is glad that "Briar Chapel had at least a more stringent approval process." They are using spray irrigation but treating wastewater to a higher standard than required. HRA pushed hard for a public hearing on Briar Chapel's 401 permit on streams and wetland impact, and better guidelines resulted.

What does Chatham County need to do differently as it develops? Something in Elaine's expression makes me add, "Your top three." One, the sedimentation control laws are not good enough. "There needs to be ground cover immediately. No waiting period."

Two, [we need] a suspended sediment standard so that you don't have to prove that sediment actually settled on the stream bottom before any action is taken.

And [we need] money put into inspecting and enforcement so that people are getting fined and forced to do it. The North Carolina turbidity standard is 50 NTUs, that measures light through water and it's a good standard. However, the home building industry got a provision so that if they are in compliance with their sedimentation control plan then they are exempt from any damages. So you have creeks that are running red for months but never with one single incident bad enough where the mud actually can be seen [deposited] all over the rocks and bottom. That's called sedimentation. You are talking about a very difficult situation. Water flows. Very rarely does sediment immediately settle. If that happens, it takes the creek a long, long time to recover. Constant suspended sediment destroys the habitat of organisms, their light is cut off, and their air and food source."

Elaine's phone goes off; she heads back to the office, but first points me to Chapel Ridge. The extra turning lane on Rt. 87 that puzzled me on the way to Pittsboro marks the spot. The pines all look young as I swirl around two rotaries, curve past sprawling club houses and mount High Ridge Road to the top, where the lots must be at the high end of the $90,000 to $150,000 range. The view stretches to a blue-gray ridge beyond the Haw. The steepness is dramatic for the Piedmont, and I see why earth bared for golf turf or home building would descend to Dry Creek.

I can't resist the sales office and a promotion packet. By early 2007, 545 of 615 lots sold, but more developments including a gated community are opening up. I am told people are coming from Massachusetts, New Jersey, Florida, even some from Cary and Apex, seeking a more rural, or less expensive, life. Big houses on lots packed with landscaping occasionally yield to ones with a blue sign announcing "Future home of Carol and Bill." The signs, all the newness, suggest an updated pioneer spirit, both fleeing and seeking. The brochure promises Chapel Ridge is both "close to everything you want; far from everything you don't." It's also like the Puritans' city on a hill, "certain to become the premier community of the Chapel Hill Area."

I cruise the streets dropping down along the golf course. Only one lonely golf cart rumbles across the bleached matt of the winter runways, but there is something cheery and promising in its beetle-like progress. Maybe some members of this Bluegreen, master-planned community will one day join the Haw River Assembly, monitoring Dry Creek and their wastewater spray irrigation system.

Back at home, I follow leads to see just how Dry Creek has been impacted. Lead one: The bible of water quality, the Cape Fear River Basin Water Quality Plan, has Dry Creek on the state's 303(d) list of seriously impaired waters; turbidity is a noted problem. That may say it all.

Just as troubling is the news that remediation plans "will be developed for identified stressors *within 8 to 13 years* of listing [italics added]."[43]

Lead two: For an eyewitness report, I call Jerry Markatos, a professional photographer who lives on Dry Creek and foretold the run-off problem. It wasn't prescience: "The situation was clearly set to happen from the very start, and the state and county were complicit. The developers cleared hundreds of acres, scraping them bare with a few silt fences down slope, pretending that would hold back the sediment. Then some of the silt and sediment barriers and ponds were prematurely removed."[44]

Markatos fumes while he zeroes in on the essence of the problem: the requirements for sediment control are inadequate: "It's specification-based rather than performance-based. When the waters fouls, they say 'Gosh darn who would have thought it,' but it was entirely clear it was going to happen."

The Cape Fear River Basin Water Quality Report agrees in more bureaucratic tones: "There are concerns that the BMPs [best management practices] are not adequate to protect water quality in Dry Creek."[45] Sadly, the speed of erosion is greater than the speed of bureaucracy. While the 8–13 years for planning go by, Markatos sees another problem: the stream of sediment "was a tracer of what will happen to the treated sewage coming from Chapel Ridge. And this is not far upstream from the water intake for Pittsboro."

Markatos sums up:

> They could have done it in steps and not left bare that many acres, but it's happening according to what's cheapest. People are moving here for just that clean environment. Developers show butterflies and salamander [in their advertising], but really are finding any avenue to cut costs. That's fair to a point, but public interest has to be reckoned in ... instead the real costs are borne by others downstream.

Lead three: I turned to the River Watch report on the HRA website, a month by month trail of decline and frustration.

> Understanding the rules and regulations that govern sediment and erosion control in North Carolina is like going down the rabbit hole in *Alice in Wonderland*. Things do not seem to be as they are, and logic seems turned upside down. Layers of rules and exemptions, lack of enforcement, and lack of common sense are the norm. We have a system where waters the color of clay after every rainstorm are considered "normal."[46]

I learn a crisis hit in August 2005. A heavy rain sent so much exposed earth sliding that Dry Creek found itself featured in the Raleigh *News and Observer*. John Holley, chief engineer of State Department of Land Quality, had some explaining to do, especially with a mandate

In 2005, Dry Creek (center right) ran into the Haw River carrying so much sediment that much aquatic life suffocated. Construction that leaves land bare leads to avoidable disasters when it rains (copyright Jerry Markatos).

on to reduce pollution in Jordan Lake. The basins dug out to catch the mud were downsized too early and overflow was the result.[47] While workers scooped up dirt in buckets to get it back on site, Holley said: "It's not likely they will get it all."[48]

Bluegreen Corp. was fined, but six inches of sediment on parts of Dry Creek meant River Watch samplers found few signs of macroinvertebrate life. Worse, orange streams continued when the adjoining development at The Parks at Meadowview was cited for sediment violation on August 31, 2006.

Holley's office seems to blame until you learn that four state inspectors cover 16 counties. This means they visit each construction site once, with additional visits, if any, averaging every 3 to 6 months. In contrast, Wake, Orange and Durham counties have their own local and much more frequent inspection, which the state encourages with grants. Holley admits that his office's work "is largely complaint-driven."[49]

Will future creeks be spared? Briar Chapel is coming on line and the earth bulldozed there could head for Pokeberry and Wilkinson Creeks.

Citizen action made some changes when Chatham voters elected candidates with a controlled growth platform — Tom Vanderbeck, George Lucier, and Carl Thompson; a new Environmental Review Board, better sediment and stormwater control, and some of the state's most stringent river buffer ordinances were put in.[50] This is significant since it is not always easy to be heard in Raleigh where the NC Realtors and the Homebuilders' Association PACs ranked first and fourth, respectively, in donations in 2006.[51]

In the meantime, those noticing a sediment disaster waiting to happen or in full progress can call 1-866-STOP MUD, or the Haw River Assembly, 919-542-5790, where they may be referred in any case.

Lead four: Finally, I check in with Allison Weakley, a biologist, who got involved when she saw "landscaping literally washing down hill" on a development on Boothe Hill. Her hope for Chapel Ridge is that when it is re-permitted, the "state will require monitoring wells on the edge of the site. Since, legally, wastewater can't leave a site, sampling four times a year would determine if it is."[52]

Weakley thinks Briar Chapel may cause less damage than Chapel Ridge because "there was more public involvement because of ordinance requirements" and it is on "relatively flatter land." Developers were required to change the planned giant pipes at stream crossings to "U shaped culverts or bank-to-bank bridges, the least harmful." What's the difference? Pipes "change stream conditions significantly; they increase velocity and produce a waterfall and bowled-out depression at the downstream end which blocks the movement of fish and stream insects."[53]

Since the brochures of Chapel Ridge include the Haw River and Jordan Lake as major attractions, it would be hard to argue against the benefits of a healthy Haw. The finger may be pointed at Chapel Ridge and Briar Chapel here because of their size and their chance to do it right, but runoff damages the Haw all over the watershed.

So what does the river need? To be spared toxins and lethal doses of sediment, fertilizer, and wastewater. We can stop the casual use of poisons for something as trivial as weed control. We have designs to control run-off, but we aren't always using them.[54] Stream buffers are basic safeguards. And we can intensify our search for alternative ways of dealing with wastewater.

Our denial that what's good for the Haw is good for our bodies is part of our disconnection from the natural world. Saying we are one with the river isn't just a romantic notion.

Assembling for the River: Chicken Bridge to Bynum

When I ask around for someone to canoe the run from Chicken Bridge to Bynum with me, it is July. The repetition of heads ducking down and peeling away could have been choreographed. I catch words: "way too low," "too dry." Then I ask Lynn Featherstone; "Sure, we can do it. I know a way through. Call me when I get back from vacation, and let's see if we don't get a little rain."

This is further evidence that Lynn Featherstone, co-founder of the Haw River Assembly, will do anything for the river. His wife, Brenda, is willing too. Lynn watches Greensboro's weather; their rain is Chatham's next day's paddling, and a very dry July yields two lush drenchings.

So one Saturday morning when a cold snap has temperatures in the low 80s, Lynn Featherstone hangs over Chicken Bridge gauging if there is any probability to this venture. The Carolina Canoe Club has cancelled; only Jim DuPree is there, his lone boat telling him he missed an e-mail.

"Care to join us?" Lynn asks him. "I think we can do it."[55] The call of the river carries the day.

At the put-in, Brenda nods to the barn swallows with mud nests in the bridge's crevices; "You see them on every bridge of the Haw." The river is full of their swooping and sharp little cries; "It's early, and they probably have young to feed."

There's no traffic on the high bridge; butterfly plants' lilac blooms light the green island that has grown from trees trapped by bridge piers. I am thrilled we are going. Just the way Lynn ambidextrously ties a left-handed bowline signals I am in the presence of a master. He started paddling when a graduate student at Chapel Hill in the 1970s, and that was it. And Brenda? "Well it was kind of a requirement," Lynn jokes. With her dark curly hair, red helmet and purple jacket, Brenda looks regal in a kayak until you see her in action. Did he make her such a demon-canoer?

"No, I just pointed her in the right direction. Before paddling, we used to be Taj Ma Tent campers; what was that first tent we had? I think it was 16 by 12 feet." They laugh too about early aluminum canoes — "rock magnets."

Jim and Brenda are quickly at the first rocky ledge, waiting for Lynn to be satisfied that I have forward, reverse, draw, and cross-draw down and can stand a rocking boat. "You know what, let's get going," he declares cheerfully, not knowing that I reverse right and left, take notes underway, and have a mind that will slip away to the whole Haw and the light bursting clouds of this Nordic summer day. I am no one's ideal paddling companion.

But Lynn will be able to contain himself. He is a great club football coach at Chapel Hill. All day, his instructions are clear and unperturbed. He's entitled to whatever he says to Brenda on the car ride home.

For all of Brenda and Lynn's addiction to whitewater, they take in the whole river, too. When a heron lifts off, they note it's a male from the blue cast of its gray feathers. Lynn points out the Mitchum and Webster sites where he talked with archaeologists. Why Native peoples chose it is clear: wide and fertile floodplain, unusual on the Haw, plus the convenience of small feeder creeks and the whole Haw for fishing.

"[Native people] would channel fish with rocky walls funneling them into a net and then head up river and come down beating the waters to get the fish moving into it. The archaeologists figure that this was probably a summer headquarters where they came together. Winters, they probably thinned out into smaller groups so they had more hunting ground."

Brenda, way ahead, signals with her yellow paddle. Three deer pick their ways across the rocks above Sawtooth Ledge, utterly lovely and jaunty as Thai puppets. Their delicate lines seem part of an enchanted world or … from another perspective, dinner. "You can see how this would have been a natural dam here." Indeed, ahead is more rock wall than river. Dark's Mill, built by Henry Lutterloh in 1790, used it to channel water to grind grains.

Some kayakers came to know its lively twists and turns as Sawtooth Rapid. One of them, "Slap in the Face," highlights the twin virtues of whitewater kayakers — daring and humility. "See the remnants of that tree trunk? It used to go across the channel. First timers who didn't scout and whipped around these rocks wouldn't know to duck under. It could give you quite a knock."

Lynn points Jim and Brenda through the bewilderment of rocks. We pass Watermelon Island, near Rock Rest, and just before Terrell's Creek, river left, Lynn tells me, "This is Bear's Amusement Park." I turn to see a thirty foot rope swing and a platform. "The only thing keeping this section of the Haw from a Wild and Scenic designation is that rope swing out over the water and Chicken Bridge; there's a mileage requirement for no manmade structures."

When the water flattens, we pull up beside Brenda, who likes "more help from the river." This section, "Second Doldrums," suits my purpose. I want to ask Lynn about the Haw River Assembly he co-founded. Lynn is so passionate about the Haw that when he found signs of beavers' return in the early eighties, he didn't care that he was in muck up to his neck. Co-founder Chuck Brady paddled the navigable Haw in 1982 to draw attention to its condition and then went on to orbit the earth 271 times in the Space Shuttle in 1996. The Haw River Assembly comes from good stock.

HRA started, I learn, in 1982 in an eddy above Gabriel's Bend on the Lower Haw (below U.S. 64). Brady and Featherstone had seen the dyes in the Haw shift from yellow to red to green over the years and knew that the $120 million Jordan Lake Project would soon fill their favorite rapids "with a bathtub of dirty water. 'Surely,' we said, 'there must be someone else out there who thinks like us and objects to this travesty,'" Featherstone recalled.

"Back on land we called Dan Besse and Jane Sharp for advice and to get the word out. Calls started coming back. The meeting in February of 1982, in Pittsboro, was a mob scene, standing room only."

The two Southern boys who started it set the tone. "Lots of people here don't get up on soap boxes, but do care about the environment, and lost jobs are a silent concern."

What approach to take? "Raise hell? Sue and raise money? Work for consensus? Southerners don't easily confront each other. We decided to do it in the least confrontational way possible."

So they worked on a name. "Haw River what? *Authority* sounded rather hard; *association* sounded bureaucratic, but *assembly* suggested getting together and working it out, like church. That's more the way people do things hereabouts, in North Carolina. And we're pretty damn religious about getting the river cleaned up." Thus the Haw River Assembly became a conduit for voices on this river.

> People came to us with their ideas, desires and concerns, to get them acted on in the least confrontational way possible. The second year our budget was $223. Now we have seven staff people. What's our budget now? Over 500 times that. That's value added. We leveraged small sums into huge gains for the river. We're working with our 2nd generation of 4th graders at the Haw River Learning Celebration. So the Haw River Assembly was never Chuck or I; it is a place where talented people can bring their ideas and act on them. A lot of the ideas heard over the years, we've got the resources to tackle now.

Jim, realizing the resource he is travelling with, asks why rapids have modern names here; on the Cape Fear they tend to be colonial ones. William Nealy, a fierce river advocate, author, and cartoonist, comes to mind.

> He's the person who brought a motorcycling-mystique to paddling and gave grim or sarcastic names to rapids in his maps and books. Paddlers, male and female, loved the macho cachet to their sport.
>
> Howard DuBose was instrumental, too. He ran River Runners Emporium, and he got people out on the river, renting canoes, mostly to Dukies who had more testosterone than sense. He would call me up when he needed some help. And I would go and help him pry the canoes off the rocks and straighten them out. Then REI came in, and he saw the handwriting on the wall and gave it up. He was one of the founding fathers of the Carolina Canoe Club. He sold us our first boat, a junker from his rental fleet, a barge, stable as hell. We'd run it in Class 4 to 5 water.

Hunger interrupts curiosity, and Jim mentions, "It's 10 of 12."

"The next rapid is Lunch Stop." Lynn can reply with double meaning. This is where the river starts to drop 15 feet a mile. Its left, right and S-turn drops yield eddy turns and peel-outs galore.

At lunch below, the talk goes back to kayaking. Jim took a sea kayaking class in 1998 and has been on a river every week since. "We used to be addicts," Brenda remembers. "We'd even come out here in summer, when it was a boneyard like this." Lynn adds, "And we ran it when it was so high there was nothing to see but water; that was generally in a raft though.

"New types of boats got surfing the rapids booming in the eighties." Lynn eyes our canoe. "It was the ABS sandwich construction. Bend or dent it? Just take a hair dryer to it, and it's back in original shape. You could bend an aluminum canoe back, but you won't have a boat that floats. Blue Hole, a company in Sun Bright, Tennessee, invented this construction in the mid-'60s."

"*Deliverance* was '71 or so." Jim orients us.

"That did a lot for white water. And there was *The Ten Who Dared* and *River of No Return*." Lynn has seen them all. They moved him from poling on the South Fork as a ten year old, playing explorer, to finding his own whitewater thrills, to bringing kids to the Haw. "By the time you get 11–14 year olds two-thirds of the way down the whitewater, they are paddling like bandits."

Not all memories are good. "I did a Boy Scout father-son paddle once with Howard Dubose, and I really had to control myself with some of the fathers. You wouldn't believe how hard dads were on their boys. You could certainly tell the sons and fathers weren't having any fun. Then we changed it up after a break, putting the kids together, and things got kicking."

Back on the river, we are cheered to hear, "All rapids from here on are Class 2." The Haw will drop 30 feet in the two miles between here and Final Solution.

As we start into the Upper Rock Gardens, Lynn leans the canoe on edge to get through while I gaze at Canada geese shifting from rock to rock and lush clumps of swamp mallow with white petals blazing light around maroon centers.

Beginner's Peril is another hubris check; at the bottom is a pillow, a rock hidden by a billow of water, that will knock you out just as you are celebrating the curving rush of your deliverance. "They think they got it, and wham! they take a little swim." Fortunately, I am with the river master.

Just below Little Nantahala, we near the piers of the bridge that kept Pace's Mill and much else going and now capture and hold trees and debris until the Haw sweeps them away.

Not all sweet spots on the river have names, and we savor several such nameless places, ferrying left and right, turning the rush of water into sideways motion. When the Haw shoots you along its rocky passages, another kind of current catches and moves through you.

We try surfing a hole. Lynn finds where the current divides in two strands, one below rushing on and one above curling back upstream into a wave that will hold you at the tip of the rapid. Find the spot and it's teasing power and tempting fate.

Next up is the Tunnel, a tree-covered island passageway until Hurricane Fran blew it open. Here, Lynn advances me to side-surfing: the wave that curves back at the ledge needs to be so wide you can turn your canoe sideways to the current and hold there, going against intuition and leaning downstream. My lean lacks belief, and Brenda laughs knowingly from an eddy; we take in some water.

Then Dry Creek comes in, river right; "This will give us more flow," Brenda reassures us. Below is another U-shaped Souse-hole Lynn savors; it's "primo surfing with more water, a real ride, trust me." Brenda leads the way through a rock garden towards the Final Solution as we hang back. "It's called the Pipeline Gravel Bar, but I call it 'pain-in-the-ass.' This delay technique is what we call going to school on your buddies. If she gets hung up, we'll try another route."

Final Solution reminds Lynn of the story the Featherstones were known by in CCC for years. On a 12-boat club venture, Brenda and Lynn were whipped from a day of hard practice in ferrying and eddy turns. Lynn, wanting to cap the day with fun, cajoled Brenda into trying the eddy turn in the middle of Final Solution. With Lynn yelling, "This will be fun," they headed in. "We hit the eddy clean, but a tree branch smacked her in the face. She turned toward me and screamed, 'Are we having fun yet, dear?'" You could hear those in other boats breaking up laughing over the roar of the rapid. That became a refrain, that greeted the Featherstones up and down the river.

I find Final Solution worthy of its heavy name; the rocky rush stays with you. A beach and swimming hole just below serve swimmers who have lost it and paddlers who want to do it over and over.

Cicadas chime about us. Batches of them shrill the air, up and down the river. At our feet in gold waters are small fish we have seen in good numbers all day.

Lynn thinks I should do more whitewater. "You're ready; you know the moves. Do you know what to do if you swim?"

"Well, I hold onto the boat."

I can feel the buzzer go off behind me. "Maybe, if you are up river of the boat, but you probably want to stay away from it. A canoe like this full of water equals a 55 gallon drum. Do the math. With water about 8 pounds a gallon; you don't want to get between it and a rock."

"So I get on my back, curl up and go feet first down river."

"That's it."

"And tuck my arms in."

"Well, you need to leave those out for balance." I still need supervised river time.

"I want to go home," says Brenda, knowing flatwater's ahead — the impounded waters of Bynum dam. A mile or so on, we pass an old ferry spot and Powell's mill which fire destroyed around 1890[56]; Bynum Dam flooded the wood and stone dam in 1922.[57]

Ahead a rope swing marks a Bynum swimming hole where Debbie Tunnell, president of the Haw River Assembly, and Ken Tunnell float in truck tubes, and a young teenager sails fifteen feet up over the Haw and crashes in.

This summer, the island opposite the take-out is home to Wolf Man, sometimes alleged to be 16 men and often held responsible for missing chickens or any other lost object. From his camp, Wolf Man obligingly returns howls from those who call out to the primitive. In former times, Bynum men enjoyed poker games here.

Brooks Creek comes in, river right, and the dam is in sight; we aim for the take out, river left. "Watch the horizon line. When you see trees coming out of the river half way up the trunk, you can be almost certain you are heading for a dam."

Bynum Dam stretches over the wide Haw just upriver of Route 15-501. The island above the dam (center) has been home to poker players and Wolf Man (Sönke Johnsen).

In the parking lot, Lynn and Brenda brush away my thanks. "Enjoyed it. You know what we say: 'A bad day on the river beats a good day at work.'"

Mill Control, River Freedom: Bynum

As the river drops in Chatham County to Bynum, its wooded shores bring to mind a time before bridges when the first few European settlers were trying for good and ill to civilize the Haw Valley. In 1746, a surveyor of Granville land grant was driven back by the wilderness, "the country being very thinly populated, nor can we be supplied either with corn for the horses or provisions for ourselves and those employed by us, there being no inhabitants that can assist us to the west of Saxapahaw [Haw] River."[58]

But just 28 years later, a court docket recorded the last payments for killing beasts — to Daniel Murphy "for five wolf scalps" and to William Murphy "for one wildcat scalp."[59]

Native peoples were sometimes seen as savage, judging from one family's records which reflect a racism that picked up where smallpox left off:

> He heard a cowbell tinkling rather near his clearing and might have gone out unarmed ... had it not occurred to him that the sound was not quite natural. He grabbed his gun and found his suspicion of an ambushed Indian well founded. After killing the savage, he summoned another pioneer family to help him keep watch during the night. Two hostile Indians that approached in the darkness were shot, and from then on the family was unmolested.[60]

In these times, the Haw was an obstacle to travel. In 1778, a permit was granted for a ferry below Bynum at Redfield ford; an 1833 map shows only two ferries on the Haw, but three on the Deep River.[61] Travel was still hard, as Francis Asbury, an evangelist connected with John Wesley, noted in 1800: "We had no small race through Chatham County; we were lost three times before we came to Charles' ferry on Haw River; and had to send a boy a mile for the ferryman."[62]

By that time, the U.S. Constitution was written, and Revolutionary loyalties were imprinted on Pittsboro and Chatham County honoring William Pitt the Younger, prime minister of England at 24, and his father, Earl of Chatham, both supporters of the colonies' cause in Parliament.[63] By then, Luke Bynum had purchased land on the Haw near Pokeberry Creek that would hold his name. On these 84 acres, his eight sons were reared, the oldest a lawyer and state legislator.[64] (Later generations would build a grist mill and then a town and cotton mill — the farthest one downstream on the Haw.)

By 1833, there were two bridges over the river in Chatham, one above Haywood and another, Lambert's bridge, east of Pittsboro.[65] A bridge here followed; those who built bridges hoped to recover the costs and more in tolls, like these: "1 horse carriage or buggy, 25 cents ... 1 man and one horse, 10 cents, Livestock — per head, 5 cents. All foot passengers, ministers of the gospel, and county officer on official duty to pass free."[66]

Sounds like a good business, but the Haw could cut into profits. In May of 1867, Carney Bynum sold the county his covered bridge, one of the Haw's longest, for five dollars.[67] And lest we romanticize the past, know that even Bynum's covered bridge could be the site of road rage. Lawrence Fooshee London tells what happened when the rights of those first on were violated:

Bynum's long covered bridge is shown looking west across the Haw. The foundation for the grist mill, right, can be seen from the current pedestrian bridge. These muddy banks are more wooded today (courtesy the North Carolina Office of Archives and History, Raleigh).

> One day when this old man, Mr. Crutchfield, was on the bridge, and this other fellow saw that he was already on the bridge, and he was in a smart buggy and horse, and he started on it, told the other man [Crutchfield] to get off, Mr. Crutchfield said, he wouldn't have no idea of getting off, so he got out of his wagon, had this big leather strap in his hand, he told that man if he didn't get off that bridge right away, he would whip him off.
>
> So he backed out pretty quickly. That was a custom that was universally accepted.[68]

The Haw had significance in the Civil War. Frank Durham of Bynum recalls the river as boundary in that war:

> We all walked up there and walked across the river on ricks and drainage chute. Of course, if the river had been up any.... They knew how to ford; I didn't. I'd hate to try it now. But that was an old Civil War house ... and he showed us how his daddy's folks hid meat and stuff in the seaming of the house ... to keep the Yankees from getting them when they were coming through. There were a bunch of them up here — they camped around Hillsborough, you know — and they were told never to cross the Haw River, but they would do it. And, of course, their house was on the other side of the river. Somebody up there killed one of them one night.... He shot him off his horse and buried him. He was a Yankee soldier stealing his horse, and he got his horse back. Said all he had to do was whistle.... But life was pretty cheap if you was a Yankee along then coming around in here.... They were told never to cross the river no how.... Well, the war was on, and they were both bad to one another, I guess.[69]

The Haw's impact as boundary in this sad war among brothers was matched by that of its lack of bridges at fighting's end. On July 4, 1876, Henry Armond London gave his still bitter analysis: "In January, 1865, a heavy freshet swept away every bridge on Haw River, which at the time was considered a great calamity, but which resulted in being a blessing, for in April following the plundering bands of bummers from the Federal Army were prevented from crossing the river, and much property thus saved."[70]

The Haw River gave its name to those fighting in Company D of the 35th North Carolina Regiment. One of the Haw River men's name is lost but not his feat; at the Battle of New Bern, he was first among those retreating riding a locomotive's cowcatcher all the way to Raleigh. Those who survived went on to the horrors of Malvern Hill, Sharpsburg, Fredericksburg, Petersburg, and Appomattox. Of 143 men, 36 did not return.[71]

After the Civil War, Luther and Carney Bynum, grandsons of Luke Bynum, sought $55,000 in capital stocks to add a cotton mill to the grist and roller mill by the Haw. Bynum Manufacturing Company was built in 1872, a three story wooden mill with 1600 spindles and production capacity of 600 pounds of thread.[72]

Local people came to live in the 14 houses that sprang up on the hill, surrendering independence for security and the stigma of doing public work as a wage-slave. Abraham Lincoln reflected widespread opinion when he argued, "Each worker [should] receive the fruits of his labor" and ask "no favors of capital on the one hand, nor of hirelings or slaves on the other."[73] At first, this new mill work was done mainly by women. Though rents were low, wages were pitiful; Flossie Moore earned 25 cents per 12 hour day in 1893.[74] She was ten, and school was out of the question.

Control was extensive as Douglas DeNatale and Brent Glass discovered; they were two among many UNC–Chapel Hill graduate students taping oral accounts in 1970s Bynum, documenting history in the swirl of many voices. Hear two of those voices on mill owner control. First, Louise Jones: "Mr. Bynum really looked after the place.... If a little something happened that it shouldn't have, he'd go and investigate about it, and tried to keep things in good order here."[75] Wesley Snipes gave his view: "If you stubbed your toe they'd fire you. They'd fire them here for not putting out the lights late at night.... Old Mr. Bynum used to go around over the hill at nine o'clock and see who was up. And if you was up he'd knock on the door and tell you to cut the lights out and get into bed."[76]

For further control, superintendents of the mill served as head of the Sunday school at the Methodist Church which stood above the mill in clear view of the superintendent's house.

As elsewhere, the mill owners did not think themselves oppressors or exploiters. A contemporary, Marjorie Potwin, praised their purpose: "Without experience as manufacturers, they blazed the trail through a maze 'of unsolved problems greater than the Anglo-Saxon race had ever faced' in a district 'the most impoverished ever occupied by an English people.'"[77]

Adding to the impoverishment caused by the Civil War, volatile cotton prices were hardest on those without money to ride out the wild fluctuations. Farmers along the Haw might rent land and buy supplies from a merchant whose "time price charge" would be 20 percent higher than for those with cash,[78] similar to credit cards today. When there was nothing left after the crop-share, debt, and surcharge, mill life was the escape. DeNatale concludes a sense of defeat made it easy for the linthead stigma to feel like a fit. Wesley Snipes would recall, "Yes sir, felt all the time like the scum of the earth."[79] But in 1900, most mill families were no longer headed by women, only eight out of 28 were.[80]

The Bynums also struggled to make their mill prosper. In 1886, the mill sold in bankruptcy to John Odell,[81] who retained Luther Bynum and then Carey Bynum as superintendents. The name changed to J.M. Odell Manufacturing Company, but the mill still struggled and was next managed by William L. London in 1902.

It helped that the Odell Company was able to buy the water rights in 1904[82] because until then the Bynums had control of water power, which was a definite problem, as an anonymous commenter said, "Bynums owned the cotton gin, and the corn mill and the flour mill, they had the water rights, and if they didn't have enough water to run all of it, they shut the mill, the cotton mill down."[83]

From 1904 to 1955, Edgar Moore was superintendent and head of the Sunday school, save for one interval.[84] A brief affair with a worker was the cause of his transfer and a sign of some worker control,[85] as Moore's nephew, Frank Durham, who blames the woman, explains:

> Yes, there used to be especially, and still is right smart of trouble with mill superintendents, supervisors and things, and oh, women give you a lot of trouble if you don't watch them sometimes.
> Yes, they sort of make a play for you a little bit when you got a little authority and think you'll favor them or not. Yes, he did. He got caught with one over there, and he had to leave here. He left here; the rest of the help wouldn't have worked for him no more, I don't reckon; they were planning to come out, I think and strike. Planning to walk out, I think.[86]

Fire was another peril for workers and owners. On July 4, 1916, a lightning strike destroyed the mill. Some young boys thrilled to the spectacle, throwing rocks at windows, but others rushed in vain for buckets to douse flames fed by oil and cotton. All pay stopped. Then families were relieved by the decision to rebuild and the chance to work in construction. Arthur H. London came to inspect the rebuilt mill in 1917 in his Ford roadster, taking over after the death of his father, who had come by horse and carriage. By 1922, the mill had 10,000 spindles operating, a new dam and more water power; electric lights and a second shift followed. Fifteen new houses accommodated workers who came from as far as Fayetteville and Gibsonville. In 1928, there was sufficient electricity to join the Haw in powering the mill until 1940, when electricity took over and river power ended.[87]

During the Depression, the Haw and the mill ran together, providing desperately needed food and money. Wesley Snipes, near Polk's Landing, struggled with other black and white tenant farmers to hang on; he moved to Bynum on Thanksgiving Day 1929:

> The last year I raised four bales of cotton; and I carried a five hundred pound bale of cotton to Chapel Hill and it brought me twenty-five dollars: five cents a pound. The boll weevil hit. And I had four or five bales at four and five cents. And I told my wife, I said, "Never will I work on the farm and spend maybe seventy-five or a hundred dollars for fertilizer, and it'd take every bit of cotton I make to pay that fertilizer and not have a dime for the whole year for my work." So I quit.[88]

The cotton mill meant no less work: "I was sweeping, twelve hours a day. We worked fifty-five hours. We worked till Saturday at twelve o'clock.... Let's see ... sixty-five hours. Well, then about the middle of the week they put my wife to work in learning to wind. She got about twelve cents an hour.... But we ate, three meals a day."[89]

The mill was suffering too from much less demand, "stayed open but didn't make any money," Lawrence Fooshee remembers.[90] Workers' jobs were split. Then came Roosevelt's NRA — the National Recovery Act, raising wages, as Vernon Durham discusses:

President Roosevelt come in and he changed it, put it over on forty hours a week, thirty cents an hour. And all over forty hours paid time and a half. So, that's what caught them napping. We was working eleven hours a day, but the depression come on, they didn't do it then, but [we] went to working eight hours a day — like you was on a vacation....

Now, if things get dull now — if they don't draw as much as twenty four dollars a week, they can draw unemployment."[91]

His wife, Eula Durham, adds that three-room houses were 50 cents every two weeks, or $1.25 for five rooms.[92]

Wesley Snipes recalls the times with a Bynum joke:

During the Depression just droves come in from Ramseur and Saxapahaw and Burlington; all of them come hunting jobs.... They said a fellow went down to the office ... and asked for a job. He said, "No, we ain't got no job for you, not unless somebody dies." He turned around and was leaving; he started on back and this fellow fell out of the window and got killed. So he was running back down to the office and said, "How about that man just fell out of the window and got killed? Can I have his job?" He said, "No." He said, "The man [who] pushed him out gets his. [laughter] They told that as a joke. But it was rough, I'm telling you right.[93]

Tense mill relations made a tough climate for union-organizing, as Wesley Snipes illustrates:

Well, you were supposed to be there when the lights blinked. At six o'clock you were supposed to be on the job. And we were so afraid we'd lose our job I'd be there an hour sitting there and it'd be dark, sitting there 'til it was time to start.[94]

One of them there organizers come by and it was whispered around all over the mill that they was going to have a meeting to organize a union up at the schoolhouse....

I was working in the mill, and I went. I didn't sign nothing. Next morning I hadn't much more than got in the mill before Mr. Edgar [Moore], the superintendent, come in. "John, I heard you went to that union meeting last night." I said, "Yes sir, I did." He says, "Do you know if I wanted to I can fire you for not walking fast down that path. I don't have to have no excuse to fire you. I can fire you for not walking fast." I said, "I realize that, Mr. Moore. I'm aware of that, very much aware of it." I said, "You can fire me just because you don't like me, or anything you want to do." I said, "But I didn't sign no paper and I didn't join the union." It scared them all to death, and a man never did come back [laughter].[95]

Efforts to organize continued on and off. In January 1934, the unionized employees at Chatham Mills struck. A meeting followed in the Bynum schoolhouse, but efforts to unionize or strike failed here, as Mary Gattis describes:

One morning, I remember, the hands walked out down there. But my mother wouldn't go.... She just didn't believe in disturbance and things like that.... But I went. And I think she understood, cause she knew I was the younger generation. But we just walked out for an hour or two. It didn't last, and it didn't amount to anything at all.... But the Londons were dead against the union. They would have shut the mill down, I reckon, before they'd ever have a union.[96]

Wesley Snipes wanted a union too:

I think that the majority of people at that time would have loved to have a union.... See, Mr. Moore's brother run that Robert Moore store over there — and if you went out of town and brought groceries, why if he didn't like it he could fire you. You'd soon know what to do, I'll tell you that. You better go there and get your groceries....

[We would] go to old Mr. Manley.... And if I had pawned my fifty dollar check to him for forty dollars cash, he took my check. He didn't give it to me; he put it in his pocket.... Well, Mr. London's labor laws or something, that went on for years. He made thousands of dollars that route.[97]

Wesley Snipes left for sawmilling, where he sometimes made thousands in one week.[98] Others at the mill struck occasionally. Eula Durham recalled, "I know one time, the spinning room went out there and wanted more money or something, and John told them he'd shut down before he'd give anymore, and they went back to work."[99]

However, spontaneous strikes could earn results. Louise Harris remembers striking when a foreman was laid off. Workers walked out and the superintendent followed, "wanting to know what was wrong." They were soon told he would be returned to work. End of strike.[100]

Inside the mill was hot and unhealthy in summer,[101] with only the sight of the Haw to cool workers: Mozelle Riddle recalled, "It used to be so hot before they put air conditioning in there. You could walk around and into the frames and burn your legs, that's how hot the heat was.... Work and sweat![102]

Add to the heat the dusty air which Wesley Snipes linked to brown lung disease: "Between you and the light ... looked like a bunch of little worms almost, with the wheels a-running and the machinery running.... Do you know all the old people that worked in the mill died of brown lung around here, breathing that cotton dust."[103] Finally in the mid-seventies, the Carolina Brown Lung Association efforts led to stricter air quality standards and worker compensation.[104]

Through these times, the river, a primary force in Bynum, also offered fish, community space, cooling waters, and freedom. It could even help to slow the pace of work, as one anonymous worker explains:

> I was young then. We would just go out to the upper end of the mill where we could see the race, you know, the spray of the water coming down from the water wheel. We'd go out there every little bit, and look and see if the water was gone down, praying that we could go home....
>
> And when the water would begin to get low, you know, the machines would really slow down. They would run it just as long as they could, you know, before they'd stop the mill off.... Well, in the summertime we wouldn't run sometime over half the time.
>
> No! we didn't get paid ... we didn't care. We just wanted to go home.[105]

The old machines belted to the Haw meant more freedom, she explains: "Back then, the work run good enough that you could leave it for a little while and come back and you wouldn't be too far behind. But just catch your work up and run to the house and eat, come back and I wouldn't be too far behind you know, I could catch back up."[106]

The river was also for dating and community gatherings — talking, singing, and cooking at Dal Johnson Spring on the bank opposite the village.[107]

> Across the river here, there used to be a spring. Well, it's still there. But then it was cleaned out, and that's where most of 'em would have their picnics and things over there ... had a table over there and all. We'd go over there and sit on it Saturday nights, and hang lanterns up in the trees, make it so we could have lighting. Have chicken stews and things, play and sing. Lord ... every boy and girl in Bynum'd be over there.[108]

When too much home brew was involved, discipline followed, as Roy Eubanks reports: "They sentenced three or four of them, had a chicken stew or something down by the river, you know, drinking and all ... to go to church for so many Sundays. They wasn't church goers, but the mill told them that they had to go to church."[109]

The Haw certainly fed the belly. In 1887, 500 pounds of fish were reported seined, or netted, in one day at Bynum.[110] Oral histories are full of comments about great fishing. There was even a fishing shack with pots and pans near the bridge across from the mill.[111]

Many women loved to fish; Louise Jones describes her mother's "baiting the place": "After dinner [they'd] go fishing. And my sister ... and Mrs. Abernathy, and a Mrs. Murphy ... they fished with cornbread bait. It was resting in a way if you could find a good place to sit, you know. They loved to go if they didn't catch a thing in the world, they loved to go fishing just the same."[112]

Sally Fowler's father used a fish basket. "I've been with him in a canoe up here and check the baskets.... Made out of some kind of fiber, you take the basket, put the food in the basket, fish would go in ... but they couldn't get back out."[113]

Men sometimes seined in the Haw for a community fish fry. The narrow 30-foot-long nets were drawn over the river, the men walking upstream.[114] J. Nathaniel Atwater recalled, "Anybody in the community, four or five different ones would get together, enough to pull the seine up the river, a few to cook ... each one would donate a little, maybe to buy the lard to fry them in, coffee and stuff like that. Just something to do when they weren't working in the mill.... They'd all donate and buy a seine, or maybe one person might buy it.... It was just something to get out and have some pleasure."[115]

These times by the Haw would come to an end. Floods threatened the mill, one in 1945 sent six feet of water into the basement,[116] but it was not the Haw that closed the mill. New management styles and synthetic threads in the early seventies drove some workers out. One complained: "They won't let you stick your head out the door now.... You can't go out of the mill, and no one can go in, you know, that's not working.[117]

By the time African Americans were allowed to work inside the mill in the 1960s, trade agreements far beyond the Haw's watershed were moving jobs overseas. For a while, two dozen workers made lamps for hotels; that ceased; fires and rubble followed, only two attached buildings remain.[118] The workers' way of making a living and a life by the Haw is gone though they might be pleased that the mill site is now North Carolina's Department of Parks and Recreation property, with paddle access and a river bank maintained for those who want to walk and fish along the Haw.

A familiar question floats through the oral histories of mill workers: Why did so much restriction of freedom generate so little protest? Rejecting the stereotype of mill workers as passive victims of paternalism, UNC–Chapel Hill graduate students stress the strengths of a continued culture and community.[119] Community there was even if it did not build to include fair wages and workers' rights. A visit to the old Bynum store, renowned chainsaw sculptor Clyde Jones, or a Haw River festival confirm that community is still there.

Jerry Partin — Bynum Store

The Bynum community fought to keep the one-lane bridge as a pedestrian path across the Haw. I walk it on a January day when the Haw threads among dirt-parched rocks, pools in smooth patches darkened by the blue dome of the sky, then breaks into mirrored shards.

The same Haw has roared over the bridge's rails, Jimmy Stubbs of Bynum tells me, remembering kayakers hung up in island trees and rescue helicopters circling. "Speed Rock" upstream of this bridge is a marker of river levels, sometimes ignored.[120]

I pop into the Haw River Assembly's old mill house office where Elaine Chiosso exhales thoughtfully and warns me about getting ribbed at "The Store," then still used for cultural research by graduate students who served as entertainment for locals enjoying morning coffee.

The store has fully earned the name "general store." It is a post office, gathering place, general store, lawn mower repair shop and music hall. "Winter and summer, there are concerts here. Bluegrass, country, old time rock 'n' roll, jazz, blues, even reggae, but we probably won't have that again. Only night we really got complaints from the neighbors, said it rattled their windows," Jerry Partin, the store's heart and soul, tells me.[121]

Partin asks me to leave the tape off, then talks affably about Bynum and the river as he cashes out customers, gets stamps, chats. No one gets perfunctory "Have a nice day" treatment.

A man comes in with half used bags of dog food. "Know any one who can use this, Jerry?" Jerry has some ideas and takes it. "Know of any houses around here for rent?" "There's one where Pokeberry Creek crosses 15-501." "Nothing closer?" Jerry shakes his head.

Jerry grew up here, left for the Navy, and came back in '89. He remembers getting bass and catfish, and gigging when the carp and suckers ran. "Course that's not legal now. When the carp are running and the river's down, it was very easy to catch them. I remember hundreds of rock holes." He "just caught the tail end" of the fish fries that brought everyone down to the river.

Jerry knows places that don't seem to be on any maps — Harkless, two-thirds of the way down to Pokeberry Creek, where the deepest hole in the river is. It's just about where the cable for measuring the river crosses. Look for steep banks with a place below to stand. "Nobody could find the bottom of it. Mr. Harkless put a pot on his head and dove down and never could bring up any bottom. It was a good fishing hole, about all filled in now. Pittsboro put a water system across there and had a temporary dam."

He points out the road the WPA put in that is now a walking path to Harkless and beyond to Redfield Ford — the shallowest point around.

> Never heard of canoes til I got back from the Navy, but we fished and gigged all up and down it. Fish fries, picnics, Fourth of July, it was a great place to play. I couldn't swim. We'd wade out and fish off the islands.
>
> Kids were raised here like one big family. Took two school buses to haul all the kids out of here to Pittsboro. Not that many now. Used to have two stops here, one of them at this store, none of this house to house. Another waste of taxpayer money. Newer people are nice but moving here for more of a bedroom place than a community. Before, 99 percent of the people who lived here worked here.

Someone pulls up in a repainted white garbage truck that catches the sun. "Hey, turn the lights off," Jerry begs him, then suggests: "Maybe you ought to get Clyde to paint something on it and then sell it and get a new one."

The mail comes and Jerry goes to put it up. I settle with a friendly woman, a man named Jack Parker, and a major teaser whose name I learn after a series of aliases, including Jack Parker, is Roy. I ask about the river.

"It runs south. That's it."

"Rockiest river I ever seen," adds Jack.

"I can't even swim. I don't know why and I was raised here. You know I am the biggest liar in Bynum."

Jack Parker tells me, "When they put dye in up at Saxapahaw, and there'd be colored foam — green, yellow, orange, up in the trees. Bundles of foam big as cars would come down the river and up over the highway. No, not when it was flooded. They turned something loose up there about Greensboro."

They help me out with directions to Clyde Jones, local sculptor of international reputation, and I climb mill hill. The dogs who run up barking look like Clyde Jones' chainsawed sculptures. A woman with a pink phone to her ear hisses "git" to one and waves. Clyde Jones critters are in every yard, but his own place stands out — penguins, crabs and dolphins move over the clapboards of his house, and the yard is a jammed corral of alligators, giraffes, dogs, horses, and lizards. Sturdy and silly, heavy-logged bodies hold expressive heads. One dog has plastic-daffodil-eyes whacked in above biting jaws. Bark hides may be sprinkled with glitter, red polka dots, or stripped raw. Just seeing them makes me take myself less seriously and get more in tune with the sun and tree shadows bouncing across the yard.

Clyde Jones welcomes me: "Anywhere you want to go, go around."[122] Before going back to argue with his cohort about how to run Christmas lights up a giraffe's neck, he talks about the Haw. "I did fish in it. I go down there and walk up and down it. I think it's pretty and it's part of nature. I make the Assembly's T-shirts for them."

One critter from among many at sculptor Clyde Jones' house eyes the camera on a sharp January day (photograph by the author).

Later, I ask about a thank you letter from Mikhail Baryshnikov, whom he first refused to sell a critter to: "Well, I didn't [sell] at that time, but it cost me two Haw River t-shirts and a little animal I sent him. It cost me for running my mouth. See I am known all over the world. It cost me. It didn't cost him."

Why not sell? "I guess I am just tied to 'em and didn't want to sell 'em. Read the letter and then you will know all about it. What I said I don't know.

"Ya, I worked in the mill here. I just quit. Got tired of that dust and stuff. I was there when they changed over to that rayon and polyester and stuff. That's when we had a mess. Cause we didn't have machinery to run that kind of stuff. Staples too long and wet. It was hateful." He warns me not to work in a mill. "There was so much dust. Rayon wasn't as bad as the cotton. People kept working; they needed the money."

And how did he switch from mill worker to sculptor? "I started ... by seeing animals in woods and bringing 'em to the yard and making them. Eugene said I was freeing the animals from the wood to nature. And I had a broken leg too, done when I was working for a logging company. It was around the '80s. I just took this for a hobby 'cause I was hurt and wanted something to do."

I tell Clyde how much I like his work, the penguins on the house, a horse's neck straining for freedom, and plastic daisy-eyed dogs splotched with orange.

His "You're welcome" is full of soul and flirtation: "Getting along with people; a pretty smile that makes the world. You make me want to do better when you talk to me. You're part of me. We are all part of each other."

He gives me a kindly warning in parting: "I wouldn't go in that store if I were you. You got what you need without going in that store. That's up to you if you want to be made fun of."

The store is gone now; last report Jerry Partin was working at Lowe's, but the community rallied to keep the concerts going.

The Haw River Assembly, that community of river lovers —1500 members and volunteers— is located in Bynum too. One May, their festival was in full swing here. It's not a state fair traffic tangle or a budget buster; once inside, the fun is free. Kids paint the plywood cutouts Clyde Jones has tossed into a pile as casually as if they were kindling. I ask for a photo. "Sure, but get the kids in it; they're what the festival is all about." A parent nudges her son to get his work signed, and Clyde swirls out his signature.

Every river-linked organization is here and much else. HRA has a basin of water with critters scooped from the Haw. "Ow, bugs," one mother says, peering at a fishfly larva in its distinctly crawling phase. "I wouldn't want to go in a river with those creatures." Then she reads the laminated sheet: the fishfly larva are only "somewhat pollution tolerant and indicate good or fair water quality." The "ow, bugs" woman stays to check out a tiny clam.

Raging Grannies and Bill Hicks send music out to the crowd seated on hay bales, and Paper Hand Puppet Theater performs *River of Life*. Clyde Jones has a log sculpture to create and sell for HRA. He roars up his saw and yells, "What'll it be?"

"Alligator."

He casts a doubtful eye at the short stump of a log. But after he has criss-crossed ridges into the back, lengthened the tail and snout with logs, and kids have hammered on legs, it's unmistakable—alligator.

Louise Kessel draws a crowd with a tale about a river. This Jewish-Japanese woman with a face both vibrant and serene started the festival to culminate weeks and weeks of the Learning Celebration — getting fourth graders out to explore the Haw. Louise brought back what she learned on another river, the Hudson, when she worked on the Clearwater with Toshi Seeger and her folksinger husband, Pete. Her idea for the festival and learning celebration evolved further when she went on a Soviet-American sail during the Reagan Cold War years. "I wanted a project that would create an experience where people would feel changed and stirred up and take that passion and do something with it to impact where we live."[123]

She mentioned it casually to an HRA board member in Bynum's General Store, and the following April it was a reality.

> I tell people they really shouldn't do feasibility studies; the festival would never have happened if we'd done that. I contacted all the school systems in the watershed. Originally I thought we would travel by canoe, and I realized pretty quickly that wasn't physically possible.
>
> We did in turn travel the entire river.... At the beginning of the festival we would have a pipe ceremony at the headwaters with Bruce Holt giving a blessing. People would say what their hopes for the river were and pour water in a jug. Then we capped that jug and took turns each day, carrying it down a stretch of river. People would come back from the day riverstruck. They just looked different especially after those upriver passages. I mean they are not easy. People would come back covered in muck and torn up, and they would have seen things and heard things. It is so amazing to be in a completely pristine swamp environment.
>
> I have plenty of stories about that first year. It was pretty nutty. The first week it rained every day; the river flooded out of its banks. Jane Sharp, who was 72, was with us, and I remember she was sort of floating in her tent. I went around with a flashlight. "Are you ok?" "I'm fine!" rang out loud and clear.
>
> I am so happy it is still going on even though I wore out as organizer.

Now every fall, over 100 volunteers bring 1500 4th graders in the six counties of the Haw to become riverstruck. Crews camp out so ten year olds can spend a day on the river, learning about stream monitoring and stewardship, net impressing clay as the Sissipahaw did, spotting animals, being outside, in and around the Haw. Eighth graders will soon have their own program.

Festival Day continues with free canoe rides, a "river of chalk" on the footpath bridge, nets for peering at real river critters, and at the end Elaine Chiosso, Cynthia Crossen, Lynn Featherstone and others sing a bluesy "Wade in the Water," and Elaine announces to cheers: "Not all of the Haw is in good shape, but I can tell you that right here, right now, the river is safe to wade in."

Whitewater Guide: Bynum to Route 64

> Trees and open spaces can't teach me a thing, whereas men in town do. — Socrates[124]

> But very few people are listening anymore [to the natural world],
> because of this strange, almost hallucinatory, notion that only humans
> have something to say. — Christopher Manes[125]

Joe Jacob stays quiet during the conversation about best paddles, lifejackets, canoes. Though he doesn't say so, for him, it is only and all about the river. A great teacher, his wooly grey beard and hair make him look the guide he is of North Carolina and Alaskan rivers. After twenty years at a desk job with Nature Conservancy, Joe now takes all sorts of folks into whitewater, keeping even those with a risky swagger deftly corralled in the safety zone. His secret? "I never raise my voice."[126]

One April evening, four of us gather by Joe at the Bynum mill for a lesson in whitewater. Pry, draw, cross-draw, sweep. When the water forms V's pointed at you, the tip indicates a rock or obstruction. When the V's point away, your path is down the center. He draws a wide S on the blackboard on his lap, then marks the zigzagging path of water: "Fastest water is generally deepest and in the center or the outside of the curve, so if you want to slow down, hug the inside curve."

We settle in to practice eddy turns, flipping into the quiet water behind a rock or peeling out. These quick, coordinated turns earn you a rest and an invaluable chance to scout downstream. Joe watches, repeating the mantra: "speed, angle and lean."

Ferrying comes next. With the right angle of the paddle, the Haw will hit the bow and stern just right to move us across the flow. After Andrew in the stern convinces me to stop paddling, we sail sideways. Greg and Steven dunk twice. "Remember SAL," Joe encourages. "It's speed, angle, lean." Six turkey vultures watch from trees; Joe has a philosophical smile, shaking his head when we perpendicular people say, "But I am leaning."

We turn in and out of eddies, make S turns, and ferry until I am working less with images of V's and S's in my mind and more with the feel of the river. "It's all about working with the river; it's got all the power. You never want to go faster than the river; you lose touch with what it is doing." Joe adds, "It's 90 percent balance," as he peels off eddy turns without a paddle.

A few hours later, the clouds are dark purple against a wan sky and Joe nudges us off the Haw. By then, the river is in us. Curving out of the parking lot, I feel myself leaving an eddy of the Haw and happiness.

One Saturday later, we re-assemble in Bynum. Joe is unperturbed when half the group is late and the fourth a no-show. He lets us, like the river, set the pace.

I paddle with Joe and learn on a "boney" Haw demanding good steering. We follow Canada geese out the island channel to the wide Haw. "They usually find the best way."

The Haw opens and then braids with islands the whole way to U.S. 64. Harkless, a fishing spot, is marked by the gauging station, river right, and cable chair wire slung across the Haw for measuring river depths. Water height and speed are recorded mechanically inside the station; the results, speed and volume of water, or cfs, are broadcast online. Lives have been saved by flood alerts and cfs reports from 7,400 gauges nationwide, but there is always a push to eliminate some and cut costs: $13,500 a year each.[127]

The Haw is a garden of rough and edgy rocks. Blown out of an island arc of volcanoes 700 million years ago when our coast and Africa were about to collide, the lava fell into a shallow sea.[128] Today, opaque haze burns an ancient sheen onto the clay-coated rocks. Joe lets everyone know that both sides of the Haw from Bynum to Jordan Lake are now public land. No future McMansions will rise over the banks.

One mile below Bynum, past Harkless and Pokeberry Creek, a ferry at Redfield was working in 1876. Another mile farther down was Brown's mill, one of two mills on the site to lose the battle to harness the river.[129] Sycamore, ironwood, swamp maple, ash, oaks, and river birch swell over the Haw, some looking like they were here in 1876. Joe agrees the trees are larger; "probably just stopped plowing earlier," he reasons. They suit the broad scale of the Haw. We pass blooms of mountain laurel, honeysuckle, and more islands, their upriver prows crisscrossed with bleached timbers that slammed them in a flood. In the shade of one island, minnows swarm through mustard-tinted but clear water. Joe tells us the Cape Fear shiner, a federally endangered species, has been found here, a good sign for its survival and ours. It is one of the 700 fish now vulnerable, threatened or endangered in North American streams.[130]

An osprey strikes a splash, circles and then flies off with a much larger than shiner catch. Before long, an eagle glides down the river. The first time I saw one I understood what all the fuss was about. Eagles command attention. Power is in the two bold colors — white head and fan tail, all else black brown. Power is in their flight, too; this eagle careens a long way downstream without a single flap of the wings. As we lunch, it or another — there are no distinctive male-female markings — sails by again and again without my seeing a wing beat. Joe has watched them dine on seagulls at Saxapahaw Lake. They also snack on killdeer and sanderlings and clean up carcasses.

Our lunch spot is a volcanic boulder split in two, overlooking a sycamore leaning like a figurehead from an island. The river is almost empty today, one fly fisherman in a rubber boat our only company. Joe wants more people out here. His way of doing conservation work is basic: get people on the river. Why aren't they here? Nature writer Richard Louv says it is because of "nature-deficit disorder" and finds less than 10 percent of youth have experiences in the natural world. The rest are more and more plugged in, "living inside media."[131]

Joe mentions David Orr's book, *Earth in Mind: Education, Environment and the Human Prospect*, which points to one reason — the Enlightenment's domination. Students are educated in the rational and compartmentalized. "Everyone's grasping and chewing on their little pieces of research and information, and passing that approach on to students," Joe objects. In schools, feeling the spirit of the whole river is not encouraged, so the whole earth is lost to peripheral vision. "This," Joe says, "is what makes us so vulnerable to development."

That word, development, brings everyone to the subject of the 8400 acres about to be developed along Robeson Creek. Joe comments:

> At a public hearing, one of the developer's representatives is telling us how it is going to be so environmentally-friendly. And the more I listen to him, the more I get this vision of the sperm fertilizing the egg, and I just can't get it out of my head. And that sperm exists for no other reason than come hell or high water, it's going to fertilize that egg, and it doesn't care what happens after that, that someone has to raise this kid.
> So I raised my hand and apologized to the audience because "some of you are going to be offended by this analogy," then I said: "You are just a sperm; you do not care about this community after that; you are out of here."
> Someone else in the audience adds, "You know, this is 20th century-thinking, and we are in the 21st century. Have you heard of global warming? You are going to take out all these trees, and you are going to have to mitigate that."
> "Well, we can't do that."
> "Well, why not?"
> "Because we have always done it this way."
> Typical developer: full of "we want all your input, our doors always open." [It's like] the mill owners and how they were going to raise people up, that same condescending thing. We

are going to bring all these jobs in and build all these houses. And we have been buying it forever. That's why I loved it when this guy said, "That's 20th-century thinking."

When we get back on the river, that current of thought leads to another:

> We are so into consumption in this country. We are so wasteful of things we don't really need, like plastic bags at the supermarket. I mean we take our resources whether they are made from oil or corn, and we throw them in landfills. But our economy is all about convincing somebody that they need or want what you have, whether it be a product or service.
>
> I don't know why we have such a severe case of testosterone poisoning in this country. I always bring it back down to the people who are the power-brokers. It is all about oil; we're in wars because of oil; we're in wars, so they can maintain their power, and that's why we haven't come up with alternative energy. It's not in their best interests. And that's treason; it's not about their best interests; it's the best interests of everyone.

I mention a report that North Carolina alone put enough taxes into the Iraq War to bring renewable energy to every household in the state. Joe responds:

> Part of the problem may be that if you can put a solar panel on your house, then they can't meter and sell it to you.
>
> What happened here along the Haw River with mills is happening in this country right now with other power brokers. We keep their engines running as they get wealthier and more powerful, and it's not in their best interest for us to be anything but their workers. Capitalism is always going to have that elite at the top controlling things. The only reason it works is because you and I have the opportunity to make some money within that system, and maybe that's not bad. Well, it is bad when you think what we have done with the rest of the life on this planet. It's not just about humans.

We paddle on as the soft heat of the day leads to meditation. And since I took the stern after lunch, shoots are full of surprises and rock mergers. I am missing the concept of "side slip."

"When I go right, you go right."

"I am going right," I tell Joe as I head us diagonally across the river.

"No, when I do this, you do this," he explains, but I don't get it. Our diagonal canoe makes a perfect shape for the current to carry — rock merger.

"When I go right, you have to go right."

"But I am going right."

"No, you're turning into the current."

Somewhere between words and rocks, my river muscles get it. Joe goes right to miss a rock, and I go right just enough to keep the boat parallel to and NOT DIAGONALLY ACROSS the current. It is easier this way.

"I thought you were caught in the coriolis effect," Joe says. "Usually, if people want to go right, I say it's the coriolis effect. Objects moving in a straight line in the Northern hemisphere appear to veer to the right; in the southern hemisphere they appear to veer left. It's because of the rotation of the earth. Airplane pilots in particular have to account for it."

The river strands around islands until we side-slip river right by the site of the many iterations of the Taylor-Henley Mill[132] to the take-out with its freshly-painted river gauge and parking lot. Too soon. I feel like running the river all over again. With its many islands and channels, there are many Haws here.

Rapid Power: Route 64 to Jordan Lake

The U.S. 64 bridge over the Haw stands where the western curve of Africa and North Carolina wrenched apart 200 million years ago as Pangaea split into continents. Stress fractures careened through rock. From South Carolina almost to Virginia, a vast rock wedge dropped over a mile down. The Triassic Basin was born.

River waters, carrying silt, sand, gravel, and cobbles into this Deep River section of the basin, turned it to a swampland where dinosaur fed on lush growth or each other. This swamp was home to tree fern, cycads, horsetails, crayfish, crane-fly and cockroaches, and, likely, two ten-foot reptiles: the crocodile-like Rutiodon and the Rauisuchid Thecodonts with a mouth like a Tyrannosaurus Rex.[133]

Fast forward to 150 years ago when one could observe the Haw and filled-in Triassic Basin from a bridge. The old crossing, Cow Ford, saw a number of bridges come and float off, starting with Griffin's Bridge.[134] Looking at the Haw then could fire this sequence of thought—rapids, power, mill. The 28.5 feet per mile drop spelled dynamic energy.

Herbert Poole and Mark Chilton have traced the old mill sites here, now kayakers' play spots, and the crossings that made them profitable. At Seven Island Falls, about a mile down, was a mill built by Gideon Kirksey in 1765; Hadley-Williams Mill followed with enough waterpower for "a summer sawmill, a grist mill, and a foundry."[135]

Moore's Mill was just above Robeson Creek, atypically river right, where it used a 600-foot natural sluice as a raceway for a "saw mill, a cotton gin and a foundry."[136] Chilton adds that George Lucas first received permission to build a mill here in 1792 and over the years business arrived via ford, ferry and six different bridges, ending with the old iron bridge that marked a kayaker's take-out.[137]

Now under Jordan Lake is the Hartsaw's Mill site, never actually built on, at the Gunter-Harris Island ford.[138] Bland's Mill, swept away twice, was where Jordan Lake Dam is now, supporting Mark Chilton's point that great mill sites don't change, not within a few centuries anyway.[139] In a fish trap above this mill, a 133-pound sturgeon, seven feet long, was caught in September 1888.[140]

This may seem like a long list of mills, but authors of an 1899 report on water saw even more potential: "Before leaving Chatham county it may be said ... that Haw river offers a large amount of power in its course through the county, very little of which is utilized.... The bed and banks are almost everywhere good, the county is hilly, but not mountainous and the climate healthy. A disadvantage is the use of small falls [and] the sudden and large rise to which the river is subject on account of the narrowness of the bottoms [floodplains]."[141]

The practicality of mills is far from what kayakers seek in the thrashings of the Haw. Four decades ago, they came here for wildness: Bob Brenner, Paul Ferguson, Howard DuBois, Brenda and Lynn Featherstone, and others including William Nealy. This man whose story we now turn to was "poet laureate of whitewater" and cult hero.[142] Nealy authored ten books on rivers and extreme sports and had a "love affair with the Haw."[143] His story exemplifies part of the Haw's history—the human dementia triggered by a river that's up: men and women running shoots, surfing holes, holding on at the edge of entrapment under merciless hydraulic pressure. Below U.S. 64, the Haw roars in Lunch Stop Rapid, Ocean Boulevard, Gabriel's Bend, Harold's Tombstone, and Moosejaw Falls, and

The entrance to the Triassic Basin below U.S. 64 provides the Haw's steepest rapids, which offer water power and extreme sport unless drought intervenes, as here. The Haw's rocky drop hampered inland migration, river transport, and trade (Sönke Johnsen).

once did in now-sunken S-turn rapid, smooth-ledged Pipeline, and, Nealy's favorite, Finder's Keepers.[144]

William Nealy throws light and humor on the people who could enjoy flatwater bliss but chose otherwise:

> Whitewater boating has been around for quite some time. Independently invented by aboriginal cultures all over the world, before the invention of the wheel and the cocktail party, negotiating rapids was a necessary evil that had to be overcome in order to use rivers as roads.... During the '50s, some river runners began discovering that people would actually pay money to get wet and scared to death. A new industry was born. Along came the baby boom babies ... masochism became really BIG BUSINESS.
> "What?!" you say, "boating is *fun*. What's all this neo–Freudian crapo about masochism anyway? Sounds like a bad attitude to me!" Maybe you're right, BUT when you get into year-round compulsive river-running, "fun" must be redefined. The idyllic summer time runs with warm water and beautiful women lounging on the rocks (hopefully "oohing" and "ahing" your surfing technique, perhaps offering you a sip of wine or a cool beer in some secluded eddy) are few and far between. Mostly it's sitting in a steamy smelly car with four other coevolutionists dressed in silly rubber suits, waiting for the snow to quit so you can do the first winter descent of some godforsaken little mountain stream. Or it's cringing as you hear your paddling buddy tell the "death run on the Suchnsuchee River" story for the fifth time this trip. I could go on. Being a whitewater addict is no bed of roses.[145]

Of course, no one did more to advance interest in whitewater and get people into holes and out of them safely than William, not Bill, Nealy. He mapped and cartooned the boaters' Haw, with militant hippies like his alter-ego Mr. December, and gave gritty, no-nonsense advice to paddlers and "worked a lot of rescues too," as Lynn Featherstone adds:

> You mean "Not Bill." We called him that because he was always having to tell people his name was William.... He gave the river a constituency and gave them an identity.
> He was an EMT; local rescue services for the most part have a dim view of people who get out on white water, since they primarily see those who get themselves in trouble. [William Nealy] developed working relations with Triangle rescue squads ... training them in the demanding art of whitewater rescue.[146]

Bob Sehlinger, president of Menasha Ridge Press, Alabama, says of its Chief Engineer of Humor:

> In my thirty years of publishing, he was one of two people I have worked with I actually consider a genius.... He could take aspects of whitewater paddling that people learned through experience from years and years and make it understandable in art and word to his readership.... He was an advocate in regard to many river issues. In his art, he integrated caricatures of himself as a rebel and a wild man, and in actuality he was a very shy intellectual, incredibly curious person who was as well read as anybody I have met in my life.[147]

William Nealy honed his own skills and river passion on the Haw, but started very early in life on an Alabama creek. "Polio Creek was a crack in suburbia, a kid Ho Chi Minh Trail, the forbidden zone. Within its sheer cool walls a youngster could walk for miles without once falling under an adult's reptilian gaze.... Naturally we were forbidden to play in the creek by the mom and dad units. 'Typhoid!' 'Rats!' 'Snakes!' and, obviously, 'Polio.'"[148]

He nearly drowned in Polio Creek, at flood, but instead climbed out on a drain pipe and went on to his life with Holly Wallace. Together they "comprised about ⅒ of Birmingham's tiny peacenik hippie population" and in 1977, they fled together after locals "strung barbed wire across the rapids" and torched Liquid Adventure Canoe Shop on Locust Fork River, leaving "KKK posters ... nailed to the ruins." Chapel Hill looked good.[149]

Nealy's job at Carrboro's River Runners' Paddle Shop, endlessly giving out directions for running the Haw, was the irritant that put him on his path in life — he produced a river map, followed by "two volumes of river maps ... 4 outdoor humor books on whitewater, skiing and mountain biking, as well as three instructional books" for kayaking, inline skating and mountain biking.[150] *Kayak* sold 250,000 copies in five languages. Menasha Ridge Press flourished, and being vice president kept Holly Wallace at the heart of a scene she would otherwise have had to abandon because of rheumatoid arthritis. She had kayaked the Haw until even duct-taping paddles to her hands did not work.[151]

William Nealy paddled on, promoting river sense and river rights and caricaturing the people he brought to the Haw and bigger rivers. Flatwater folk like myself he cartooned paddling up a hill. He defined others: "Despite the charming egalitarian aspects of the 'sport,' inner-tubing is proof that natural selection is still hard at work."[152]

Cool, authentic Nealy opened rivers up to those who wanted in and let them know he was no member of the elite: "Probably the most important thing to understand is that it's pretty much a warrior society, especially at the high end. It's a pure meritocracy kind of thing. The big bad warriors operate up here in their big bad universe, and then there's all the rest of us down here, which is good because it pushes the limits of the sport and the learning curve."[153]

William Nealy's suicide in 2001 left his many friends and readers with a huge gap. Chronic back problems, dependence on pain killers, and depression may be explanations. We can't know; he lived his life and chose his death.

Many tributes followed: Mike Holland opened the William Nealy Memorial Haw River Trail; mountain sports festivals and Kate Geis's film, *Riversense*, were dedicated to him.

His books still entertain kayakers in rubber suits in steamy cars waiting to hit the Haw. Perhaps someone will make a movie of Holly's and his life to bring major attention to this river he worked hard to protect.

Protectors: Robeson Creek and Haw Watershed

While paddlers like William Nealy fought for the Haw in their ways, others inventoried Robeson Creek, documenting the case for its protection. Its ravine community is now sheltered in the Jordan Lake Natural Area. Red-back salamanders scamper here isolated from their mountain roots and unaware of other red-backs at White Pines Promontory. Even some Carolina anoles have migrated in from the coast. Mussels, otters, and Cape Fear shiners live below deep-cut ravines that cross the Triassic Basin fault line. Once you hear the mesodron albolabris lives here, you will yearn to see the white-lipped forest snail for yourself.[154]

Hank's Chapel Road takes me over ridges to a put-in by a green Haw, rimmed by hollies and river oats. Catherine Deininger, stream steward coordinator for the Haw River Assembly, meets me for a paddle to Robeson Creek. We canoe over the Haw's lost rapids; ahead, deep-down, lie unexplored Paleo-Indian sites.

Since 2000, this young mother and scientist has worked on Robeson Creek with NC State University faculty and creek residents, saving the Haw one creek at a time.

William Nealy's cartoons have amused, galvanized and educated generations of paddlers on the Haw and other rivers. These are from Kayak: The New Frontier *(William Nealy/Menasha Ridge Press).*

Why Robeson Creek? "HRA picked Robeson Creek for the first stream project because it was an urban watershed in their own back yard. North Carolina listed it as impaired in 2000 because of chlorophyll a, from too much algae, and impaired aquatic habitat, not finding the aquatic bugs a healthy creek has. Also Town Lake [created by a dam on Robeson] is impaired for aquatic weed parrot feather, which is usually released from home aquariums."[155]

In fact, Catherine tells me, in 2003 the town was under a mandate to cut phosphorus by 71 percent to reduce chlorophyll a levels and algae from their wastewater treatment plant and septic leaks and runoff above and below the plant.

With a situation ripe for change, Catherine Deininger went at it:

> We got grants for a Stream Steward Campaign, wrote a handbook, talked to landowners and businesses and started an awards program. Meanwhile, NC State Water Quality Group, the Town of Pittsboro got involved, along with NC Extension Services and Chatham County Soil and Water Conservation office to work on stormwater runoff to address chlorophyll a levels. More initiatives got started.
>
> NC State, led by Karen Hall and Dan Line, started getting grants and leading the effort to get stormwater management best practices going. Soon, the Robeson Creek Watershed Council was formed. In 2006, Robeson Creek was chosen by the state and the EPA as one of seven priority watersheds for North Carolina. They were impressed by the active Robeson Creek Watershed Council, so we got more attention from the state and the EPA.

We turn into Robeson Cove and spot a rope swing, but an ice-skim with broken edges that clink like wooden chimes keeps us from temptation. Two jon boats have lines out for crappie.

The sight of Robeson Creek, a cobbled ramp under Hank's Chapel Road bridge, reminds Catherine of the urgency of protecting Robeson from the imminent development on its whole eastern edge.

> Land [8,400 acres] along Hank's Chapel Road has been bought by Jim Goodnight of SAS, Inc., but Preston, Inc., is developing it. They may develop land that is four times the size of the Town of Pittsboro. That's why strong ordinances on development are so important now and we are working on that.
>
> The problem is that low impact development sometimes runs counter to old subdivision rules. For instance, now you are required to have a stormwater pond if you have a certain amount of runoff, but in low impact development, you want home cisterns and rain gardens to filter and slow stormwater as close to the source as possible. And instead of increasing the impervious surfaces by using curb and gutter, you want grassy swales that act as an absorbent ditch to slow runoff.
>
> Now, if a development tries to reduce impervious surface by using narrower roads, they run up against all kinds of requirements for fire trucks and emergency vehicles. These rules need rethinking so you have needed access without excess asphalt.

At the mouth of Robeson Creek, water slips by cobbles in a tree-arched streambed — a compelling invitation to climb its tunneled world. A kingfisher rattles from a stone perch. But Robeson Creek is far too dry for anything but a hike up it, so we let the wind push us back to the access. Catherine Deininger continues,

> The project has worked out so well because we have had so much involvement from so many different people and organizations.... It's so great to see so much going on. The state is talking to the town and the town is talking to the people and the people are talking to the EPA. It's just that whole dance, and getting people to realize they can make a difference.

Robeson Creek is the right size watershed to figure out what is going on. It's not like taking on the whole Haw River.

Clean Water Management Trust Fund

Nancy Guthrie and the Clean Water Management Trust Fund (CWMTF) are taking on the whole Haw River and more. The trust fund buys lands along state streams and rivers to guard water quality. Spearheaded by Senator Marc Basnight in 1997 after massive fish kills occurred, especially in the Neuse River, CWMTF has a budget of $100 million annually, but only if the legislature is pushed to approve it.

When Nancy became one of CWMTF's five state representatives, she could work to improve conditions: "I had been gathering data on water quality. I love that job; it got me in every stream and river in the state, but I sometimes felt I was documenting decline. When I did GIS mapping of the impaired streams of the state, you could make out where I-40 was, even with no roads on the map."[156] Nancy now helps people apply for funding that may buy easements to create wooded stream-buffers, restore streams, reduce stormwater, improve wastewater treatment, or study the river. "Individuals must partner with a local government or an organization like a land trust or conservation group."

The CWMTF works with other state agencies funding different aspects of a project in an intricate mix Southerners manage well.[157] At Glencoe, Nancy Guthrie said, they gave funds for a buffer along the Haw. Other projects included

> a take-out site at the Bynum dam ... protection of the headwaters of the Haw ... the park at Swepsonville ... the river frontage easement of Jane Iseley's farm ... and just recently the Conrad tract on Mear's Fork. Paul Kron and Janet MacFall have headed up studies on the Haw we funded ... and Mebane's study on handling stormwater, without piping it into a stream ... bought a large piece of land in the Cane Creek [river left] watershed.... There's a lot of work going on.

"I keep looking at North Buffalo and South Buffalo," Guthrie nods to the map before us: "On South Buffalo is one of our more interesting projects. Right near I-40, we have the only inflatable dam I know of in North Carolina. When it rains, the level and pressure of the water trigger and actually inflate this dam. It just stands up and moves the flood water into a wetlands area. It deflates after a day or so; you don't want the water staying there and mosquitoes breeding."

The list could go on, but Nancy has a meeting at Soil and Water Conservation — time for two last questions.

> What's hard? It's having millions of dollars and realizing it is not nearly enough for all of the problems we still have and all of the good ideas that we don't fund.
>
> Satisfying? I am a little biased, being a paddler, but what is most satisfying is when the state has put money into protecting a river in a way that means people can get out and enjoy that area. We do favor projects that are on the ground.

5

Jordan Lake
Chatham County

> I looked over Jordan and what did I see...
> A band of angels coming after me,
> Coming for to carry me home.

Looking over Jordan, what do we see? Triassic Basin, swamp, river bottom, hunting grounds, farmland, whitewater ... lake. Long after dinosaurs roamed a swamp here, in Paleo times (12,000–10,000 B.P.) people hunted the land where the Haw is joined by its major tributary, New Hope Creek (which runs by Chapel Hill).

Two archaeological sites on the Haw (past Robeson Creek), known as 31Ch29 and 31Ch8, go back to the Paleo Era and are among the most important sites on the East Coast. Stephen Claggert and John Cable, documenting the excavation, describe how "professional archaeologists and federal agency planners" rushed a "salvage-oriented effort," 25 people working steadily for four months, to excavate the sites systematically before lake water rose over them.[1]

The sites 31Ch29 and 31Ch8 provided "a rare occasion" and a "record ... that promised to span the entire Holocene epoch," layered in time-marking strata.[2] The riverside location itself denotes a river people. Hardaway-Dalton projectile points were found here[3]; their characteristic bifurcated base revealing the presence of Native peoples by the Haw earlier than ever before.

Thicker sediment deposits suggest a colder, wetter time. As the last Ice Age warmed, pine and spruce weakened their hold, and oak, beech and hickory grew. "The earliest Indians to occupy the Haw River sites ... lived within a very different environment from that of today.... The Paleo Indians ... lived in small family groups and lived by hunting wild game, particularly whitetail deer, and gathering available nuts (hickory, acorn), roots and other plant foots."[4] They likely ate turkey, bear, elk, deer and perhaps even caribou as well.[5] There may even have been a 1000-year overlap of human and mammoth and mastodons in early Paleo times.[6]

The changing style of stone tools at the Haw River sites support Lewis Binford's theory that "hunter-gatherers" changed from collectors to foragers because the climate shifted. Collectors lived where seasonal changes were sharp, following herds or streams for foods, probably using the same places repeatedly. Their tools would be few, high quality, hunting-related tools made for their long term use; such a people would be grouped by tasks.[7]

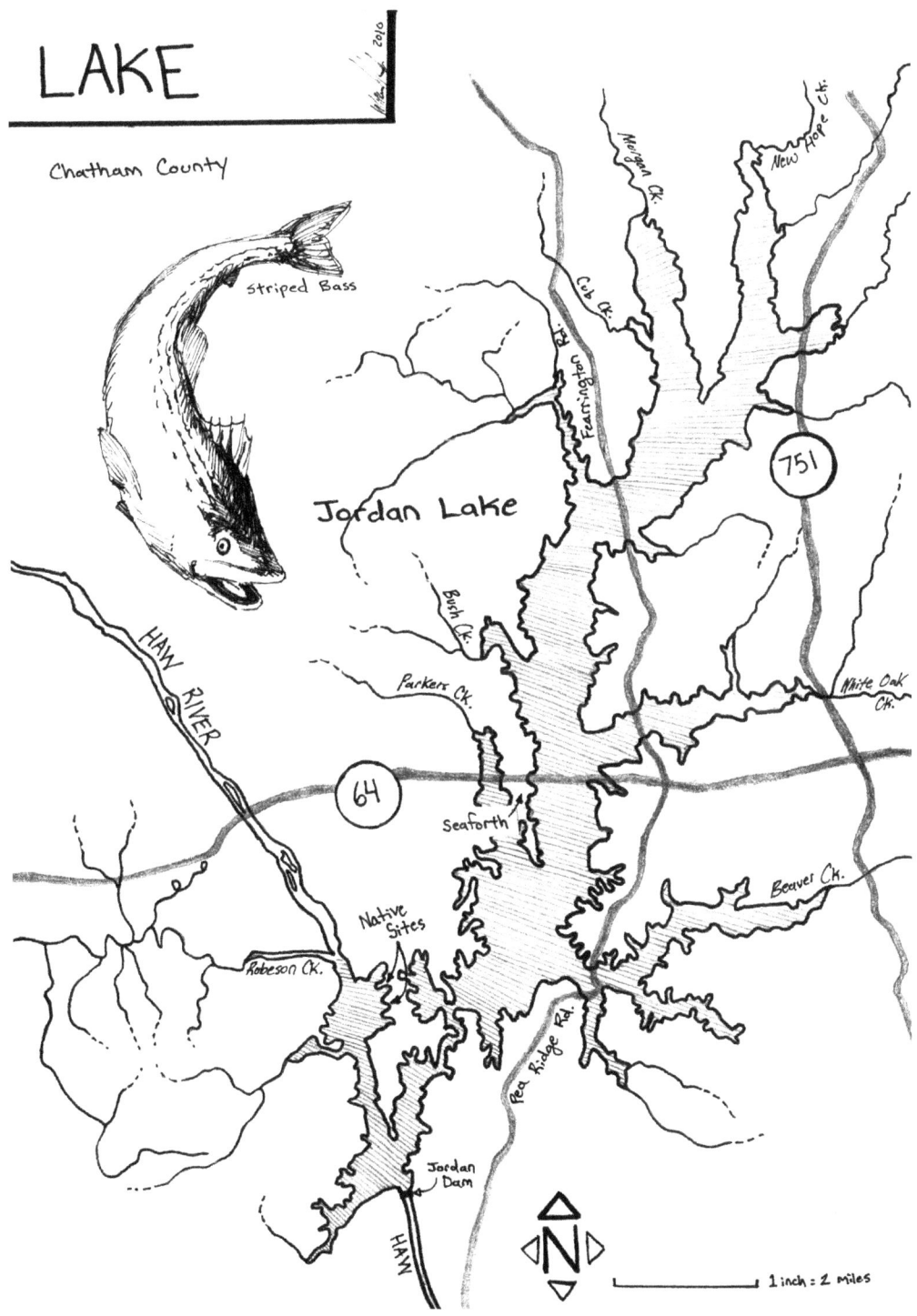

Lake (William Rusch).

B. Everett Jordan Lake, viewed from the visitors' center above the dam, covers where the Haw flows in, left, over a now submerged Gunter-Harris Island. New Hope Creek once looped in from between the shores in the background to converge with the Haw just before the dam, at the photograph's right edge (photograph by the author).

Nearer the surface, thinner bands of sediment show the climate of the Haw drying and warming, as the Paleo turned into the early Archaic period (10,000–8,000 B.P.). Instead of a few high quality tools of collectors, living on the move, a wider variety are found here, less specialized, more plant-related, suggesting a forager way of life with people roaming around their settlements until, resources exhausted, their whole group moved on.[8]

The arrowheads uncovered also sustain the idea that the Archaic and Woodland periods were distinct, though not sudden, shifts. Fragments of pottery confirm the sequence of development archaeologists predicted. Earliest pottery was crumbly and sandy, likely made from nearby sources, while later Woodland ceramics suggest a deliberate search for materials to produce a tougher and finer ceramic, and pots without evident coil marks.[9]

There may well have been more discoveries to come, even Clovis artifacts, in the fertile triangle formed by the New Hope and Haw. Recent findings of Albert Goodyear on South Carolina's southern border suggest that the crossing over the frozen land of the Bering Strait, Beringea, could have occurred 50,000 to 40,000 years ago rather than from 28,000 to 10,000.[10] But the dam-gate closed; the lake flooded; boats glide over remaining clues.

Salvage Archaeologist

Crist Holden is a freelance salvage archaeologist who has also worked under Al Goodyear and Dennis Stanford on digs that probe deep into the past. In his hands, the napped stones of the Haw, ancient tools of survival and of art, connect him and the thousands who have attended his shows to the peoples of the Haw. Holden's mission is to preserve these relics so their reminder of other ways of living and surviving is not lost. His license plate reads CLOVIS, the name of the people of the Paleo period whom he most admires.

Crist Holden holds a case of Clovis points from the Paleo era, 12,000–10,000 B.P. (photograph by the author).

I've been consumed by my passion, my love of Native Americans since I was very young.[11] It's the rush you get knowing you are holding basically the earliest form of art and the fundamental basis for survival on the late Pleistocene and Holocene landscape. The stone itself represents only 5 percent of what they owned; in the South, the rest is lost to humidity, acid soils, and construction, so everything else is subject to conjecture. It's a story told in stone, and that's why I call my show "Shadows from the Past."[12] People see it and get an idea or figure out for themselves how these people lived and survived in a much harsher environment. Everything these days is so convenient. I like to see people pick something up and see how it fits in their hand, and figure out for themselves how it was used; some do, very quickly.

When I was seven years old, I found my first piece, a Guilford ax, on my uncle's farm in Merry Oaks. I grew up on a Chatham County tobacco farm that, like most Piedmont farms, was camped on at some point. So I grew up on a site. Steve Claggert said, "You are in one of the richest areas in the nation for artifacts." I have found stuff everywhere I have ever walked.

The Haw has a rugged landscape. You don't have the floodplains you do on other rivers, but all the hills above the Haw were occupied, the little streams, all the stream confluences. Every confluence I have been to on the Haw, I found archaeological sites, and on the adjacent hills. They were everywhere.

Jordan Lake? I was old enough to go out and look when they were building it. You may not often hear this, but a hurricane flooded Fort Bragg. You can't have a military base flooded.[13] So that's why they built it as well as water supply and recreation. Jordan Lake at one point was the largest salvage archeological project in the United States; that means it had already been disturbed by de-forestation and plowing.

Oh yes, we were bitter 'cause I grew up out there on the Haw, and a lot of beautiful farmland and historical houses were destroyed, a lot of lives affected. I heard of one person who had a heart attack and another who went crazy. A lot of people lost friends and family. Whole communities were destroyed, like Seaforth, now a recreational area, largest community on the lake and the best farm land, and they destroyed that. But that's the philosophy of modern man; we destroy to re-create.

I remember walking out there as a kid. The confluence of the Haw looked like a war zone of artifacts; everywhere the ground was basically covered with debutage, flakes and artifacts. It was a twilight zone, once a thriving village and suddenly everyone's gone.

My classroom was the outdoors. I was put in touch with the three brothers who were the pioneers of the lower Haw — George, Royce and Jimmy Reeves of Pittsboro — who had been out on the Haw River for decades. They actually found a Clovis campsite between the New Hope and the Haw.

But at the Jordan Lake sites, they couldn't go beyond that and look for pre–Clovis for two reasons. Back a few years ago, if you talked about pre–Clovis, it was like talking about a ghost. It was almost taboo. People held to the Clovis First model that no one could make it over here before ice formed a landbridge between Siberia and Alaska. It was a barrier to research.[14] I think it was also political. If you found an actual Clovis site undisturbed, they would have to spend a lot of time and money and possibly postpone the [Jordan Lake] project.

When they were building Jordan, they had been out there two years, and they were actually digging in Clovis-level sands and finding artifacts, and they shut the entire operation down because water was filling in the holes faster than they could get the water out.

What was it like in Paleo times? the Clovis period? A lot like modern Canadian forests today, and it was a hell of a lot cleaner and teeming with wildlife, not like today when we have so much development. It was basically a cornucopia of wildlife for Native Americans, and their most precious commodity, stone, was abundant. Around the Silk Hope and Bennett, NC, area is the largest crop of rhyolite I have seen aside from Morrow Mountain in the Uwharries. And wherever your stone sources were is where you had your largest concentration of Native Americans. A lot of their villages were based on these floodplains, basically during the summer months, and off-season when it started to get cooler, they would move away from the floodplains and go up into seasonal mobility camps. They harvested plants and nuts, acorn and hickory nuts, all kinds of wild grapes, you know, the muscadine. I still live off a lot of this stuff.

In the Paleo period, they would have used fire to locate others or tell them they had seen incoming game and for cooking. I have always heard that there were not that many Clovis or Paleo Indians or pre–Clovis people, but I disagree. The accumulation I have gathered says otherwise.

The Paleo-Indian people only used the highest quality raw material, and a lot of times they would take a blade, and because this stuff is so sharp, they could open up an entire hide very quickly. They were very big on blades, micro-blades or pressure flake tools; they could use these for cutting, scraping, carving; that's why I call them multi-purpose tools.

The further you go back in time the higher the quality and technology. In the Contact period, they were introduced to metal, copper and rifles, but even long before that in the Archaic period with the mega-fauna gone, the quality is not as good. In fact, they still consider the Clovis to be the most technologically-advanced hunters to ever walk the face of the earth.

What is the difference in quality? The stone, the craftsmanship. These Clovis points are thin. They would do overshot flaking that would extend basically across the entire bi-face. All Paleo points have the concave base. A prominent flute made it easier to mount. They would grind these base edges. It might take three or four days to make one of these arrow points while one of these Woodland pieces might take just 30 to 45 minutes to make out a pre-form and another 30 to 45 minutes to finish it.

The most unusual artifact, a Clovis one, I've ever seen in the United States came off of Weaver Creek, a mile and a half from the Haw. When I found it on a sandbar about fifteen minutes before dark, I'd seen the color and thought it was a flake. It was so close to dark I wasn't sure. I'd walked past it three times before I decided to go back and pick it up. I went down to the water to see. I thought "Oh, my God, I picked up a Clovis that was picked up about six thousand years later by this Stanly culture, and they just put their signature on top of this." Some other collectors were telling me it was like an edge-filed scraper or this or that, and I am saying, "No, they did it basically as an artistic expression." They probably could have used it, but I don't think they did. I've seen lots of flakes that were made into an effigy or ceremony artifacts. That is my prize. I was very lucky. I have a friend who been doing this for eight years and has never come close to these findings. He thinks I am connected at some spiritual level. A Clovis piece is like the Holy Grail for artifact hunters.

I found a cache of Hardaway blades this year, four miles from the Haw on a construction site. This blew everybody away. We were walking around a bulldozer, and I found one piece first lying up on its side; then I dug into the sand with an expresso spoon and found these others. There have only been about twelve of these this age found in the United States. They are just after the Clovis; the Hardaway is the final frontier of Paleo.

I usually hunt construction sites these days. I always get permission

Three artifacts Crist Holden found by the Haw: from left, a Paleo-era Clovis point, an Archaic Kirk point, and a Woodland Yadkin arrowhead. The Clovis point is high quality rhyolite that fractures predictably; it has overshot flaking, parallel to the blade. Its thinness and removed channel flake make it aerodynamic and able to penetrate. By comparison, both the thicker Kirk and Yadkin points show random percussion and pressure flaking. The Kirk point's medial ridge would cause more damage; the Yadkin arrowhead is even rougher and less uniform (photograph by the author).

from the land owner; they usually don't care. I come in right when they cut the trees, after a rain. This is about preserving these artifacts for future educational purposes. That's why I do workshops and take people out to find and identify artifacts.

No one really cares about these disturbed sites or disturbed archeology, but then they are paved over forever and lost. Someday, someone will be digging in our ashes trying to put the pieces back together and understand a lost culture.

Well, you are a scavenger archaeologist, I say. "Yes, salvage, scavenger — I am the vulture that sits in the tree up there waiting for the road kill."

George Moses Horton

At the beginning of the 19th century, George Moses Horton, poet and slave, bearing his owner's name, was moved to land between the Haw and New Hope where he was "a cow-boy." As his brief autobiography relates, "In the course of this disagreeable occupation, I became fond of hearing people read." Horton began teaching himself to read using white school children and their discarded spelling books.[15] It may have been on the banks of the Haw that he "retire[d] away in the summer season to some shady and lonely recess, where I could stammer over the dim and promiscuous syllables in my old black and tattered spelling book, sometimes a piece of one, and then of another."[16]

When George Horton's master moved to Pittsboro, he squeezed in Sunday trips to Chapel Hill to sell vegetables and fruit to the hundred or so students there.[17] There his passionate love of words found an appreciative audience ... after a while. Horton tells us:

> The collegians who, for their diversion, were fond of pranking with the country servants who resorted there for the same purpose I did, began also to prank with me. But somehow or other they discovered a spark of genius in me, either by discourse or by other means, which excited their curiosity, and they often eagerly insisted on me to spout, as they called it.... But I soon found it an object of aversion, and considered myself nothing but a public ignoramus. Hence, I abandoned my foolish harangues, and began to speak of poetry, which lifted these still higher on the wing of astonishment; all eyes were on me, and all ears were open. Many were at first incredulous; but the experiment of acrostics established it as an incontestable fact. Hence my fame soon circulated like a stream throughout the college. Many of these acrostics I composed at the handle of the plough, and retained them in my head, (being unable to write,) until an opportunity offered, when I dictated whilst one of the gentlemen would serve as my emanuensis. I have composed love pieces in verse for courtiers from all parts.[18]

Horton, who was paid a quarter or more for these love poems,[19] went on to be the first African American to publish a book in the South,[20] and the third to write a book, after Jupiter Hammon and Phillis Wheatley.[21]

Both words and freedom compelled Horton to write. Poetry was liberty of mind, and his earnings amassed for buying freedom. Caroline Lee Hentz, though no critic of slavery, helped him publish a book of poems, *Hope of Liberty*, in 1829, before he could write.[22]

University of North Carolina President Joseph Caldwell was also a supporter, but his successor, former governor David L. Swain, proved treacherous, possibly delaying Horton's freedom by twenty years. In 1844 and again in 1852, Horton entrusted Swain with letters to key abolitionists. These pleas for help in his liberation were not naive gestures. Indeed, William Lloyd Garrison had published his poems, and Horace Greeley knew of him and might well have helped him raise the $250 he needed. However, Horton's letters were found still among Swain's papers many years later.[23]

While Horton waited for replies, he continued to write. When Chapel Hill students were replaced with Union soldiers, he wrote on[24] until he was finally freed by the words of the Emancipation Proclamation.

In his verse are metaphors of water, inspired likely by his streams, the Haw and New Hope.

> Wedlock is woe
> Without the stream of love[25]
>
> On fertile borders, near the stream,
> Now gaze with pleasure and delight
> See loaded vines with melons teem
> It is paradise to human sight.[26]

After the Civil War, other former slaves would struggle alongside white tenant farmers battling pests, market prices, and depleted soil. The local population dropped in the years from 1890 to 1920. But those who stayed later found practices that revived their farms, and they fought back when faced with land seizures by eminent domain for the Jordan Lake project.[27] One hundred and fifty-two families' homeplaces and 40 cemeteries with 2000 graves were at stake, as well as miles of the Haw's deep pounding rapids.[28]

Busloads traveled to Washington to protest in 1963, when 35,000 acres of the best land were set to go underwater for "New Hope Lake."[29] Environmental groups and kayakers joined the protests, and the courts delayed the project. Trees and bushes grew up on bulldozed land. But the dam project prevailed. In 1982, the flood gates were lowered; the rains came, and the land quickly became B. Everett Jordan Lake, the only aspect of the project completed ahead of schedule.

The River Hits the Wall: Jordan Lake and Dam

> Nature to be commanded must be obeyed.— Francis Bacon, 1620

Though some think of lost homes, lost rapids or a river's health and wish the Haw still ran full force, undammed to join the Deep, others find Jordan Lake brings pleasures: floating under the sky, canoeing toward osprey, reeling in bass. However, when the water hits the wall of Jordan Dam, it becomes impossible to ignore poor water quality and a 303(d) listing. The river is a major reservoir and recreation area.

Lest you think the Haw is not really part of Jordan Lake, let the numbers reassure you; the Haw's waters range from being 70 to 90 percent of the lake.[30] When the muddy Haw is slowed, sediment drops, and algae thrive in the nutrient rich waters, turning the lake green. It is time to explore this green. How worried should we be? What can be done?

Right now, fishermen's worries are future tense. The fish are biting; algae, fertilized by the excess nitrogen and phosphorus, make abundant fish food. The booming fish population is part of the reason Jordan is home to the most bald eagles on the East Coast in late spring. The Jordan Lake Striper Club is open to newcomers and full of enthusiastic fisherfolk with online postings by Reel Deal, Happy Hooker, and Fin-addict.

Troy Roberson, a club leader, finds fishing in Jordan fine.

> The fish are doing extremely well. Ninety-nine percent of the time I fish for striper, or striped bass; they're aggressive, hard-hitting, fighting fish as big as 18 pounds and lots of fun. They roam more and are out in the main body of the lake. Striped bass are stocked in Jordan because the river is not long enough to allow their eggs to float for 72 hours and incubate without getting covered up with silt. But they are otherwise flexible and can live in fresh or salt water.

But Troy Roberson adds: "I have seen some large algae blooms in the north part of the lake and heard of them in the Haw region."[31]

Corey Oakley, who monitors game fish for the North Carolina Wildlife Resources Commission, agrees on the fishing:

> It is certainly one of the better fisheries. For crappie, it's one of the top reservoirs in the Southeast and maybe the top in the state. Fish grow very rapidly here. Crappie reach 10 inches at two and a half years of age. That is extremely fast growth and would take longer in other reservoirs, if they even did reach that size.
>
> Nutrients do provide the source for the large crappie population in Jordan Reservoir. But there are problems that can occur from excess nutrients. Rapid development, especially in the Piedmont, can degrade the water systems. Without the proper stream buffers, nutrients and sediment will run off into the state's rivers. Alamance, which the Haw runs through, is a rapidly developing county. There has to be a balance. It's a hot button issue.
>
> I hope we can handle it with regulations as we did the over-fishing of crappie. In 2003, the data indicated the crappie population was getting down to two year olds; they were 87 percent of the population, and we were relying almost solely on them to reproduce. The other 13 percent were one year olds. This can lead to a crash in reproduction and the overall population. Therefore a regulation was implemented; currently, it is a minimum length of 10 inches and a creel limit of 20 fish per day. Before there had been reports of anglers harvesting 200 to 300 crappies a day. Since the regulation, the crappie population has continued to grow rapidly with greater numbers of 3, 4, 5 year old fish. The lake is really outstanding. It's a fish-growing machine.
>
> Yes, it is eutrophic, and that's great for growing fish. Right now there are plenty of fish out there, but there has to be a balance. Too many nutrients in the reservoir lead to massive algal blooms and fish kills, whereas too few nutrients slow fish growth and cause the fishery to decline.[32]

Great fishing is the temporary silver lining of a looming crisis. The Division of Water Quality has been worried since listing Jordan Lake as nutrient sensitive in 1983, just one year after the floodgates shut. Nutrient sensitive waters (NSW) have a harmful overdose of two otherwise good ingredients — nitrogen and phosphorus — and are called eutrophic. In 1997, it got worse, and the upper arm of Jordan Lake fed by New Hope Creek was impaired and 303(d) listed, and in 2006, all of Jordan Lake was on the 303(d) list. This requires the state to develop a plan to correct the problem. It's like being put on a diet or, in this case, TMDL, Total Maximum Daily Load limiting phosphorus and nitrogen. Farmers, developers, municipalities, the 65 wastewater treatment plants, and everybody in the Haw watershed have to reduce the nitrogen and phosphorus sent into it. Some may think, "But I don't live by a stream," but the water that leaves their roofs, driveways, lawns, and pipes is pulled by gravity to streams and river.

Protest and finger pointing followed the reduction mandate. Many people don't understand runoff or realize that the proposed rules for cleaning up Jordan will clean up the creek near them. Most of us don't talk about eutrophic conditions or even notice a problem unless fish die, swimming spots close, tap water tastes funny, or children get rashes. And there is dispute about how bad it is and who's responsible. Add the pressure of rapid development (North Carolina is the fifth fastest developing state), and the controversy heated up.

The Jordan Lake Rules define how different stakeholders will meet the TMDL and how much they will have to reduce their nutrient contribution to the Haw. Hearings on the rules were lively. Developers and realtors predicted economic ruin. One mayor claimed the costs "would put us out of business."

Farmers agreed to the rules but asked for aid in meeting them. Rule supporters pointed out that similar regulations in the Neuse, Tar and Pamlico area had not brought economic doom. They also noted we are losing land — 383 acres a day — to development, a major river stressor.

Part of the debate is the question: How bad is Jordan? The visible signs of trouble are a fish kill, masses of floating fish, and tiny algae plants that green the water on hot, light-filled days.

The algae which fish thrive on can turn and kill them when they compete for oxygen. These microscopic plants create dissolved oxygen during the day, but without the sun to drive their growth at night, may suck so much oxygen out that fish can't breathe. The result is a fish kill; one hit on March 21, 2006.[33] Test results for dissolved oxygen and chlorophyll a, a substance given off by algae, confirm the problem.

A high pH/alkali factor shows lots of plant photosynthesis, too. A healthy balance of acid and alkaline is seven, but in the summer of 2006, pH was 9 and over, almost like a weak bleach solution.[34] Results like these led the Cape Fear River Basinwide Water Quality Plan 2005 to state: "The Haw River and New Hope River Arm will be added to the 303(d) list. TMDLs are currently being developed to address the Impairment in Jordan Reservoir."[35]

For Haw River Assembly members, these bland words sound a trumpet blast of hope. All the Haw's streams will be helped through buffers, stormwater controls, sediment controls, best management practices in agriculture, as well as reduced nutrients in wastewater outflow.

If you work for a city in wastewater management and the finger points most easily at you, the report's words sound more like a dentist's drill. Steven Shoaf, director of utilities, disputed that the lake was that troubled: "It's not a crisis, not even close. The problem is non-point sources [NPS]." (NPS means run-off from roofs, roads, lawns, developments, and farms that does not come out of a pipe.)[36]

Eric Davis, water and sewer operations manager, concurs: "It's a non-point source issue. They are regulating us to death, pushing us to go lower; it's almost like a physical workout. If we were seeing wholesale fish kills, we would understand. They are constantly changing the rules as you go" over what seem like "vague worries."[37] The big need is maintenance, new sewer pipes, for Burlington, so "it's like painting your house when the foundation is crumbling."

As upset as utilities managers of the upper Haw are, they are on the easy "south end" of the TMDL plan where the requirement is only to reduce nitrogen by 8 percent and phosphorus by 5 percent. Chapel Hill and Durham on New Hope Creek must meet a 35 percent nitrogen reduction goal along with 5 percent for phosphorus.[38]

The demands are different partly because the water in the Haw's end of Jordan Lake moves through in a week or two, and motion is health to a river. New Hope Creek pools at the upper, shallow end, where it is blocked once by the dam-like effect of Fearrington Road, then is blocked again by the faster-moving waters of the Haw. Waters in New Hope Creek may take a year or more to pass the dam.[39]

Though the municipalities on the Haw are already permitted to discharge over 1,500,000 pounds of nitrogen and almost 200,000 pounds of phosphorus a year, they protest because further reduction is expensive.[40] The cost of having healthy water and rapid growth is one taxpayers generally prefer not to calculate.

Steve Shoaf and the Haw River Assembly folks are in agreement on one point: the majority of nutrient load, 64 percent of the nitrogen and 83 percent of the phosphorus, is coming from non-point sources, not the "point sources" of wastewater treatment or industrial pipes.[41]

Point sources were regulated by the Clean Water Act signed by Richard Nixon in 1972. The act declared that all waters be fishable and swimmable. This crucial act passed in the days when Dr. Don Francisco found only sterile muck at Haw River and when 11 industries and 4 municipalities were discharging directly into the Haw,[42] as in many cities. Pollution was highly visible when colored foam rode the waters of the Haw. Industrial chemicals were burning on the surface of Lake Erie.

Without the Clean Water Act of 1972, nobody would be kayaking or swimming in the Haw today. The Clean Water Act has more than doubled the miles of U.S. rivers that are fishable and swimmable.[43] But the act went after point source polluters only. Now both sides agree non-point sources, runoff, have to be trapped and filtered or channeled for treatment before joining the Haw's streams.

I check with Peter Caldwell of the Division of Water Quality to get the official word on how bad conditions are in Jordan Lake and the Haw watershed. He re-forms the question: "Is Jordan currently supporting its intended uses for recreation and drinking water?" then adds, "The lake regularly exceeds the state's chlorophyll a standard, enough to list it on the 303(d) list of impaired waters."

Are the algal blooms a problem?

"They can be, but so far aquatic life, recreation, and drinking water have not been affected to a significant extent. There have been some fish kills but nothing massive. The system may be on the edge of having a problem." And that, Caldwell considers "an important reality. The population growth is virtually out of control."[44] Indeed, Guilford, Alamance, and Chatham counties all grew 17 to 20 percent in the 1990s; the projected growth from 2000 to 2020 is 25 to 30 percent, though an economic downturn may slow that.[45]

When I ask again how bad the algal blooms in Jordan Lake are, Peter Caldwell points me to a phycologist.

Haw Under the Microscope

In Mark Vander Borgh's Raleigh office, slime-filled fish tanks are stacked on the window ledge. This is, after all, the office of a man who studies algae for the Division of Water Quality. My "How bad is Jordan?" question turns out to be way too broad for a scholar of microscopic single-cell wonders like the cyanobacteria, blue-greens, dominating Jordan. Mark sticks to known particulars, blue-greens among them in all their 200 variations, which he recognizes at a glance.

When I ask just how bad blue-greens are, I find we owe our life to them.

> Blue-greens are some of the oldest organisms and the ones responsible for giving us oxygen. They are really like bacteria in their ability to break down different things. They started making oxygen, so the first organisms to change the atmosphere were blue-greens. They created these coral reef like things called stromatolites in the early oceans, and as they started to produce more and more oxygen, plants started taking off through evolution and creating an atmosphere. It's really sort of funny. We owe them a lot, and yet they are one of our pseudo-enemies in some ways.[46]

The window Mark Vander Borgh opens on the blue-greens' world makes me want to cheer them on. They are single celled; they get swept downstream and land in the sediment and hang out, unable to bloom, until favorable conditions for growing. "Blue-greens don't have flagella, so they can't swim. They are primitive organisms, and they can't tolerate low light like diatoms."

So how do they survive at all? "Your light and warmth is up here, but your nutrients are down here. How are you going to get both? You have to have some kind of mobility mechanism. And they do. It's only a floating mechanism."

They operate like hot air balloons? "Yes, they have gas vacuoles. Instead of pushing their oxygen or gases out, they can put it into vacuoles [pouches] that are like a submarine ballast."

And the dam at Jordan gives them their break?

> This [floating mechanism] is only effective if the water isn't churning around and sweeping them downstream. You would not have had filamentous blue-greens or these other ones without the dam. They like warm water, slow flow, high light, and lots of nutrients. They want to be in the most optimal position in the water column; the blue-greens will be at the bottom in the afternoon, evening where the nutrients are, and they can go to the sunlight other times.
>
> Alga are really good and the basis of our food web; blue-greens are unique in regard to other forms of algae; they can "fix nitrogen" [break it down to a form fish and humans can use]. Only blue-greens can do it. That's a competitive advantage. They take nutrients and create carbohydrate structures and feed the fish and then us, as it goes up the web.[47]

Their pluck is inspiring; however, I learn blue-greens take over; they are the dominant alga in Jordan from April to December.

> Blue-green is one of the smallest types. To put it in perspective, if a blue-green is the size of a BB pellet, others can be the size of a softball. They can have the lowest density by count and turn around in a month and have the highest density by count, as high as 90 percent.
>
> And fish are caught in a system where organisms are producing oxygen to a concentration of 130 to 140 percent saturation and hyperventilating then six to eight hours later lower the saturation to 40 to 50 percent; fish gasp for air because now algae are utilizing it. It's not just the suffocation, but the extreme change.

Time to ask for a verdict on the state of the lake. "Sure, Jordan is very eutrophic. Is that serious? What is important to whom? What values do you want to put on the system? It's a resource and a highly managed resource. Every drop of water in Jordan has probably been used once or twice by someone or some animal upstream. Add to that it's dammed; it's highly urbanized."

I want a prediction. "At what point do fish kills take over? There's going to be an optimum point for trout or bass. For which fish? At what level is that? I don't know."

A guess? "That's what I can't give you. That's what the modelers do. My knowledge is on a microscopic level; I am the person who looks at algae under the microscope and gives an analysis of what I see. As water stagnates, health concerns become great; blue-greens can produce toxins. Have they been shown to produce toxins in Jordan Lake? We don't have any evidence for that."

How serious is the health threat?

> What one person is allergic to, another one isn't. I am not a health expert. If someone offered me catfish fillets from Jordan, I'd eat 'em. And swimming? There are times of the year I'd prefer not to go in. The greener the water, the more organisms growing in it, and even though I'm not afraid of my algae coming to get me....
>
> The important thing to know about blue-green toxins is that we don't know. And it's not like we're not trying. We just found out they produced toxins. Well, what kind of toxins? How much is actually toxic?

That's as close to an answer as I may get, so I just enjoy Vander Borgh's Powerpoint images of algae sparkling under the microscope like jewels — opals, diamonds, pearls. There's even beauty in the high-density blooms when filamentous blue-greens dash in lines, with the playful drive of a Paul Klee drawing. Unfortunately, this visual spice comes under the heading "extreme" algae bloom. "I had to come up with this 'extreme' designation when explaining Jordan algal assemblages. In the 160 samples I analyzed last year from around the Piedmont, only Jordan Lake had extreme blooms — over 100,000 units per milliliter [units can be a filament, colony or single cell]."

I just captured part of my answer; Jordan is in worse shape than any other lake around. So why isn't the lake covered with algae? Vander Borgh responds, "Jordan doesn't get the nasty surface blooms that create public outcry. Jordan has blooms throughout the water column that turn the water green."

These more pervasive but less obvious algal blooms were so extreme in Jordan that Mark Vander Borgh also had to come up with a new way of photographing them — in layers. In the blooms, not only does one kind of algae dominate — blue-greens — but one blue-green in particular, Pseudanabaena. It can form chains 15 cells long that shimmer like strips of opals in photos. The take-home message Vander Borgh ends his Powerpoint with is: "If you drink a half shot glass of raw water from Jordan, you could be ingesting over a million units of algae."

Is this a problem? I am not even going to ask. But Vander Borgh senses my question. "Is there a problem? Well, the normal person is sure with 50 percent or less proof. Scientists are not satisfied even if they are 99.9 percent sure. This means it can take them a long time to say there is a problem.

Teetering on a Crisis

Amy Pickle, a lawyer with the Southern Environmental Law Center, does not hesitate to say that Jordan Lake is near a crisis point.

> No one wants to swim through algae or dead fish, of course, or in water that is so basic

> [alkali] it is similar to bleach. Some people have already reported that Jordan water affects, dries their skin. The cost of treating drinking water increases. Aesthetics, recreation are affected.
>
> Yes, the condition of Jordan is a problem. It is a rare instance where some one breathes the air and falls dead. It is about an increased risk of health problems.
>
> The huge side benefits [of reducing nitrogen and phosphorus in Jordan] is to the tributaries, keeping those waters cleaner, healthier. We're just figuring out how to have the development and growth people would like to see. It's not easily regulated.
>
> We are working to get retrofits for stormwater, to change agriculture, change new developments. It will take decades for the lake to be clean.... It's a scientific fact the lake is polluted, and everyone has to pitch in and do it. We've known for decades, the limits set in the Clean Water Responsibility Act of 1997 are not a surprise.[48]

Cam McNutt of the Division of Water Quality thinks its urgent too; he points out that of the 28,775 impaired acres of water in the state, over one-third, 10,571, are in the Haw Valley which has nearly all of the 10,834 acres impaired in the Cape Fear: "So the lesson we want to teach is ... if [water quality] is fair, it's time to fix it while you still can. Don't let it get worse because every time you take a drop down, it takes more money to fix it."[49]

Michelle Woolfolk did modeling and made predictions for Jordan Lake for the Division of Water Quality for many years. That may explain why she sounds like someone accustomed to a maelstrom.

> Jordan is highly eutrophic. It is not unexpected for what it is because it's downstream of major cities. With Jordan there is nothing they can do; it is not naturally a lake, only the Carolina Bays are. The dam will keep it from being a wetland with artificially high water levels. It will just be nice and green.
>
> Nothing bad has happened, but the state's job is to be the steward of water. People don't complain until we are at a point where there is nothing we can do. If the state allows things to get that bad, then we made a mistake. And the state is not doing its job. So far there have been almost no fish kills, so that shows that our early work did help.

Can we afford the costs?

> You can just say that recovery is more expensive and politically harder even and still does not guarantee it will succeed. I like to compare it to being sick. Eating green vegetables all along or getting cancer at 50, with chemo and surgery. Was that less expensive than eating green vegetables all those years?
>
> I would love to see people take more interest in the river, what the Haw River Assembly [Learning Celebration] does. I like to get people out there seeing what the river looks like and should look like. It's a shame people don't value their resources until something goes wrong with them.[50]

The Jordan Lake Rules did pass into law in August 2009. A 5 percent phosphorus reduction in wastewater outflow had a year's deadline, but others were extended. Janet Mac-Fall notes two big wins: "Unincorporated areas, like Alamance County, will be held accountable for the nitrogen and phosphorus they contribute to the river. Stormwater rules in the past have only been applied to towns and cities."[51] And secondly and perhaps most significant, municipalities are mandated to improve stormwater control in already existing development (i.e., put in stormwater retrofits) if testing in 2017 reveals that other changes have not improved conditions in the lake. Grady McCallie of North Carolina Conservation Network says it is one of the first times nationally this has happened; "The North Carolina program is an important precedent."[52]

All sides might have been in closer agreement if the public could understand the costs of clean water and that run-off belongs to all of us. Right now many assume the Haw can be used for disposal, yet clean water will flow from the tap.

Stephen Shoaf had to deal with this lack of awareness as Burlington's director of utilities, so he videotaped people's answers to questions like "Where does your water come from?" "How does it get to your house?" The guesses were funny and painful; one in sixty knew the answers.

"They buy fertilizer at K-Mart. When I ask, 'What's in your fertilizer?' They can't tell you what the N P K stands for [nitrogen, phosphorus, potassium]. Why did people settle next to rivers?" Shoaf asks rhetorically. "So their garbage was gone. Take antibiotic pills, for example, what did the instructions used to tell you? Flush down the toilet. This is the most responsible job in the world — taking care of the drinking water."[53] (Instructions for drugs may now say "dispose of properly," but few places have take-back programs.)

This problem isn't limited to the Haw. The nutrients from artificial fertilizer have been a mixed blessing, increasing food production and population, but causing seasonal dead spots at river deltas all along the world's coasts where algal blooms suck out oxygen. Near summer's end, the lifeless zone at the Mississippi Delta is about as large as New Jersey.[54] That's a major loss of sea life (and seafood).

A river runs through us. Can we feel it? If we find that connection between our health and the Haw's, we will get to solutions. I am feeling I need some river time; perhaps we all do.

> Who hears the rippling of the rivers will not
> utterly despair of anything. — Henry David Thoreau[55]

Drinking the Haw: Cary, Apex and Global Fresh Water

Who drinks the Haw? More people than ever intended when Jordan Lake was first bulldozed for flood control in 1967. However, soon after the floodgates closed, the pooled waters of the Haw and New Hope Creek were destined for drinking in Northern Chatham County, Research Triangle Park, Apex, Morrisville, Holly Springs, Wake Forest, Cary, and some parts of Orange County and Durham.[56]

Cross over Jordan Lake on U.S. 64, then look north to see the station that pumps Jordan water into people's bodies, bathtubs and businesses. Continue six miles east and you drive between two water pipes headed for treatment, duplicated so a breakdown in one doesn't stop the flow. The pipes turn north on the American Tobacco Trail and swing under pine woods before entering Cary-Apex Water Treatment Facility.

Kelvin Creech, manager, and I stand atop one of its brick buildings surrounded by blue roofs in a vast chain-linked park, with very little water in sight. Some of it does lie olive gray below us in shallow square pools. Creech is responsible for that water leaving here fit to drink.

Healthy water is a human struggle. A breakthrough came in 1855 when Dr. John Snow showed that cholera was waterborne; turbid waters' links to typhoid and dysentery would follow and bring complex water treatment for some. However, worldwide an estimated 1.7 million people die a year from preventable waterborne disease.[57]

Unlike wastewater treatment, which largely replicates natural processes, the magic here is chemical. A high voltage electrical jolt to oxygen turns it to ozone, O_3. When this enters the water, a strong oxidation begins that can remove petroleum, herbicides, pesticides, and reduce iron, manganese and sulfur. Using ozone is expensive, energy-intensive, and state of the art. Ozone is one of the strongest disinfectants out there, killing bacteria and viruses. It also aids coagulation; add alum and polymers, and sediment adsorbs or clumps together. "It's really working through electro chemistry," Creech explains, "alum has a positive charge."[58]

Amazingly, this clumping action is taking place through a grate below our feet where grey wisps of pollutants form. These floc will sink to the bottom and be skimmed off, then channeled to a building two hundred yards off to be centrifugally spun out. The resulting dry cake is composted and recycled as fertilizer since alum is a micronutrient.

The clearer water drops through holes in pipes near the surface. We climb down stairs leading into a vast room housing three ozone generators. There's only a hum in the room, but inside the dark blue cylinders, a high-voltage electrical storm is converting O_2 to O_3. Below the cement floor where the water flows unseen, the ozone bubbles in. Creech detects its smell, like the air after a thunder storm. Only one ozonator is needed for winter demand, 12 million gallons a day, and two for summer demand which peaks at 25 million, but averages 20 million. The third ozonator is backup; tap water must flow.

Filters come next. We walk out to where clearer water makes a second and final appearance in rectangular pools. The flat bottom is really seven feet of filters: four feet of anthracite coal, two feet of sand and one foot of gravel. Sand and gravel filters are not new. Egyptians used them in 2400 B.P. Filtering through charcoal is mentioned in early Sanskrit and Greek writings and may go back to 6000 B.P.[59] Downward, the Haw's waters sink to greater purity.

Moving a few hundred yards underground, it pools under two huge low domes to undergo final chemical treatment. The pH is adjusted to 7.5–8.0 to control corrosion; lead or copper pipes would leach at more acidic levels. Orthophosphate is used to guard against the same effect.

Chlorine joins the mix to control the growth of organisms since ozone is unstable, throws off an oxygen molecule and ceases to work as a disinfectant. The chlorine will go in with ammonia as chloramines; alone it can be a carcinogen.

After a dose of fluoride, that controversial tooth protector, the water flows into the pipes of Cary. The new, improved Haw also pours luxuriously from ten faucets in a sink the length of a whole wall, for endless testing for health, taste, odor, and color. One test is for manganese, which has to be kept low so laundry won't stain.

Safe drinking water is a chemical miracle hard to maintain. Even small leaks of ozone could be a health hazard to workers. And ozone may react to certain organics in ways that are unhealthy.[60]

Now that a patient Kelvin Creech has sketched out the water treatment system, we gaze off at the domes and discuss waters' path to residences (75 percent), irrigation (2 percent), commercial use (21 percent) and industry (2 percent),[61] then on to wastewater treatment and back to the river.[62]

"We only use 40 percent of our allocation now. At this point in time I have no concerns. It needs to be well managed, of course." Creech acknowledges that global warming—"a big shift in climate patterns—could change that."

Cary and Apex, which share this plant 77/23 percent, are, in fact, permitted up to 32 million gallons a day — the lion's share of the lake. The waters of Jordan may swirl in currents and roll in waves, but they are divided up with straight lines on the Division of Water Resources graphs. A bottom layer is reserved for the inevitable sediment; another line designates the volume for flowing water, dubbed "water quality pool"; a third section is the 100 million gallon reservoir where imaginary lines grant Cary-Apex 32 percent, Durham 10 percent, Chatham County 6 percent, Morrisville, 3½ percent, and so on.[63]

Perhaps part of Creech's confidence about supply comes from Cary's conservation program, which started in 1997 to reduce consumption by 20 percent. First, officials reach out to school children to teach home conservation — tight faucets, shorter showers, and the right appliances. The big users are toilets (27 percent of home water used), washers (22 percent) and showers (17 percent). A front-loading washing machine can cut electricity and water use by over half. Shorter showers save 2.2 gallons a minute off the 17 gallon average.[64] If the younger generation tunes in, the savings will be dramatic.

The conservation program also addressed the 60 percent of water lavished on Americans' obsession, our beloved monoculture, the lawn. Cary's sprinklers were reined in with alternate day water restrictions and something like a tuna can. Lawns get shallow-rooted with over an inch of rain a week, and the can helps residents measure that inch.

Creech takes pride in Cary's proactive measures. "Some see the solution as selling more water, making more revenue ... but Cary is really a model in North Carolina, maybe even on the East Coast. Even so, the West is decades ahead of us in water conservation; we've learned a lot from Arizona, Colorado, California."

A final question: Does Creech drink bottled water? He shakes his head no, amused, but he doesn't proselytize. "It's a personal choice. Some people call to ask us about that. We don't recommend or discourage. In most cases it's bottled tap water; it's not coming from the top of a glacier somewhere."

Americans assume water will flow from the tap, but for many people on our planet, there is no tap. One billion do not have a reliable source of clean water. Two thirds of the world has relatively insufficient to scarce water.[65] With global warming threatening to immerse barrier islands and coasts, salt water, 97.5 percent of all water, will be even more abundant. However, the shortage is fresh water, most of which is frozen, leaving 1 percent of the world's water drinkable if unpolluted. World use has tripled in the last 50 years,[66] compounded by a similar increase in population.

For the Haw watershed and the world, questions arise about best use of water. Lawns hardly qualify. Even some productive uses like for orange groves in California or cattle and wheat in arid climates are unsustainable. It takes 900 liters of water to grow a kilogram of wheat and 15,000 liters for a pound of beef.[67]

The Colorado River no longer pours into Mexico; its drastically reduced flow now has to pass through the Yuma Desalting Plant. Crops by the Nile require such heavy irrigation that the Nile does not reach the Mediterranean all year.[68] Water re-use, de-salination, conservation, ending water-wasteful subsidies could all ease the problem.

We cannot devolve into a world where water goes to those who can pay. Right now, the U.S. tops the world with per person water use; by one comparison, people of Masai are using 5 liters of water a day, while residents of Los Angeles average 500 liters a day.[69] Corporations attempting to privatize water are meeting resistance. In 2000 in Bolivia, the people of Cochabamba fought for their right to water and won. In Grenoble, France, the same struggle went on.[70] In Kerala, India, farmers in the village of Plachimada found their water level drained by a new Coca-Cola plant. Their protest brought an initial victory when the plant shut in 2004.[71]

Will rivers remain part of the great public commons? Or will Pepsi's Aquafina and Coca-Cola's Dasani replace tap water? Global Water Corporation of Canada contracted to ship 58 billion liters of glacier water to China for bottling, luring investors to "harvest the accelerating opportunity as traditional sources of water around the world become progressively degraded and depleted."[72]

River scholar Luna Leopold reminds us not to take our river bounty and our celestial location for granted. Closer to the sun, our rivers would be vapor. Farther out, they'd be ice. Though rivers came to us slowly, born of volcanoes' vapors,[73] we can lose them quickly.

6

Coastal River to the Sea
Chatham, Lee, Harnett, Cumberland, Bladen, Columbus, Pender, Brunswick and New Hanover Counties

River Travel and Trade: Jordan Lake to Mermaid's Point

Hot March days bring a flash-spring; John and I drive through tunnels of redbud and unclenching dogwood, canoe on top, to Jordan Dam. The feeling switches to industrial at the massive dam, 113 feet high and 1330 wide, a far cry from the milldam here in 1868.[1] High above the dam, outside the Visitors' Center, we watch the Haw spread over the old Triassic Basin. Cloud shadows move across the lake under the vast sky. Inside, photographs show the old confluence of Haw and New Hope and humans' power over the earth. Hydro Matrix will soon install turbines to capture energy at the base, the most dynamic point. When the lake is full, the release spews and boils like Big Sur in high winds.

Today, the release is mild though the flow still marbles and roils. Headed to the rocky put-in, I pass Latino families fishing from the rocks; my informal demographic study tells me these newest immigrants, who come community by community like those of old, love the river more than any other group, finding what is vital in their new land.

Soon I am propelled by dam release along the Haw's narrowed canal until backwater from Buckhorn Dam slows me. Ahead or behind me are Matt Steible and Nolan Wildfire, two Elon students on a spring-break trip down the Haw. Matt, the organizer, long had this trip in his imagination. As "Phoenix Paddler," his mission was to know the Haw by running with it and to advance its rebirth:

> I'm not looking to set the world on fire; I just want people to know what they have in their backyard.[2]
> The Haw is a very personable river. In the west of North Carolina, there may be other rivers with more white water. To the east, there might be rivers with wider banks. But smack dab in the center of Carolina, there's a gem that cuts through our daily lives.[3]

As Matt made plans to "play Huck Finn for ten days,"[4] his zeal was fired by discovering he was duplicating Chuck Brady's run down the Haw exactly twenty-five years earlier. Brady, who died in 2006, was the stuff of legend, Haw River Assembly's co-founder, doctor, Blue Angels pilot, astronaut. He had brought attention to the Haw's pollution by talking to people and reporters along the way and toting a jug of water from the source, proclaiming:

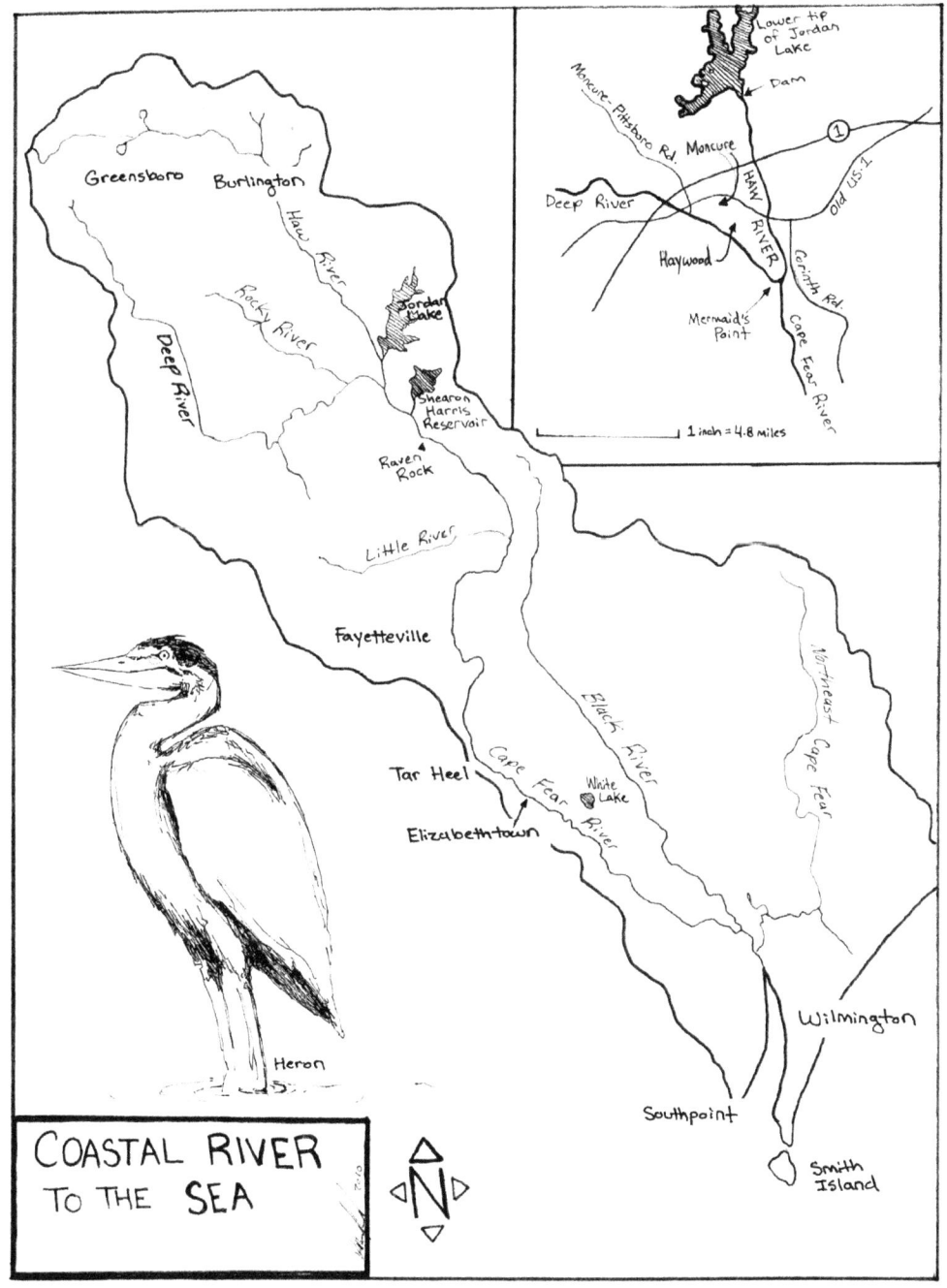

Coastal River to the Sea (William Rusch).

"The only way you'll get clean water at the end of the Haw River is if you carry it down from the start in a jug."[5] Today, Matt has that very jug as he and Nolan head to meet Mark Garner, commissioner from Brady's hometown of Robbins, NC, and Brady's friends, William and Lynn McDuffie. They traveled down Bear Creek and the Deep River to join Matt and Nolan at Mermaid's Point and pour their jugs into much cleaner waters than Brady could.

The Haw sprays out of B. Everett Jordan Dam and roils on toward Mermaid's Point in a rock-lined channel. A ramp for observation, lower right, is sometimes under water (David Eilers).

I paddle on; the Haw will channel us together sooner or later. A heron cruises the air above the river, level as a ship on seas. The tops of oaks are thinnest green and golden rain. A bold sycamore stands out; this hollow tree, heartless and pale, was the name some Native peoples gave to Europeans. This one splays out like Virgo, as a ghostly half-moon passes through its branches.

In no time, I am at Rt. 1, checking in with John and Joe Jacob, who was the anchor of Matt's trip and advised a day off when the river rose six feet after two inches of rain. That was after Matt went through the Guilford and Rockingham County swamps in the rain, but still arrived on time when a worried Joe went to meet him. Joe was still smiling about Matt's send-off at Haw River State Park, where a church youth group, Happening, lifted him above the cold rain singing "Row, row your boat" and hoisting banners, "Water is the MATTer" and "Have an Hawsome trip." This was a well-planned venture with potlucks scheduled, where Matt and Nolan impressed with their stories and appetites.

Joe tells me the guys are ahead, so I move on, glad to have my body back in the river, and pull against it. There is no more solid feeling than water holding you. Two men catching crappie tell me, "Two guys are just ahead."

The water is thick olive, over-rich with the nutrients and algae of Jordan, the blue-green jewels under Mark Vander Borgh's microscope shining in it.

River right is Haywood, where Archibald Murphey and other speculators once owned parcels in the early 1800s, reckoning river traffic would make it a fulcrum of trade. Haywood's

On their way to Mermaid's Point, Nolan Wildfire and Matt Steible (top, from left) pose with other Haw River advocates, Elaine Chiosso, Cynthia Crossen and Joe Jacob (front, from left). Dry Creek is in the background (David Peterson/Haw River Assembly).

still quiet ways, softened further by the spring day, appeal to me more than the anticipated boom; houses mellow into the land rather than mushrooming up. But this place of Deep and Haw River convergence was always in the mix, once considered as a site for the University of North Carolina and the state capitol (after the Raleigh capitol burned).[6]

Earlier, in the Revolution, both Cornwallis and Greene took comfort at Ambrose Ramsey's Tavern here and likely drained supplies from his mill.[7] Cornwallis waited for a bridge to be constructed over the Deep as he retreated to the coast; this time Greene was in pursuit.[8]

In the 1820s, a land boom was spurred by the Cape Fear Navigation Company's attempts to turn the Cape Fear, Haw, and Deep into lanes of commerce. If Smilie's Falls and Buckhorn below could be by-passed then the Rip Van Winkle state could shake off its dependence on Virginia and South Carolina for trade routes. The Deep offered access to veins of coal and black oxide.

When you hear that they worked to blast the granite of Smilie's Falls with hot vinegar and gunpowder, you know what is meant by determination.[9] This was before dynamite. A canal could bypass Buckhorn Falls, but the connecting lock proved flimsy. William Thompson's, later the successful engineer, gave this sputtering denunciation of that attempt:

> The simplest laws of hydraulic pressure were totally disregarded. The sides of the locks were merely upright posts braced at each alternate post, with single lining of one inch plank,

and with no other support whatever. As might have been expected, they were forced open soon after being subjected to the head necessary to pass a boat through them. They were miserable attempts at lock building and it would have been far better had the money been thrown into the river; as the effect of their failure, although some twenty years in the past, is still found in the minds of many enterprising and intelligent gentlemen, who look upon any further attempt at improving this noble river as utopian.[10]

In the 1850s, Haywood speculation reboomed; fancy summer residences, Bob Brown's 50 room hotel, and Miles Stephens' bar business rested on hopes of river transport. This time, William Thompson headed up the Cape Fear and Deep River Navigation Company's efforts; digging to enlarge a slough around Buckhorn Falls started in December 1849. On March 11, 1856, the *John H. Haughton,* "the first paddle-wheeler to trail pitch pine smoke across the Chatham County treetops," arrived.[11] On the return trip, a rowdy crew of Haywood men launched chairs and a few of themselves overboard, but Captain Matt Brady could still rejoice; along with some drunks headed for the Fayetteville jail, 135 cotton bales were delivered: proof Chatham goods could get to Fayetteville, Wilmington and beyond.[12]

Alas, in only three years, floods knocked out the engineering that tamed the Cape Fear and a bit of the Haw for transport. Confederates made repairs to keep the iron ore coming. Then Sherman's Freshet carried off every lock and dam from Haywood to Buckhorn in 1865. One river streamer was stranded at Haywood and served briefly as a cobbler's shop.[13]

A piercing smell comes in from river left. Plywood glue and raw wood fill the air. Moncure Plywood and Arclin, maker of synthetic resins, line up with Progress Energy along the eastern bank. The smell dominates even when my eyes rest on oaks whose curving branches hold clumps of mistletoe.

Then a bend dissipates it. And I see two rivers come together a long way from their start in Forsyth County. An ancient capture by a tributary of the Haw brought the Deep and Rocky rivers over here.[14]

Four canoes bank at Mermaid's Point, the now dam-flooded spit of sand where mermaids once came to wash the salt from their hair and sing into the night.[15] In their place, Matt and Nolan huddle with three others, in obvious good cheer and full-throttle story swapping. Nolan spins around in surprise; Matt runs to clear a place for my canoe, but I am just passing to salute them. When I tell them I am taking notes for the end of my book, Matt and Nolan leap into high fives. I circle out of the Haw, across the Deep, and into the Cape Fear. On the banks, the limey light of new leaves fuses down the dark trunks. I paddle back against the river's ever on-coming waters; the Haw is in good hands; Matt and Nolan are already talking about next year's run.

> The rivers flow not past, but through us. — John Muir

Cape Fear Rolls On: Mermaid's Point to Atlantic Ocean

When I head back, the river heads on; gravity makes the Haw's choices. The earth's dense core pulls the water to sea level at Southport. The Haw, which first seeped out of the earth at almost 1000 feet, has come 110 miles and dropped 842 feet; the river, now at 158 feet above sea level, will run 202 miles to the Atlantic.[16]

When the Haw joins the Deep, the name "Cape Fear" takes over. "Haw" is lost, not to a mightier river, as the Ohio becomes the Mississippi, but to the flexing power of coastal Plantation owners and a name for the river that simply dug in earlier. Once the Haw or Saxapahaw River held its name all the way to Little River (above Fayetteville).[17] The Cape Fear's name underwent many changes including River Jordan on a 1539 map[18] to Charles River on a 1662 map,[19] and a 1730 map has it as Clarendon River.[20] The latter two are savvy nods to the power of a king and his first minister, also a lord proprietor. It was also Cape Faire,[21] but the name for the shoals that wreck many a sailor is the one that held. Those who want the Haw's name to continue to the crossing of ocean and river currents must hope patiently for another change.

The Cape Fear River comes with a movie title to its credit, a psychic thriller now in two versions, and some impressive chroniclers: John Hairr's concise and fascinating *From Mermaid's Point to Raccoon Falls*; John Sprunt's fine *Chronicles of the Cape Fear*, and Malcolm Ross's engaging, loosely spun history, *The Cape Fear*.

As the Cape Fear River, the hydro powerhouse of the Haw gives way to a river that serves coal and nuclear power: coal fired Carolina Power and Light's Cape Fear Plant below Haywood, Shearon Harris Nuclear Plant near Buckhorn Dam, and Brunswick I and II Nuclear Plants past Wilmington. Buckhorn Dam (removed in 2010) harnessed power from 1908 to 1962[22] and stopped shad, but did not pollute with mercury or radioactivity.

The Upper Cape Fear is rocky like the Haw, more impediment to travel and transport than help. Downstream was different. Many ferries lessened the divide; one remains in Bladen County, a two-car ferry operating on a steel cable.[23] Below Fayetteville, the river was the road. Shallow-draft riverboats stopped at dozens of landings sites.[24] There were grander riverboats too, like one of the last, *City of Fayetteville*, that sank with a load of heavy cotton in 1913.[25] Today Rt. 87 and 421 run with the Cape Fear to the sea, replacing the railroad, the river's one-time competitor in transport.

In 1856, there were 13 dams and 14 locks on the Cape Fear, just above Fayetteville.[26] Today three locks operate on the whole river. Whether they still should is in question; state and national fishery and wildlife people want them taken out so the river can once again be a major food source, with shad and herring able to migrate to their spawning waters.

The river is recreation and reprieve. At Raven Rock State Park, people hike and camp by beautiful 200-foot quartzite cliffs, sculpted by the river. Lillington too has a

A steamer boat, H.H. Brimley, *is about to be loaded at one of the many landings on the Cape Fear River in 1910. Smaller versions of this vessel operated on the Haw and Deep rivers in the few years they were open for transport and locks and dams were not wrecked by floods (courtesy North Carolina Office of Archives and History, Raleigh).*

broad field for camping by the river and an easy put-in and fun rapids. Below, almost all of the Cape Fear is undeveloped; in some counties only two bridges cross it. Bear and bobcat, though rare, are around. The trip down river that the boys of Haw River took is still a great one. The Black River, a tributary above Wilmington, offers paddlers "Outstanding Resource Waters" with river otters and 1700-year-old bald cypress.

Cape Fear supported industry and sometimes suffered for it. At Erwin, "the world's finest denim" was spun until a bitter closing in 2000 took 1400 jobs to Mexico.[27] Much earlier, longleaf pine provided another industry, along Little River on the Harnett-Cumberland County line. This tree, with foot-long needles, has sap for rosin, tar, turpentine, pitch — all needed on wooden ships to treat wood and ropes. Naval stores were a powerful economic force from the 1700s to the Civil War, but the waning of wooden ships," General Sherman's pyromania, and lumbering reduced the long leaf's domain by 90 percent.[28] The nickname "Tar Heels" still remains on the state and a basketball team in Chapel Hill.

At Fayetteville, contamination from the DuPont factory's use of C-8, perfluorooctanoic acid, for Teflon and food packaging materials, has put a planned water intake in Bladen County in doubt.[29] The EPA Science Advisory Board called C-8 "a likely human carcinogen."[30]

As the Cape Fear runs on to Elizabethtown, it takes in the nitrogen soaked waters from 199,783, more or less, swine. East is hog farm county; tributaries running through parts of Sampson and Bladen counties bring some of the waste of 1,743,669 more pigs to the river.[31] (Some are working to convert this manure to energy or sell it as fertilizer.)[32] Below, at Tar Heel, river right, the long grey buildings of the largest slaughter and hog packing plant in the world spread for acres: Smithfield Foods slaughters 32,000 hogs a day there.[33]

The Cape Fear seeps on through the splendid Green Swamp, past International Paper's Riegelwood Mill, built by John Riegel in 1951, and now churning out 2400 tons of paperboard per day. Its polluted discharges were reduced when International Paper took over.[34] Dioxin-producing chlorine dioxide was then phased out of the bleaching process, but weak EPA standards still allow other toxic emissions, including 986,376 pounds of carbon disulfide, a neurotoxin.[35]

Gaelic-speaking Highland Scots founded Cross Creek (1756) and Campbellton (1762), Fayetteville's forerunners, beginning this river-, then railroad-, now army-town. In 1776, a sluggish ferryman at Dawson's Ferry was blamed for the capture of 300 Scottish loyalists retreating from the Battle of Moore's Creek.[36]

The Atlantic sculpted this land. Fayetteville rises on the Orangeburg escarpment, the first of five rises formed by ocean batterings as waters advanced and withdrew, even dropping hundreds of feet, during the last 1.3 million years of Ice Ages.[37] In this century, global warming may bring a meter rise and obliterate barrier islands.

A few million years ago, the Deep River may have recaptured the river here. The indication is the abrupt left turn that the Cape Fear takes from Erwin to Fayetteville, resuming the path the Deep River was on before its capture in Chatham County.

In the swampy lands above Elizabethtown, east of the river, are more geologic wonders: the mysterious ovals of the Carolina bays, the only natural lakes in North Carolina, but part of a million elliptical, sand-rimmed holes from New Jersey to Florida. One theory holds that a meteorite exploding over Canada strewed them here about 12,900 years ago (and triggered a 1000 year ice age).[38] Jones Lake, formerly Black Lake, and White Lake, are bays with refreshing waters and sandy bottoms for all to swim in; the harsh lines of racism have diminished with the name change.

Botanists John and William Bartram, father and son, worked from Ashwood Plantation, midway to Elizabethtown on the right bluff. Their kinsman's house was their base from 1765 on as they collected plant specimens and wrote letters about the natural wonders of the colonies, and William wrote his famous *Travels in North and South Carolina*.

All the way to the sea, the river runs over land created by rivers' vast work, endlessly dropping its earthy load to form ancient deltas including the Black Creek Formation.[39] At Phoebus Landing, below Elizabethtown, the river excavated fossils from land, once ocean, then swamp. Among the wonders of the Cretaceous era (144 to 66 million BP) uncovered were 13-foot ghost sharks; a 30-foot sharp-toothed marine lizard, mosasaur; and a 40-foot dinosaur in crocodile form, Deinosuchus rugosus.[40]

The river rolls on down the Brunswick and New Hanover County line and is joined by the Black River above Wilmington. The last 35 miles, where river meets ocean water, is the Cape Fear estuary, tide-driven and salty; if the river is healthy, this is a vast nursery for fish, crabs, and shrimp.

At Wilmington, the last major tributary, the Northeast Branch of the Cape Fear pours in. It is here, far in from the shoals and the Atlantic roar, that a settlement took hold in 1733 after several tries further down river. The town boomed into a port city and major commercial center of North Carolina from 1840 to 1900. Still a thriving city, its Cape Fear Museum tells the river's story.

Wilmington's riverfront is pictured on a postcard sent in 1906 which shows docks, buildings, and a factory smokestack in the background. Foreground, a sailboat crosses the Cape Fear River near a ferry transporting a horse-drawn cart (North Carolina Collection, University of North Carolina at Chapel Hill).

Some of that history, retold in Philip Gerard's *Cape Fear Rising*, is harsh. In the 19th century, the river carried to death or to freedom African Americans who could be slain without penalty. The 1898 uprising in Wilmington is now seen as a coup. It blocked the power of the Fusion Party of impoverished whites and blacks who had overcome prejudice to see their common interests. In this city, "a symbol of Black hope," Black male literacy was higher than Whites, and Blacks owned a dozen restaurants, and one, Alexander Manly, owned the *Daily Record*.[41] That was until Alfred Waddle burned the newspaper and replaced the mayor at gunpoint, declaring: "We will never surrender to a ragged raffle of Negroes, even if we have to choke the Cape Fear River with carcasses."[42]

Today, at the edge of downtown, almost under the bridge at Sutton Street, is Cape Fear River Watch, working for the river along with the North Carolina Coastal Federation. Like the Haw, the Cape Fear is now drawing people to it, but mishandled development is a river hazard.

Past Wilmington, Orton Plantation, one of the few remaining plantation houses, reigns over the bank. In 1904, James Sprunt bought Orton and wrote *Chronicles of the Cape Fear* here.[43] Earliest Europeans had simpler dwellings in settlements at Charles Towne near Town Creek and at Brunswick, river right, which aborted quickly. The third and last group left in 1668, having fallen foul after "seizing and sending away [Cape Fear Indians'] children under the pretense of having them educated."[44] The Cape Fear Indians, who were numbered at 206 in 1715, later re-settled west of Charleston, South Carolina,[45] before vanishing.

Native peoples were once all along the river, and inland peoples came here too for fish and Yaupon, ilex vomitoria, a holly used for purging rituals. They greeted the first European explorer of North Carolina's coast, Giovanni de Verrazzano, sponsored by Francis I of France to find a route to China. In the spring of 1524, Verrazzano saw fires here and found people who fed him, until he seized a Native child farther up the coast to display to Europeans. He would die four years later, for similar behavior to the Caribs, still thinking Pamlico Sound stretched to the Pacific.[46]

The Cape Fear, in full tidal motion, passes Fort Fisher ferry and the fort crucial in the Civil War to Confederate blockade runners and the last open supply line. Its fall on January 12, 1865, after massive bombardment, made the Confederacy's surrender inevitable.

The island known as both Smith and Baldhead looms against the Atlantic. Sea turtles nest here; a few months later their 100 or so hatchlings will stream out together in a perilous shamble to the sea. Ghost crabs and gulls are only their first predators.

The open sea is ahead. The Cape Fear is the only coastal river in North Carolina not hemmed in by barrier islands; the freer flow brings health and is likely why it escaped the pfisteria outbreaks especially bad on the Neuse in the 1990s.

At the Cape Fear's mouth is Southport, site of Fort Johnston in 1754, and home to many river pilots and fishermen. In the early 1700s, Blackbeard prowled these waters, a terrifying sight with smoking hemp-rope entwined in his matted locks and pistols and cutlasses strung about. He had a lucrative business, with the governor of North Carolina turning a blind eye, until he was captured and beheaded in 1718, age 38. The same year, Stede Bonnet, the Gentlemen's Pirate, was caught in the mouth of the Cape Fear and executed along with 49 other pirates, male and female.[47]

Knowing the infamous Cape Fear shoals at the river's mouth was the pirates' advantage. Three lighthouses run today to protect boats from the shifting sea floor, known as the

Graveyard of the Atlantic. This treacherous spot is the final meander of the Haw. Gravity's work on the waters is now done by the pull of the moon on the turning earth's tides.

Whatever we have passed on to the Haw and all the 6049 miles of Cape Fear streams head on to sustain or weaken ocean life. Strands of the Haw's swirling waters will be far out, shining back the lights of the Milky Way and all the stars undimmed.

Future Haw

What is the future of the Haw? The answer lets us know our own. Here is the realistic hope of Horizon 2100, a coalition with a vision for North Carolina:

> They are in the river before the shadbush blooms on the banks of the Haw, before the dogwood leaves are the size of a squirrel's ear: that's what the old fishermen would say. The fish come with the earliest hint of spring — pushing up the Cape Fear and the New, the Neuse, Tar, Pamlico, Roanoke, Cashie and Chowan....
>
> A century of fixing North Carolina rivers has brought the river migrants home. For the state's eight species of anadromous fish — Atlantic and shortnose sturgeon, hickory and American shad, striped bass sea lamprey, blueback herring and alewife —... they gush ... through rapids where the rivers roil over the Piedmont fall line, nosing ever upstream to spawning grounds tens, scores, hundreds of miles from the sea....
>
> In the Cape Fear River, the big shad — some grow to eight pounds or more — skirt Lock and Dam No. 1 through a half-mile long canal designed specifically for fish.... The spawning grounds of the Deep and the Haw lie just above....
>
> They're in the river again ... and so are the fishermen ... those who fish for a living, who fish for fun, who fish for tomorrow's supper.[48]

Let's hope.

Dams will be gone from the Haw; new ways of capturing water's energy will not block fish spawning. Coal and nuclear power with their massive use of water and toxic pollution will be read about in history lessons, along with those rare diseases, asthma and cancer. Solar, wind, geo-thermal, and new forms of hydro-power will fuel the Haw Valley.

All streams will have wide wooded buffers. Cattle will be fenced out. Children won't believe at first that our waste used to go in the river. We will separate out tub and sink drains for re-use and have new ways of treating waste water.

Marking storm drains with the name of the river basin will be unnecessary; people will spend so much time walking trails, canoeing, fishing, and swimming that they will know their own river address. Putting paint or oil down sewer drains will be as outrageous as putting it in one of those chlorinated swimming pools people used to have. Storm water will seep through gravel driveways and rain gardens and collect in containers for outdoor watering.

Tax breaks and taxes will forward the cultural revolution of the river, channeling us all to better practices. We will have changed not only what we buy and build, but what we want to own and live in, what we see as beautiful. Landscaping will feature native plants, so lawn chemicals and herbicides will seem like poisons of the Dark Ages; organic farming will have changed the health of the land and Haw as well. Wetlands will be valued for their work, and we will squish into them, fascinated.

Environmental Protection Agency standards will protect both the river and its people. Chemicals will have to test safe before they are marketed.

The 6300-mile web of streams that drains the Cape Fear basin, an area just larger than New Hampshire, will run clear.

Can we do it? Did the whitetail deer, almost extinct in 1925, come back? Did the beaver and wild turkey come back?[49] Did the eagle come back? Did the Haw come back? It was on its way when the 21st century began.

Opposite: *The Haw and Cape Fear River Basins are shown in a 1775 map, Henry Mouzon's "An Accurate Map of North and South Carolina" (courtesy North Carolina Office of Archives and History, Raleigh).*

6. Coastal River to the Sea

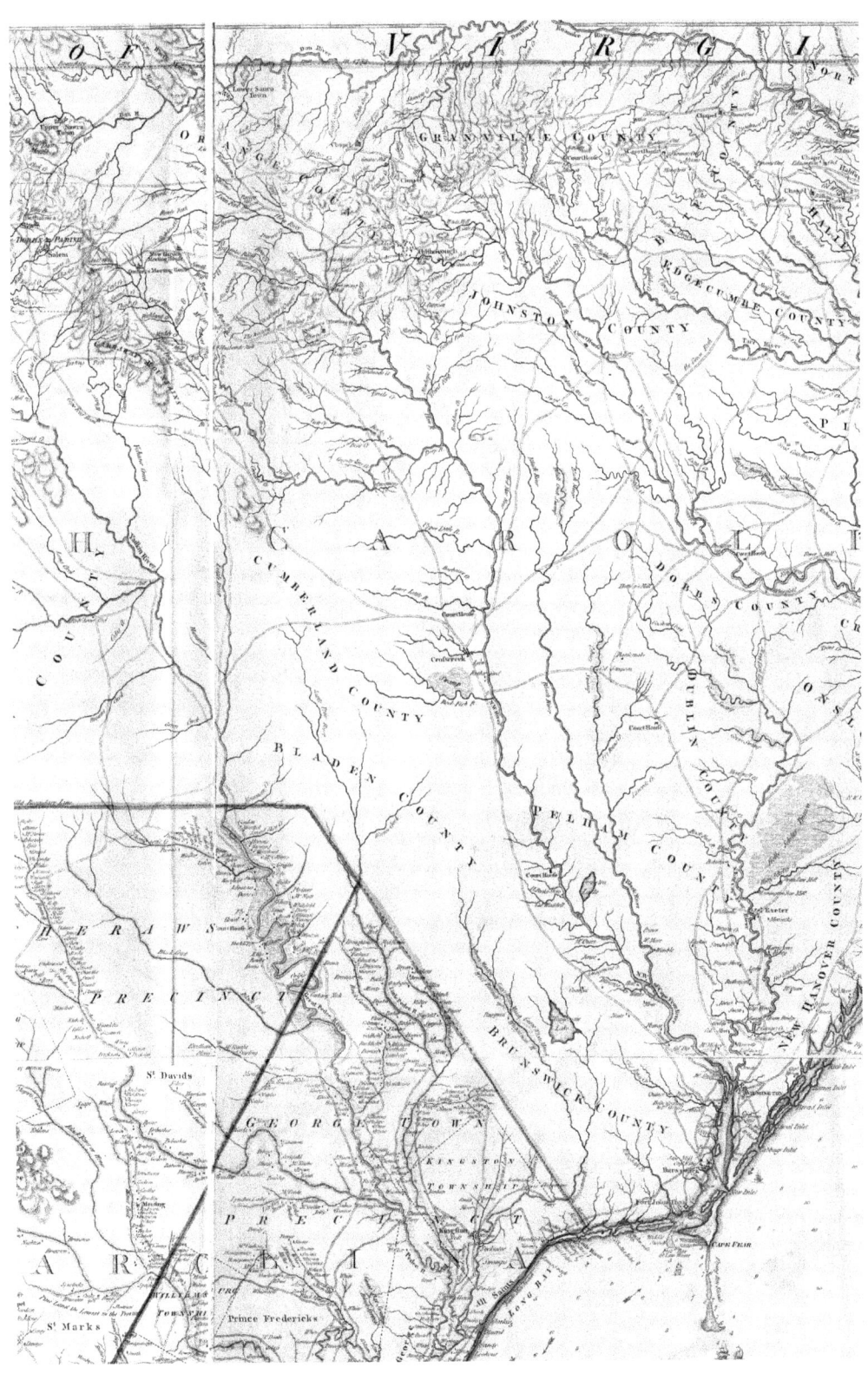

Appendix

Citizen Organizations Working for the Health of the Haw

Blue Ridge Environmental Defense League
P.O. Box 88
Glendale Springs, NC 28629
(336) 982-2691
http://bredl.org

Cape Fear River Watch
617 Surry St.
Wilmington, NC 28401
www.cfrw.org

CERES: Coalition for Environmental Responsibility and Education through Synergy
dr_mike@mindspring.com

Chatham Conservation Partnership
http://chathamconservation.wikispaces.com

Clean Jordan Lake
2912 Enson Place
Raleigh, NC 27603
www.cleanjordanlake.org

Clean Water for North Carolina
2009 Chapel Hill Road
Durham, NC 27707
(919) 401-9600
www.cwfnc.org

Conservation Council of North Carolina
P.O. Box 12671
Raleigh, NC 27605
www.conservationcouncilnc.org

Deep River Clean Water Society
c/o The Abundance Foundation

Friends of Bolin Creek
P.O. Box 234
Carrboro, NC 27510
www.bolincreek.org

Friends of the Deep River
P.O. Box 624
Jamestown, NC 27282
(336) 337-0811
www.bredl.org/FODR/index.htm

Friends of the Rocky River & Rocky River Heritage Foundation
(292) 777-9750
www.rockyriverchatham.org

Haw River Assembly
P.O. Box 187
Bynum, NC 27244
www.hawriver.org
info@hawriver.org
(919) 542-5790

Haw River Trail Partnership
info@thehaw.org
(336) 229-2229
www.thehaw.org

Morgan Creek Valley Alliance
c/o J. Randall/Botanical Gardens

UNC–Chapel Hill
C.B. 3375, Totten Center
Chapel Hill, NC 27599-3375

New Hope Creek Corridor Advisory Committee
34 Pond View Court
Durham, NC 27705
www.newhopecreek.org

North Carolina Conservation Network
19 E. Martin St., Suite 300
Raleigh, NC 27601
(919) 857-4699
www.ncconservationnetwork.org

Northeast Creek Streamwatch
www.northeastcreek.org

Piedmont Environmental Center
1220 Penny Road
High Point, NC 27265-9182
(336) 883-8531
www.piedmontenvironmental.org

Piedmont Land Conservancy
1515 West Cornwallis Drive
Greensboro, NC 27408-6334
(336) 691-0088
www.piedmontland.org

Preserve Rural Orange
P.O. Box 1314
Carrboro, NC 27510
www.preserveruralorange.org

Robeson Creek Watershed Council
http://www.bae.ncsu.edu/programs/extension/wqg/srp/robeson.html

Southern Environmental Law Center
200 West Franklin St., Suite 330
Chapel Hill, NC 27516-2559
(919) 967-1450
www.southernenvironment.org

Triangle Land Conservancy
1101 Haynes Street, Suite 205
Raleigh, NC 27604
(919) 833-3662
www.triangleland.org

West End Revitalization Association — WERA
P.O. Box 661
Mebane, NC 27302
www.wera-nc.org

Chapter Notes

Chapter 1

1. Those involved include Lynn Featherstone, Haw River Assembly, Piedmont Land Conservancy, U.S. Fish and Wildlife, and the NC Clean Water Management Trust Fund.
2. E.C. Pielou, *Fresh Water* (Chicago: University of Chicago Press, 1998), 5.
3. In the Piedmont, underground water does not open into vast underground lakes, like the Midwestern Ogallala Aquifer.
4. Wallace Stevens, "Sunday Morning" poem.
5. Kevin Moorhead, "Wetland Conversions: Counting the Losses," *Wildlife in North Carolina*, March 1995, 27–31.
6. Alice Outwater, *Water: A Natural History* (New York: HarperCollins, 1996), 142. This book is highly recommended reading for all those interested in the ecology of water, and a resource I relied on heavily.
7. As William Schlesinger pointed out in a talk at Elon University, September 16, 2009, wetlands denitrify with minimal output of nitrous oxide (N2O), a by-product of manufactured fertilizers which produce greenhouse gases at a rate 200 times that of carbon dioxide.
8. Outwater, *Water*, 32–33.
9. Mark Vander Borgh, "Questions and Answers about Algae," *Haw River Assembly*, http://www.hawriver.org/index.php?contentid=39 (accessed June 25, 2007).
10. "North Carolina Navigability Report," American Whitewater, http://www.americanwhitewater.org/content/Wiki/access:nc (accessed January 20, 2010).
11. Pielou, *Fresh Water*, 138.
12. Outwater, *Water*, 23.
13. Ibid., 4–5, 24. Outwater reports that European beavers have no gift for dam construction; it may have been lost when they were almost exterminated around 1300.
14. Ibid., 25.
15. Ibid., 24 and 32.
16. Joel Vance, "A Good Problem to Have," *Wildlife in North Carolina*, August 2009, 5.
17. Alice Outwater, "A River Runs Through It," *Haw River Assembly*, http://www.hawriver.org/library/publications/Alice%20Outwater%20Talk%20on%20Haw%20River.pdf, 14 (accessed June 21, 2009).
18. Outwater, *Water*, 4–5, 20.
19. Kim Willis, "Peaceful Coexistence with Beavers," *Haw River Assembly*, www.hawriver.org/index.php?contentid=38 (accessed May 3, 2006).

Chapter 2

1. Laura C. Martin, *The Folklore of Trees and Shrubs* (Chester, CT: Globe Pequot Press, 1992), 76–78.
2. Fred Hughes, *Guilford County, North Carolina Historical Documentation Map*, 1980. Surprisingly, the section of the Haw from the source to Eversfield Road was placed on the 303(d) list of impaired waters in 2000 because the aquatic life expected at the Eversfield testing site only received a fair rating. Possible sources of the problem were reported as runoff from roads and yards, agriculture, and treated wastewater. Erick Fleek, senior biologist, Bioassessment Unit, Environmental Sciences Section, NC Division of Water Quality, e-mail message, February 8, 2010.
3. Parks, Richard. "Atlantic and Yadkin Railway." A Southeastern Regional Railway—1930's to 1940's. http://www.r2parks.net/A&Y.html. (accessed June 9, 2012).
4. Julian Crandell Hollick, *Ganga: A Journey Down the Ganges River* (Washington, DC: Island Press, 2007), 7 and 103.
5. Luna Leopold, *A View of the River* (Cambridge: Harvard University Press, 1994), 58 and 65.
6. Kent Bowker, "Albert Einstein and Meandering Rivers," *Search and Discovery Net*, http://www.searchanddiscovery.net/documents/Einstein/albert.htm (accessed January 20, 2010).
7. Pielou, *Fresh Water*, 97–99, 119.
8. Alan Barry Nelson, "A Brief Introduction to the Geology, Hydrology, and Geomorphology of the Haw River Basin, North Carolina, from Its Source to the Headwaters of the Cape Fear River," unpublished manuscript, 2002, 35.
9. Leopold, *View*, 74–77.
10. Clyde Sorenson, "What Do Deer Markings Mean?" *Wildlife in North Carolina*, November 2005, 39.
11. Jelaluddin Rumi, "Where Everything Is Music," Coleman Barks, translator, *The Essential Rumi* (San Francisco: HarperCollins, 1995), 34.
12. Harry Watson, quoted in *The Way We Lived in North Carolina*, ed. Joe Mobley (Chapel Hill: University of North Carolina Press, 2003), 111.
13. Ibid.
14. Elizabeth A. Fenn and Peter H. Wood, *Natives and Newcomers: The Way We Lived in North Carolina Before 1770* (Chapel Hill: University of North Carolina Press, 1983), 70–71. The authors add that the Scots-Irish came here early after having migrated to the Ulster area of Ire-

land and competed so well with England that they were pushed on to America.

15. Watson quoted in Mobley, *The Way We Lived*, 132.

16. Lindley S. Butler, *Rockingham County: A Brief History* (Raleigh: North Carolina Department of Cultural Resources, Division of Archives and History, 1982), 13.

17. Bill Morris, "Buffalo Chronicles," *Wildlife in North Carolina*, July 2005, 20–25.

18. Fred Hughes, *Guilford County, North Carolina: A Map Supplement* (Jamestown, NC: Custom House, 1988), 64.

19. Ibid., 65–67.

20. Ibid., 64. Fred Hughes also suggests that because they were removed from the coast with interchanges of goods and diseases like "typhoid, malaria, yellow fevers," life was healthier here (66). Pellagra would come later when abrasive farming sent the topsoil into the Haw and sharecropping forced farmers into cash crops.

21. Hughes, *Historical Documentation Map*.

22. Hughes, *Map Supplement*, 7, 105.

23. Robinson and Stoesin, *History of Guilford*, 64.

24. Ibid., 64.

25. Jack L. Perdue, "Old Mill of Guilford" (pamphlet), 1. The mill is now at 1340 NC 68 North, Oak Ridge, NC.

26. Ibid., 2.

27. Ibid., 3–4.

28. Robinson and Stoesen, *History of Guilford*, 136.

29. Hughes, *Map Supplement*, 29.

30. Ibid., 27, 29 and 47.

31. Sallie Stockard, *History of Guilford County* (Knoxville: Gaut-Ogden Printers, 1902), 49.

32. Hal Sieber, *Holy Ground: Significant Events in the Civil Rights–Related History of the African-American Communities of Guilford County, North Carolina, 1771–1995* (Greensboro: Tutor, 1995), 19.

33. Ibid., 21–23.

34. Ibid., 23.

35. George Hendrick and Wilene Hendrick, *Fleeing for Freedom: Stories of the Underground Railroad as Told by Levi Coffin and William Still* (Chicago: Ivan R. Dee, 2004), 10.

36. Elaine Chiosso, map, unpublished.

37. Dr. Bryan Walls, 700 Club interview, http://www.cbn.com/spirituallife/ChurchAndMinistry/ChurchHistory/underground_railroad (accessed May 15, 2006).

38. Ibid.

39. Dr. Bryan E. Walls, *The Road That Led to Somewhere: A Documented Novel About the Underground Railroad* (Windsor, Ontario: Olive, 1980), 68–69.

40. McKinley Collins, interview, August 3, 2006.

41. Thomas Berry, *The Great Work* (New York: Harmony/Bell Tower, 2000), 12–13.

42. Ibid., 17.

43. Thomas Berry quoted in Derrick Jensen, *Listening to the Land: Conversations about Nature, Culture and Eros* (White River Junction, VT: Chelsea Green, 2004), 2.

44. Ibid., 18.

45. Kingsbury explains that conservation easements block future owners from developing. Owners enjoy the land while they get cash or tax benefits for their donation. It can be a sweet deal. Kalen Kingsbury, interview, May 9, 2008.

46. The park hopes to buy up more land along the Haw to extend its camping, hiking and educational possibilities.

47. Kathryn Royall, interview, January 13, 2008.

48. Quoted in Taft Wireback, "Park's on New Course," *News and Record*, January 11, 2008, A1.

49. Dan Tenaglia, interview, May 8, 2002. His Web site, www.missouriplants.com, has information on North Carolina plants.

50. Stanley Trimble, *Man Induced Soil Erosion on the Southern Piedmont, 1700–1970* (Ankeny, IA: Soil Conservation Society of America, 1974), 22.

51. Ibid., 23.

52. Ibid., 52.

53. Ibid., 54.

54. Ibid., 157, citing A.O. Craven.

55. Ibid., 70.

56. Ibid., 97.

57. Ibid., 104, 106, 115.

58. Ibid., 132.

59. Robert Walters and Dorothy Merritt have recently presented a new theory of erosion suggesting that Piedmont rivers like the Haw were often braided until greatly changed by mills. They argue the deep cut rivers of the Piedmont were caused by 10,000 mill dams turning rivers into "series of linked mill ponds." When the Haw blew out dams, the river carried the soft, looser sediments trapped behind the dam away, scouring the banks and hollowing out river channels, thus explaining why the Haw flows low in its banks. Robert Walter and Dorothy Merritt, "Natural Streams and the Legacy of Water-Powered Mills," *Science,* January 18, 2008, Vol. 319, 299–305. However, Dan Royall, Department of Geography, UNC–Greensboro, notes there is not yet evidence to support the braided river theory. Phone conversation, May 7, 2008. Janet MacFall, Elon University, found the most erodable soils to be where agricultural fields deposited dirt on low slopes, and the river is in a narrow channel. Talk at Elon University, April 28, 2009.

60. Charles Dyson Rodenburgh (ed.), *The Heritage of Rockingham County, North Carolina* (Wentworth, NC: Rockingham County Historical Society, 1983), 23.

61. Butler, *Brief History*, 11.

62. Rodenburgh, *Heritage*, 2 and 18.

63. Robinson and Stoesen, *History of Guilford*, 43–48. I relied on Robinson and Stoesen for the details in the next four paragraphs.

64. Ibid., 47.

65. Ibid., 48.

66. Butler, *Brief History*, 16, 23.

67. Ibid., 24.

68. Rodenburgh, *Heritage*, 19.

69. Lindley Butler, quoted by Brent Lancaster, "The Haw's Got Some Problems," *Times-News*, March 26, 2000.

70. Kevin Moore, district watershed conservationist, Rockingham County, phone interview, May 17, 2005.

71. John J.W. Rogers, *History and Environment of North Carolina's Piedmont: Evolution of a Value-added Society* (1999), 41. Rogers adds that 3 percent of Piedmont water is groundwater supplied via wells.

72. Greg Feller, OWASA public affairs, interview, June 18, 2008.

73. Cat Warren, "Jonesing for Jordan," *The Independent Weekly*, March 5, 2008, 13.

74. Cat Warren, "Thirsty? Dirty? Sorry," *The Independent Weekly*, January 30, 2008, 14.

75. Kelvin Creech, Cary-Apex Water Treatment Plant, interview, March 23, 2006.

76. Mark Binker, "Candidates Offer Ideas to Save Water," *News and Record*, January 21, 2008, A7.

77. Gary Chamberlain, *Troubled Waters: Religion,*

Ethics and the Global Water Crisis (New York: Rowman and Littlefield, 2008), 104.

78. Samuel Solomon, *Water: The Epic Struggle for Wealth, Power and Civilization* (New York: HarperCollins, 2010), 373.

79. Julia Milstead, site communications leader, Progress Energy, phone interview, May 15, 2009.

80. Rogers, *History and Environment*, 74. He adds that 80 percent of all water in rivers in the state is cycled through coal and nuclear plants for cooling.

81. Grady McCallie, NC Conservation Network, e-mail message, January 15, 2010.

82. Fred Pearce, *Keepers of the Spring: Reclaiming our Water in an Age of Globalization* (Washington, DC: Island Press, 2004), 29.

83. Chamberlain, *Troubled Waters*, 186.

84. Allan Williams, phone interview, May 12, 2008.

85. Allan Williams adds: "And there is an economic fairness, the irrigation water used in July and August is some of the most expensive water that we produce because we are only selling it for the two or three months out of the year. As a utility, most of our costs are fixed. Whether our plant treats 10 million gallons of water a day or 30 million gallons, we still have the plant and the operators there, so when we cut back during a drought, all we are really saving is electricity and chemicals which is probably only 10 percent of what our charges represent."

86. Rodenburgh, *Heritage*, 34.

87. Outwater, *Water*, 57–59.

88. Hughes, *Historical Documentation Map*.

89. Hughes, *Map Supplement*, 13.

90. Butler, *Brief History*, 16.

91. Kevin Moore, phone interview, May 17, 2005.

92. Patrick Beggs, project manager, Watershed Education for Communities and Officials, NC State University, phone interview, May 5, 2005.

93. Jason Doll, senior scientist, program manager at Santech Consulting Co., phone interview, May 17, 2005. Low oxygen had to be addressed because too much manure, animals' and humans', causes algae to thrive and suck so much oxygen out that bugs, fish, and mussels may suffocate. This is the problem in Jordan Lake, delivered from the creeks up river. Holding the line here on BOD, biological oxygen demand, gave the local creek, Haw, and Jordan all protection.

94. Jason Doll, phone interview, May 17, 2005.

95. Henry David Thoreau, *Walden and Civil Disobedience* (Boston: Houghton Mifflin, 2000), 107.

96. Solomon, *Epic Struggle*, 100–101.

97. Tempie H. Prince and Margie P. Greene, "High Rock Mill and High Rock House," *Journal of Rockingham County History and Genealogy*, December 1979, 54.

98. Tom Magnuson, interview, January 14, 2005.

99. Tom Magnuson adds: "Indians don't want to get their feet or their moccasins wet. You want to preserve both. Cutting your foot could be a death sentence. If you were injured, you were left. You don't risk the group for the individual."

100. Prince and Greene, "High Rock," 51–52.

101. Ibid., 47.

102. Tom Magnuson surmises east as well.

103. Robinson and Stoesen, *Heritage*, 43–44.

104. Butler, *Brief History*, 17 and 24.

105. Herbert S. Turner, "Affluent Tar Heels Resorted in Summer to Lenox Castle," *The State*, September 1, 1969, 12–13, 24.

106. Butler, *Brief History*, 4.

107. Prince and Greene, "High Rock," 54.

108. Grimsley T. Hobbs, *Exploring the Old Mills of North Carolina* (Chapel Hill: Provincial Press, 1985), 7.

109. Ibid., 7.

110. Prince and Greene, "High Rock," 58.

111. Nelson, *Geology of the Haw River Basin*. I am indebted to this work and conversations with the author throughout this section.

112. Fred Beyer, *North Carolina: The Years Before Man—A Geologic History* (Durham, NC: Carolina Academic Press, 1991), 5. This book was also an essential resource.

113. Ibid., 6–7.

114. Ibid., 9–13.

115. Ibid., 17–20. Beyer makes clear that this current theory is based on available but limited evidence; much has been eroded away.

116. Rogers, *History and Environment*, 16. The vast erosion may be hard to fathom, but as Rogers writes, "People who live in the eastern half of North Carolina Piedmont live on rocks that were once part of South America."

117. John McPhee, *Annals of the Former World* (New York: Farrar, Straus and Giroux, 1998), 48.

118. Beyer, *Years Before Man*, 38–39.

119. Ibid., 42 and 58. The river starts on the Milton Belt, created by the volcanic outbursts that erupted later when the continental plate pushed under the west side of Avalon Terrane.

120. Ibid., 32 and McPhee, *Annals*, 33.

121. Beyer, *Years Before Man*, 40–41.

122. Ibid., 79.

123. Ibid., 80.

124. Nelson, *Geology of the Haw River Basin*, 5.

125. McPhee, *Annals*, 147–148, quoted in *Geology of Haw River Basin*, 8.

126. Ibid., 36–38.

127. George L. Bain and Harvey W. Bruce, eds., *Field Guide to the Geology of the Durham Triassic Basin* (Durham: Carolina Geological Society, 1977). The Haw leaves the Triassic Basin as the Cape Fear one mile above Buckhorn Dam at the Jonesboro Fault. Durham Triassic Basin is part of the larger Deep River Triassic Basin that stretches from South Carolina almost to Virginia.

128. Rogers, *History and Environment*, 19.

129. Nelson, *Geology of the Haw River Basin*, 15 and 19.

130. Beyer, *Years Before Man*, 119.

131. Ibid., 110 and 168.

132. Nelson, *Geology of the Haw River Basin*, 33.

133. Beyer, *Years Before Man*, 168.

134. Ibid., 137–140.

135. Nelson, *Geology of the Haw River Basin*, 31.

136. Rogers, *History and Environment*, 18–19.

137. Nelson, *Geology of the Haw River Basin*, 44.

138. McPhee, *Annals*, 89–91.

139. Mark Chilton, *An Historical Atlas of the Haw River* (Carrboro, NC: Self-published, 2008), 61.1.

140. Ibid., 59.7.

141. A put-in here has long enjoyed public use while people like Mike Holland and Cary Allred worked for more official status with the Gants. That has come about and a deluxe put-in is now in place—thanks to the Haw River Trail Association, Alamance County Parks and Recreation, Eagle Scout candidate Baden Piland, Burlington Parks and Recreation Department, and the NC Department of Transportation, which moved the bridge, and, of course, the Gant family, who own this land. The rebirth of the Haw in Alamance takes a whole county.

CHAPTER 3

1. Joe Jacob of Haw River Canoe and Kayak Co. and Susanne Gomolski of Kayak Adventures are two qualified local guides; the Haw River Trail Web site, www.thehaw.org, is a source for other new enterprises.
2. Paul Ferguson, "Accident of the Haw River: A Near Death Experience," http://www.rockfamilytree.com/hawrivertrail/accident%202004.htm (accessed December 15, 2006).
3. Ibid.
4. Outwater, "A River Runs," 24.
5. Ibid., 24–25.
6. Robin Clarke and Jannet King, *The Water Atlas* (New York: New Press, 2004), 42–44.
7. Outwater, "A River Runs," 24.
8. John Jordan, interview, January 31, 2008.
9. "Energy efficiency and renewable energy," Department of Energy, https://appsl.eer4e.energy.gov/states/energy_summary.cfm/state=nc (accessed April 13, 2009).
10. Outwater, *Water*, 106.
11. Outwater, "A River Runs," 26–27.
12. Taft Wireback, "Dam Removal Good News for Endangered Fish," *News and Observer*, September 10, 2007, A1.
13. Jim Wilson, "Fish of Yesterday, Fish of Tomorrow," *Wildlife in North Carolina*, October 2007, 22 and 28.
14. Wade Hadley, "Fish in Streams in Chatham Over 100 Years Ago," *Chatham Historical Journal*, No. 2, October 1989, quoting the *Chatham Record* of May 8, 1884.
15. Ibid. Summarizing the *Chatham Record* of May 15, 1884.
16. Outwater, "A River Runs," 27.
17. Martin Doyle, associate professor of geography at UNC–Chapel Hill, quoted by Colie Hoffman, "Dam Nation," *Endeavors Magazine*, winter 2007, January 17, 2007, http://research.unc.edu/endeavors/win2007/damnation.php (accessed February 7, 2010).
18. Ibid. Deadly PCBs were released when a Hudson River dam was let loose in 1972.
19. Martin Doyle quoted, ibid.
20. American Rivers, cited by Matthew Preusch, "As Dams Go Down, Kayaking Rises," *New York Times*, August 9, 2009, http://travel.nytimes.com/2009/08/09/travel/09explorer.html?8dpc=andpagewanted=print (accessed February 7, 2010).
21. "Beaten Paths: Some 17th Century Southeastern Trading Path Maps," Trading Path Association, http://www.tradingpath.org/content/view/44/51 (accessed April 9, 2010).
22. Tom Magnuson, presentation to Elon University class, June 9, 2002.
23. Carole Watterson Troxler, "Along the Trading Path: The Fluidity of Racial Status in the Early Eighteenth Century Southern Piedmont," in *Voices from Within the Veil: African Americans and the Experience of Democracy*, ed. W.H. Alexander, et al. (Newcastle upon Tyne: Cambridge Scholars Publishing, 2008), 397–399.
24. Ibid. Laurel Crone Sneed's unpublished work on Thomas Day, noted craftsman, suggests that successful free blacks might own slaves to avoid the appearance of challenging racist culture, as Day did even while he secretly supported and funded abolitionists in the North. Interview, September 10, 2009.
25. Ibid. In fact, the North Carolina Provincial Congress of 1776 gave *all* freemen the right to vote for representatives and those with property could elect senators, but free blacks would lose the vote in 1835. Powell, *Four Centuries*, 187 and 282.
26. Adolphus Teague, "Altamahaw," *Alamance Gleaner*, March 18, 1886, 3, quoted in Margaret Elizabeth Gant, *The Raven's Story* (Burlington: Counterpoint Graphic Productions, 1979), 33.
27. Julian Hughes, *Development of the Textile Industry in Alamance County: Evolution of Warp and Welt in Alamance* (Burlington: Burlington Letter Shop, 1965), 105.
28. William Murray Vincent, *Shuttle and Plow: A History of Alamance County, North Carolina, Part II: Recovery and Renewal* (Alamance County Historical Association, 1999), 370.
29. Roger Gant, presentation to Elon University class, June 5, 2002.
30. Vincent, *Shuttle and Plow*, 360.
31. Hughes, *Development*, 104.
32. As Mark Chilton explains, mill ponds and races protected the mill from the direct force of the Haw; the race could be short here because the drop was so great.
33. Gant, *Raven's Story*, 35.
34. Vincent, *Shuttle and Plow*, 361. He adds: "The first cloth, a type of coarse osnaburg [coarse cotton used for sacking and upholstery], was loomed in 1887. By 1888, cloth from the Altamahaw Mill and the nearby Ossippee Mill ... was shipped from a railroad station, known as Mill Point (now Elon College).... "The mill office, built between 1889 and 1890, was said to be the most modern commercial building in Alamance County, with central heat, hot and cold running water, and carbide light. By 1895, the Altamahaw Mill operated 6500 spindles, 300 looms, and employed 275 people."
35. Bureau of Labor and Printing, Annual Report ... 1901, 420 and 422, quoted in Jacqueline Dowd Hall et al., *Like a Family: The Making of a Southern Cotton Mill World* (Chapel Hill: University of North Carolina Press, 1987), 102 and 378.
36. Vincent, *Shuttle and Plow*, 361.
37. Carole Watterson Troxler, *Shuttle and Plow: A History of Alamance County, North Carolina, Part I: Old Allemance* (Alamance County Historical Association, 1999), 123.
38. Outwater, "A River Runs," 23.
39. Ibid., 22 and 23.
40. Ibid.
41. Jane Iseley, interview with Victoria Strange and Martina Donati, in *Haw River Oral History Project*, Elon University, Janet MacFall, ed., fall 2006, DVD.
42. Chilton, *Historical Atlas*, 50.3.
43. Billie W. Phillips, *Growing Up in Glencoe* (Burlington, NC: Nall Printing, 2005), 9.
44. Vincent, *Shuttle and Plow*, 362.
45. Phillips, *Growing Up*, 1.
46. Sydney Nathans, *The Quest for Progress: The Way We Lived in North Carolina, 1870–1920* (Chapel Hill: University of North Carolina Press, 1983), 31.
47. Dave McCarns, quoted in Hall et al., *Like a Family*, 39.
48. Helen Walton, interview, April 2003.
49. Geo. F. Swain, J.A. Holmes, and E.W. Myers, *Papers on the Waterpower in North Carolina: A Preliminary Report* (Raleigh: Guy V. Barnes, Public Printer, 1899), 153.
50. Nathans, *Quest*, 31–32.
51. The list includes Alamance County, the City of Burlington, Clean Water Management Trust, and Natural Heritage Trust.
52. Lynn Cowan, interview, July 1, 2003.

53. Hall et al., *Like a Family*, 151.
54. Jerrie Nall, quoted by Adrienne Owens, "Spinning a New Yarn," *Burlington Times-News*, December 1, 2002, A2.
55. "25 Years of Cotton Mill Progress: Wonderful Transformation Which Has Been Wrought in Mill Communities of North Carolina," Textile Progress edition, *Charlotte Observer*, 1918.
56. "Compulsory Education Inaugurated at the Glencoe Mills Forty Years Ago — Clean Community of Splendid People — This Alamance Mill Another Plant Which Has Done Much for Its Employees," ibid., 34.
57. Nathans, *Quest*, 36. This appeal is also the opening for a letter from Thomas M. Holt to "all the mills at Haw River" dated January 9, 1896, and titled "A Word with My Operatives," which adds in place of the last sentence: "and in case of panic, which is threatening the whole world, the more I make the longer I can give you employment. There is one way in which you can help me and thus benefit yourselves that is, do your trading in our store." Gail and Bob Knauff, *Fabric of a Community: The Story of Haw River, North Carolina* (Haw River Historical Association, 1996), 82.
58. Nathans, *Quest*, 36.
59. Kathy Barry, personal correspondence, June 26, 2009, and Kathy Barry and Jerrie Nall, "Glencoe Cotton Mill and Village: 1880–1954; Quick Facts," Textile Heritage Museum, leaflet.
60. Nathans, *Quest*, 34.
61. Allen Tullos, *The Habits of Industry: White Culture and the Transformation of the Carolina Piedmont* (Chapel Hill: University of North Carolina, 1989), 18.
62. Ibid., 16–17, Ethel Faucette is quoted.
63. Ibid., Don Faucette is quoted.
64. Ibid., Paul Faucette is quoted. There are many statements about what brought ruin to the river, most often suggesting it is something that happened upriver, but the earliest state monitoring did not occur until 1968, so information is scarce.
65. Ibid., 18, Ethel Faucette is quoted.
66. Ibid., 24.
67. Don, Thelma, and Garland Massey, interview, July 4, 2003.
68. Chilton, *Historical Atlas*, 45.6.
69. For comparison, the first cotton mill worldwide is thought to have been set up in 1771 in Cromford, England, by Richard Arkwright. In 1801, it was copied (stolen) by Samuel Slater who memorized the design for a mill on the Blackstone River in Pawtucket, Massachusetts. Solomon, *Water*, 221.
70. Hughes, *Development*, 46.
71. Vincent, *Shuttle and Plow*, 358.
72. Mark Andrews, interview, January 14, 2004.
73. Jason Copland, phone conversation, May 5, 2009.
74. Hall et al., *Like a Family*, 265.
75. Ibid., 264.
76. James Copland, interview, January 14, 2004.
77. Edmund Andrews, "Textile Quotas to End," *New York Times*, November 2, 2004, A4.
78. Ibid.
79. "Chinese Imports Worry Textile Leaders," *Times-News*, January 11, 2009, B2.
80. Eric Davis, interview, March 7, 2006.
81. Clarence Sell, interview, March 7, 2006.
82. Terri Buckner, post to sludge@lists: subject [Sludge] Safer Sewage Sludge, May 10, 2008.
83. Sim Van der Ryn, *Toilets Papers* (Santa Barbara, CA: Capra Press, 1978). Van der Ryn argues in his introduction: "The soil is starved for the natural benefits of human manure, garbage and organic materials that go down the drain and to the dump. So agribusiness shoots it up with artificial fertilizers made largely from petroleum. These synthetics are not absorbed by the soil and leach out to pollute rivers and oceans. We each use eight to ten thousand gallons of fresh water to flush away materials that could be returned to the earth to maintain its fertility. Our excreta — not wastes but misplaced resources — end up destroying food chains, food supply and water quality in rivers and oceans."
84. Rogers maintains that even towns that collected waste in honey wagons often dumped it downstream. *History and Environment*, 78.
85. Clarke and King, *Water Atlas*, 52–3.
86. Mike Magee, *Healthy Waters: What Every Health Professional Should Know about Water* (Bronxville, NY: Spencer Books, 2005), 111.
87. Cornelia Dean, "Drugs Are in the Water. Does It Matter?" *New York Times*, April 3, 2001, D1.
88. Sandra Steingraber, *Living Downstream: A Scientist's Personal Investigation of Cancer and the Environment* (New York: Vintage Books, 1997), 99 and 190–193.
89. Ibid., 174–7 and 53.
90. "Lawn and Garden Pesticides: Hazards and Alternatives." Fact sheet from Toxic Free North Carolina, *Toxic Free*, http://www.toxicfreenc.org/informed/pdfs/Garden_pesticides.pdf (accessed February 7, 2010).
91. Steingraber, *Living*, 218.
92. Ibid., 201. Citing J.S. Osborne et al., "Epidemiologic Analysis of a Reported Cancer Cluster in a Small Rural Population," *AJE*, 132, Suppl. 1 (1980) 87–95.
93. Ibid., 165–66, 179 and 270.
94. "Non-Hodgkin Lymphoma Linked to Herbicides in Two Studies," NCAP: Northwest Coalition for Alternatives to Pesticides, November 2008, http://www.pesticide.org/hhg/herbicidesNHL.html (accessed April 9, 2010).
95. Mark Shapiro, *Exposed: The Toxic Chemistry of Everyday Products and What's at Stake for American Power* (White River Junction, VT: Chelsea Green, 2008), 58.
96. Steingraber, *Living*, 271.
97. Theo Colborn, Dianne Dumanoski and John Petterson Myers, *Our Stolen Future: Are We Threatening Our Fertility, Intelligence, and Survival? A Scientific Detective Story* (New York: Dutton, 1996), 138. Some pesticides, the authors state, are 10 times more potent today. They also note that golf course pesticide drenchings may be as high as four times that of farm fields. Ibid., 218.
98. Ibid., 9, 11.
99. Stuart Levy, "The Challenge of Antibiotic Resistance," *Scientific American*, No. 3, March 1998, 278, 5–6, Web source, Janet MacFall https://blackboard.elon.edu/courses/1BIO215A0703/content/_318842_1/AntibioticResis.htm (accessed May 12, 2007).
100. Senior seminar, Professor Janet MacFall, environmental studies, Elon University, class discussion, May 3, 2007.
101. Levy, "Challenge," 2.
102. Henry David Thoreau, *Journal XI*, 4–5, quoted in Robert Lawrence France (ed.), *Thoreau on Water: Reflecting Heaven* (Boston: Houghton Mifflin, 2001), 19.
103. Troxler, *Shuttle and Plow*, 29.
104. Knauffs, *Fabric*, 8.
105. Tom Magnuson, e-mail message, April 7, 2000.
106. Knauffs, *Fabric*, 8.
107. Mark Chilton, talk at the Chapel Hill Historical Association, January 27, 2008.

108. Hugh Lefler and Paul Wager, eds. *Orange County: 1752–1952* (Chapel Hill: Orange Printshop, 1953), 15.
109. Troxler, *Shuttle and Plow*, 56.
110. Knauffs, *Fabric*, 9–10.
111. Beyer, *Years Before Man*, 38–39.
112. Knauffs, *Fabric*, 10.
113. Knauff, *Fabric*, 13, and Brent Glass, *The Textile Industry in North Carolina: A History* (Raleigh: Division of Archives and History, NC Department of Cultural Resources, 1992), 23.
114. Knauffs, *Fabric*, 12. Gail Knauffs speculates that the gardener was Rudolf Topel, who went on to design Airlie Gardens, now a state park in Wilmington. Interview, January 18, 2010.
115. Knauffs, *Fabric*, 13–15.
116. Troxler, *Shuttle and Plow*, 517–519. In Geo. S. Mabry's "Sketch of Alamance County," this black principal of Graham Colored School notes as a sign of progress in 1895, "They were employing colored labor entirely to do the brick work on this new mill [Cora Manufacturing] which will be three stories high ... under the supervision of a skilled colored workman, Wm. Rogers of Graham."
117. Knauffs, *Fabric*, 15 and 19.
118. Hughes, *Development*, 19.
119. Vincent, *Shuttle and Plow*, 371.
120. Ibid.
121. Hall et al., *Like a Family*, 104.
122. Ibid., quoting Raleigh *News and Observer*, Oct. 18, 1900, p. 1, and October 10, 1900, p. 2; and North Carolina Bureau of Labor and Printing, *Annual Report ... 1901*, 416.
123. Vincent, *Shuttle and Plow*, 272.
124. Hall et al., *Like a Family*, 105, quoting *Alamance Gleaner*, November 1, 1900, p. 3.
125. Knauffs, *Fabric*, 21.
126. Barry and Nall, "Village Quick Facts."
127. Tullos, *Habits*, 24, citing Melton Alonza McLaurin, *Paternalism and Protest: Southern Cotton Mill Workers and Organized Labor, 1875–1905* (Westport, CT: Greenwood Press, 1971), 156–160.
128. Knauffs, *Fabric*, 31.
129. Hall et al., *Like a Family*, 260.
130. Henry "Hank" Sapoznik, "You Ain't Talking to Me: Charlie Poole and the Roots of Country Music," Columbia Records and Legacy Recording, booklet, 4.
131. Knauffs, *Fabric*, 68 and 21.
132. Ibid., 23 and 27.
133. Ibid., 38–39.
134. Don Francisco quoted in Tom Dillon, "River's Cleanup Effort Highlights Concerns Over Erosion, Runoff," *Times-News*, September 13, 2001, C1.
135. Richard Jarrett, interview, June 7, 2002.
136. Mike and Spencer Trollinger, interview, March 24, 2003.
137. Haw River Trail Association brochure.
138. Hughes, *Development*, 29.
139. In 1867, John Trollinger died, and E.M Holt, having insisted on payment in gold rather than Confederate dollars for cloth for soldiers' uniforms, was able to buy his land at Stony Creek and the Haw and build Carolina Mill in 1869. Knauffs, *Fabric*, 13.
140. Hughes, *Development*, 29.
141. Gant, *Raven's Story*, 8–10, 18–19.
142. Troxler, *Shuttle and Plow*, 241–243.
143. Gant, *Raven's Story*, 17–18.
144. Vincent, *Shuttle and Plow*, 329.
145. Alan Sanders, phone interview, April 13, 2002.
146. Tony Laws, interview, May 1 2008.
147. Swain et al., *Waterpower*, 151–154.
148. The new put-in is at Red Slide Park; heading east to Haw River, take the last left before the bridge.
149. Knauffs, *Fabric*, 92.
150. Isaac Groves, "Haw River Ponders Downtown Vision," *Times-News*, February 12, 2008, B1.
151. Mark Chilton, talk at River Landing Inn, May 2, 2010.
152. Tom Magnuson, interview, January 20, 2004.
153. "Alamance Environmentalist has a Big Day in April 1955," *Daily Times-News*, February 25, 1987.
154. Mark Kemp, "Does the Haw Cause Cancer?" *Daily Times-News*, February 26, 1987, 6A.
155. Scott Henderson, phone conversation, July 23, 2008, and Mark Kemp, "Legislators Push Haw Cleanup," *Daily Times-News*, February 25, 1987, A1.
156. John F.D. Smyth, *A Tour in the United States of America*, Vol. 1 (London: 1784) reprinted (New York: Arno Press, 1968), 168–170.
157. Ibid., 223–24.
158. Sue Sturgis, "Panel Rejects Broad Cuts in Air Pollution," *The Independent Weekly*, February 15, 2006, 15.
159. "Stericycle Awareness Campaign for Clean Air," Blue Ridge Environmental Defense League, http://bredl-medwaste.org/fromwhere 0809.htm (accessed June 1, 2010).
160. Sam Kiser, member of GASP, phone interview, March 2, 2006.
161. "Mercury Fact Sheet," *Health Care Without Harm*, www.noharm.org (accessed March 26, 2006).
162. Margaret Wooster, *Living Waters: Reading the Rivers of the Great Lakes* (Albany: State University of New York Press, 2009), 114.
163. Amount confirmed by Ray Stewart, environmental engineer, North Carolina Department of Environment and Natural Resources, e-mail message, July 6, 2009.
164. "Fish Consumption Advisories," Epidemiology Department of Health and Public Services, http://www.epi.state.nc.us/epi/fish/safefish.html (accessed February 7, 2010).
165. World Health Organization, "Dioxins and Their Effects on Human Health," http://who.int/mediacentre/factsheets/fs225/en (accessed January 20, 2010).
166. Ray Stewart, e-mail message, July 6, 2009.
167. David Mickey of BREDL, Blue Ridge Environmental Defense League, helped to get GASP going and linked to a nationwide movement to switch incineration of medical wastes to safer methods, like autoclaving.
168. Martha Hamblin, phone interview, March 1, 2006.
169. Lee Barnes, "Should They Know What They Burn?" *Times-News*, January 2, 2005, A6.
170. "Stericycle Fined for Excess Mercury Emissions," *Health Care Without Harm*, http://www.noharm.org/details.cfm?type=documentandID=989 (accessed April 9, 2005).
171. Ray Stewart, phone interview, March 20, 2006.
172. Lois Marie Gibbs, *Dying from Dioxin: A Citizen's Guide to Reclaiming our Health and Rebuilding Democracy* (Brooklyn: South End Press, 1995), 270.
173. As noted by Charles Duhigg, for broader solutions, EPA regulations need to change. For example, scrubbers in coal burning plants reduce airborne pollution but transmit the problem to water, where EPA regulations have not kept up with that pollution. Charles Duhigg, "Cleansing the Air at the Expense of Waterways," *New York Times*, October 13, 2009, A1 and A20.
174. "Environmental Responsibility," Stericycle,

http://www.stericycle.com/environmental-responsibility.html (accessed February 7, 2010). The Web site further states: "The process does not create any regulated air or water emissions, and it enhances the ability to recycle plastic materials." And this: "Currently, about 10 percent of Stericycle's U.S. treatment capacity includes incineration, down from more than 30 percent in 2001." "Alternative Technologies and the Promise of ETD," and "Environmental Responsibility," Stericycle, http://www.stericycle.com/environmental-responsibility.html (accessed February 7, 2010).

175. Ray Stewart, e-mail message, July 15, 2009.

176. "American Lung Association Annual State of the Air Report," cited in "Triad No Longer Home to Worst Pollution," *Times-News*, April 30, 2005, C6.

177. In 2010, Carolyn Cole and others in Clean Air Now (CAN) organized and got a state hearing to appeal to the Environmental Management Commission to implement the EPA's new higher standards for air pollution at Stericycle in 2012 and not 2014. Those present also argued for monthly unannounced inspections. Resolutions signed by four counties and seven municipalities supported CAN's efforts. In January 2011, the EMC decided to implement new standards in 2013.

178. Troxler, *Shuttle and Plow*, 52–54.

179. Knauffs, *Fabric*, 12.

180. "Demography of the Occaneechi Band of the Saponi Nation," information sheets, Tribal Office of the Occaneechi Band, Mebane, NC.

181. Forest Hazel, interview, January 9, 2008, and phone interview, January 24, 2008.

182. Forest Hazel with Lawrence A. Dunmore, "A Brief History of the Occaneechi Band of the Saponi Nation," 1995, unpublished manuscript.

183. Omega Wilson, interview, January 23, 2008.

184. Quoted in Ethel Stephens Arnett, *The Saura and Keyauwee on the Land that Became Guilford: Randolph and Rockingham County* (Greensboro, NC: Media Printers, 1975), 89.

185. James Mooney, *The Siouan Tribes of the East* (Washington, DC: Government Printing Office, 1894), 65.

186. T. Trawick Ward and R.P. Stephen Davis, Jr., *Indian Communities on the North Carolina Piedmont A.D. 1000 to 1700* (Chapel Hill: Research Laboratories of Anthropology, University of North Carolina, 1993), 428.

187. Ibid., 143.

188. Chester B. DePratter, Charles M. Hudson and Marvin T. Smith, "The route of Juan Pardo's explorations in the interior Southeast, 1566–68," *Florida Historical Quarterly*, 62, 1983, 125–158.

189. John Lederer, *The Discoveries of John Lederer*, William P. Cumming and Douglas Rights, eds. (Charlottesville: University of Virginia Press, 1958), 72.

190. R.P. Stephen Davis, Jr., "John Lawson and the Native Peoples of Carolina," in *Lawson's Legacy: Nature Writing and North Carolina, 1701–2001*, ed. Robert Anthony, North Caroliniana Society (in press).

191. Stephen Davis, e-mail message, February 22, 2010. Stephen Davis' help, clarifications, and writings were invaluable to me throughout this section.

192. The Oenuchs were likely the Enos, according to Stephen Davis, ibid.

193. Lederer, *Discoveries*, 70 and 121.

194. Jerry Mander, *In the Absence of the Sacred: The Failure of Technology and the Survival of the Indian Nations* (San Francisco: Sierra Club Books, 1991), 216–219.

195. John Lawson, *A New Voyage to Carolina*, ed. Hugh Talmadge (Chapel Hill: University of North Carolina Press, 1967), xv. Lawson's book has the very complete title *The History of Carolina Containing the Exact Description and Natural History of that Country Together with the Present State Thereof and a Journal of a Thousand Miles, Travel'd thro' several Nations of Indians, Giving a Particular Account of their Customs, Manners, Etc.*

196. Ibid., 60. William Byrd would echo these sentiments in his *History of the Dividing Line Betwixt Virginia and North Carolina, 1733*. "Between Eno and Saxapahaw [Haw] Rivers are the Haw Old Fields, which have the reputation of containing the most fertile high land in this part of the World, lying in a body of about 50,000 acres." Mark Chilton, "The Hawfields: An Approach to Title History," unpublished draft manuscript, 2.

197. Ibid., 61 and 64. Lawson also praises Eno Will, his Shakori guide whom he writes owned a Sissipahaw slave.

198. Davis, interview.

199. Lawson, *New Voyage*, 174 and 50–51.

200. Ibid., 200.

201. Ibid., 196.

202. Ibid., 218.

203. Ibid., 34.

204. Ibid., 232.

205. Ibid., 190 and 195.

206. Ibid., 177–78.

207. Ibid., 240.

208. Ibid., 175.

209. Ibid., 184–85.

210. Ibid., 184.

211. Ibid., 246.

212. Ibid., 243.

213. Ibid., 4.

214. Ibid., xxxvi.

215. Ibid., 207.

216. Davis, "John Lawson," 2.

217. Ibid., 8.

218. In 1697, three "Sax:a:pax" at the coast are alleged to have joined with Sugaree and Keyauwee to kill a plantation owner's son. Individuals were both with the Tuscarora on the lower Neuse River in 1711 and with John Barnwell fighting the Tuscarora in 1712. Jack Wilson (thesis, 1983), cited in Daniel Simpkins with Gary Petherick and Roy Dickens, "First Phase Investigations of Late Aboriginal Settlements Systems in the Eno, Haw and Dan River Drainages" (Chapel Hill: University of North Carolina Press, Research Laboratories of Anthropology), October 1985, 39–40. In 1713–1717, some Sissipahaw may have been with the Saponi at Fort Christanna. Heriberto Dixon, "A Saponi by Any Other Name Is Still a Siouan." *American Indian Culture and Research Journal*, 26: 3, 2002, 65–84.

219. R.P. Stephen Davis, e-mail message, July 14, 2005, and February 22, 2010.

220. Hazel, interview.

221. T. Trawick Ward and R.P. Stephen Davis, Jr., *Time Before History: The Archeology of North Carolina* (Chapel Hill: University of North Carolina Press, 1999), 276.

222. Ward and Davis, *Indian Communities*, 143.

223. Ibid.

224. Kristen Gremillion, "Botanical Remains," in Ward and Davis, *Indian Communities*, 143.

225. Ward and Davis, *Indian Communities*, 109, 120, 141–140.

226. Davis, interview, July 6, 2005. Davis states: "The Occaneechis' stronghold on trade likely dictated the arms trade; the Sissipahaw received only novelties."

227. R.P. Stephen Davis, "Pottery from the Fredricks, Wall and Mitchum Sites," in Roy Dickens, Trawick Ward, and Stephen Davis, *The Siouan Project: Seasons I and II* (Chapel Hill: Research Laboratories of Anthropology, University of North Carolina, 1987), 185 and 284.

228. Gary Petherick, "Architecture and features at the Fredricks, Wall and Mitchum Sites," in Dickens et al., *Siouan Project*, 75–76. Lawson writes: "These Savages live in Wigwams, or Cabins built of Bark, which are made round like an Oven, to prevent any Damage by hard Gales of Wind. They make the Fire in the middle of the House, and have a Hole at the Top of the Roof right above the Fire, to let out the Smoke. These Dwellings area are as hot as Stoves, where the Indians sleep and seat all Night. The Floors thereof are never paved nor swept, so that they have always a loose Earth on them.... The Bark they make their Cabins withal, is generally Cyprus, or red or white Cedar.... In building these Fabricks, they get very long Poles, of Pine, Cedar, Hiccory, or any Wood that will bend; these are the Thickness of the Small of a Man's Leg, at the thickest end, which they generally strip of the Bark, and warm them well in the Fire, which makes them tough and fit to bend; afterwards, they stick the thickest ends of them in the Ground, about two Yards asunder, in a Circular Form, the distance they design the Cabin to be (which is not always round, but sometimes oval) then they bend the Tops and bring them together, and bind their ends with Bark of Trees, that is proper for that use, as Elm is, or sometimes the Moss that grows on the Trees, and is a Yard or two long, and never rots; then they brace them with other Poles, to make them strong; afterwards, cover them all over with Bark, so that they are very warm and tight, and will keep up firm against all the Weathers that blow. Lawson, *New Voyages*, 180 and 182.

229. Ibid., 78.
230. Davis, interview, July 6, 2005.
231. Chilton, *Historical Atlas*, 37.6.
232. There were three professors at UNC–Chapel Hill at that time. Carole Troxler, talk at Haw River Historical Association, April 28, 2009.
233. Herbert S. Turner, *The Dreamer: Archibald Debow Murphey* (Verona, VA: McClure Press, 1971), 19. Herbert Turner, whose work was relied on in this chapter, puts the Hermitage near the confluence with Alamance Creek.
234. Ibid., 24.
235. Troxler, *Shuttle and Plow*, 238.
236. Turner, *Dreamer*, 25.
237. Ibid., 24.
238. Turner, *Dreamer*, 220.
239. Rachel Osborn and Ruth Selden-Sturgill, *Architectural Heritage of Chatham County* (Pittsboro, NC: Chatham County Historical Association, 1991), 10.
240. Turner, *Dreamer*, 30–33.
241. Ibid., 26.
242. Ibid., 213, citing the Hoyt Papers.
243. John A. McGeachy, "A Dreamer's Speculations: The Financial Plight of Archibald D. Murphey," paper, History 561, North Carolina State University, 2002, http.www.lib.ncsu.edu/staff/mcgeachy/admurphy.htm (accessed June 1, 2004), 14.
244. Troxler, *Shuttle and Plow*, 239.
245. Turner, *Dreamer*, 208, 214, 204. Turner further explains that had any prisoner escaped, the jailor would have been liable for his debts.
246. Ibid., 211.
247. Troxler, *Shuttle and Plow*, 259.
248. Vincent, *Shuttle and Plow*, 346.
249. Troxler, *Shuttle and Plow*, 258.
250. Vincent, *Shuttle and Plow*, 380. Vincent adds that in 1851, when slaves numbered 3,196 and whites 7,924, Ruffin topped the list of those 14 people who owned more than 20 slaves.
251. Sallie Stockard, "Daughter of the Piedmont," n.d., quoted in Troxler, *Shuttle and Plow*, 262–63.
252. Troxler, *Shuttle and Plow*, 264.
253. Ibid., 269 and 286. Troxler states Judge Ruffin's vast connections got privileged friends leave at critical harvest or family times. Gratitude and sometimes payment followed.
254. Ibid., 308.
255. Ibid.
256. Vincent, *Shuttle and Plow*, 357.
257. Turner, *Dreamer*, 211 and 227.
258. Vincent, *Shuttle and Plow*, 357.
259. Hughes, *Development*, 41.
260. Horace W. Raper, *William W. Holden* (Chapel Hill: University of North Carolina Press, 1985), 128.
261. Robert Wyllie, "Swepson," in *Dictionary of North Carolina Biography*, ed. Dr. William S. Powell (Chapel Hill: University of North Carolina Press, 1994) Vol. 4, 490.
262. Vincent, *Shuttle and Plow*, 357.
263. Hughes, *Development*, 44. Julian Hughes, a local historian, was at one time manager of Virginia Mills.
264. Knauffs, *Fabric*, 15 and 81.
265. Vincent, *Shuttle and Plow*, 358.
266. Ibid., 359.
267. Ibid.
268. Ibid.
269. Raymond Herring, interview, July 8, 2003.
270. Hughes, *Development*, 40. Hughes writes: "The Saxapahaw Nine was an arch rival of the Swepsonville team, and many a time players of the two teams entertained the fans with a free for all when a close decision was called. Umpires calling the game in Saxapahaw or Swepsonville soon learned that it was most expedient to give the home team the benefit of the doubt when the play was close."
271. Duard Farrell, interview, July 21, 2003.
272. Ibid.
273. Farrell gives further detail: "Down from the dam is where the stage coach crossed past a rock ridge. A rut marks the old stage coach road. You can follow it along behind the tee box of number 4 and the number 3 fairway."
274. Carole Troxler, *Pyle's Defeat: Deception at the Racepath* (Graham, NC: Alamance County Historical Association, 2003), 43–44. Troxler notes the present day location now—Loomcraft Textiles on I-85 in Burlington.
275. Howard Zinn, *A People's History of the United States: 1492–Present* (New York: HarperPerennial, 1995), 64.
276. Troxler, *Shuttle and Plow*, 62 and 67.
277. Walter Whitaker, Staley Cook, and A. Howard White, *Centennial History of Alamance County, 1849–1949* (Burlington, NC: Chamber of Commerce, 1949), 55.
278. Troxler, *Shuttle and Plow*, 64.
279. Fenn and Wood, *Natives*, 90–91.
280. Ibid.
281. Martin (1829, 2: 277–278) quoted in Troxler, *Shuttle and Plow*, 77.
282. Hughes, *Development*, 32.
283. Magnuson, presentation to Elon University class, June 9, 2002.
284. Troxler, *Pyle*, 31 and 14. Carole Troxler, building

on the work of George Troxler and with the help of Tom Magnuson and others in the Trading Path Association, has worked through to the exact location of those events. In this section, I rely on her *Pyle's Defeat: Deception at the Racepath.*

285. Troxler, *Pyle*, 26, 38–9 and 44. Troxler adds that the area was then the plantation of an unpopular Whig, William O'Neal, who had allegedly taken horses for the Revolutionary cause, then sold them for profit.

286. Ibid., 29 and 52. Some accounts suggest he was asking for surrender.

287. Ibid., 64, 55, 65–66, quoting Moses Hall. Troxler also writes that the fury continued that night and Hall saw six prisoners "hewed to pieces with broadswords." Pyle survived by hiding underwater in a pond.

288. Ibid., 68 and back cover.

289. Troxler, *Shuttle and Plow*, 121. Jeff Bright and Steward Dunaway have located the likely spot of Clapps Mill—now under Lake MacIntosh not far from the marina off of Huffman Mill Road. Michael Abernethy, "Historians Pinpoint Location of Battle," *Times-News*, March 30, 2008, B1–2.

290. Vincent, *Shuttle and Plow*, 346 and 351.

291. William Vincent, quoted in Keren Rivas, "Mill Memories," *Times-News*, January 16, 2005, B1–4.

292. Vincent, *Shuttle and Plow*, 348.

293. Ibid., 351.

294. Hughes, *Development*, 9.

295. Wilbur Cash paraphrased in Cathy McHugh, *Mill Family: The Labor System in the Southern Cotton Textile Industry, 1880–1915* (New York: Oxford University Press, 1988), 99.

296. Wages there remained low and hours long; in 1929, one worker, Fern Edwards, earned $8 starting for a six day week with 10–12 hour shifts. Rivas, "Mill Memories," B4.

297. William Vincent, phone interview, July 15, 2003.

298. Hughes, *Development*, 128–9.

299. Chilton, talk to Chapel Hill Historical Society, January 27, 2004.

300. Janet MacFall, interview, March 15, 2005.

301. Chilton, *Historical Atlas*, 35.6.

302. Fenn and Wood in Mosley, *The Way We Lived*, 165.

303. Chilton, *Historical Atlas*, 33.6.

304. Tom Magnuson, talk on Trading Path hike, November 4, 2007.

305. Janet MacFall, interview, March 15, 2005.

306. John K. Terres, *From Laurel Hill to Siler's Bog* (Chapel Hill: University of North Carolina Press, 1993), 47–50.

307. Donald W. Stokes, *A Guide to Bird Behavior*, Vol. III (New York: Little, Brown, 1979), 28–30.

308. Mark Chilton, "The Hawfields," 4. I am grateful to Mark Chilton for the opportunity to read a draft of his unpublished article, which provided all the information here on rival claims to the land.

309. Bulla, Ben, interview, June 24, 2004.

310. Hoover Dixon, interview, July 8, 2003.

311. Nannie Dixon McBane, conversation at River Inn meeting of Haw River Land Owners, April 19, 2008. McBane's brother Hoover Dixon did play in the river: "We'd swim down there, have our weekly bath and fish in the river for catfish, bream, and carp. Course that stopped with the pollution after the war when all the industries started back up. Burlington and all the cities were probably dumping pure raw sewage, Swepsonville, Haw River all these dye plants. In Saxapahaw, I know company houses where commodes run straight to the river bank. You could smell it all the way up here." Dixon, interview.

312. Ben Bulla, "Saxapahaw Cotton Mills: The First Hundred Years," *The Cotton History Review*, Vol. 2, 1961, 135.

313. Chilton, *Historical Atlas*, 31.8.

314. Ben Bulla, interview, June 24, 2004.

315. Bulla, "Cotton Mills," 132–133.

316. Ben Bulla, "History of Saxapahaw," http://www.saxapahaw.com/history6.html, as published in Centennial Edition of the Burlington *Daily Times-News*, May 1949, 2 (accessed June 29, 2004).

317. Bulla, "Cotton Mills," 133.

318. Hughes, *Development*, 36.

319. Troxler, *Shuttle and Plow*, 231–234. Troxler, the key source here, further explains that freeing slaves was not easy after 1831 and the Nat Turner Rebellion. A $1,000 bond had to be posted for each slave to guarantee that person would not return to North Carolina.

320. Bulla, "Cotton Mills," 135.

321. Bulla, interview.

322. Bulla, "Cotton Mills," 136–7.

323. William Friday, "Foreword" in Ben Bulla, *Textiles and Politics: The Life of B. Everett Jordan* (Durham, NC: Carolina Academic Press, 1992), ix.

324. Ibid., 186.

325. Ibid., 154.

326. Ibid., 141. Gordon Marlette remembers: "We rode out to Saxapahaw and the National Guard was out there in their tents.... The thing that I really recall is that there were machine guns set up just outside the entrance to the bridge." As quoted in: Joseph Cigna, "City's 1934 Labor Riots Inspire New Novel," *Times-News*, November 11, 2001, A1 and A9.

327. Leon Madden, interview, July 12, 2007.

328. Bulla, *Textiles and Politics*, 202.

329. John Jordan, interviews with Ben and John Jordan, January 31, 2008, and February 7, 2008.

330. Leon Madden, interview.

331. Bulla, *Textiles and Politics*, 361. (Figure is for 1969.)

332. Dixon, interview, July 8, 2003.

333. Hoover Dixon adds: "Turnover here? It was big. They're a lot of mills. You take the best employee you had, and he would work for maybe six months then he would leave and go to Bellemont or leave and go to Gastonia. I've hired the same person at least eight times. I needed the employee, and he was trained, but I knew he wouldn't be here in six months."

334. Ben and John Jordan, interviews, January 31, 2008, and February 7, 2008.

335. Dr. Mike Holland, interview, July 21, 2007.

336. Quoted in Amy Jo Jenkins, "Activist Seeks a Cleaner, Healthier Haw," *Times-News*, March 29 2004, C2.

337. Bruce Holt quoted in *Erasing the End*, Tony White and Ben Kivlan, unpublished student manuscript, Elon University.

338. Bruce Holt, interviews, May 11 and 18, 2005.

339. Cane Creek Mountains are known as Bass Mountain to local people.

340. Beyer, *Years Before Man*, 168, citing R.H. Kesel (1974). Beyer adds that the Uwharries are not the "oldest mountains in North America. While the rocks of the Uwharries are pre–Cambrian in age, the erosion that created today's land surface is very young in geologic terms."

341. Outwater, *Water*, 55 and 63.

342. Pielou, *Fresh Water*, 69.

343. Rusty Rozzelle, program manager, Mecklenburg County Water Quality, presentation, Sustainable Communities Conference, Elon University, Elon, NC, September 19, 2008.
344. Leopold, *View*, 188.
345. Christine Tilburg and Merryl Alber, "Impervious Surface Review of Recent Literature," Georgia Coastal Research Council, http://crd.dnr.state.ga.us/assets/documents/jrgcrddnr/ImperviousLitReview_Final.pdf (accessed August 4, 2008).
346. Patrick Beggs, interview.
347. Rusty Rozzelle, Sustainable Communities Conference, 2008.
348. Elaine Chiosso, executive director, Haw River Assembly, interview, January 4, 2007.
349. Clarke and King, *Water Atlas*, 67.
350. Chilton, "Hawfields," 11.
351. Osborn, *Architectural Heritage*, 8–9.
352. Ben Jordan fills in some informed speculation about the date of the bankruptcy: "It may have been way before 1927, 1921 or 22 or 23. There was a severe recession that affected cotton mills all over the country. Now this is just speculation, I wasn't taught this in my journeys through Duke. They were in heavy war production, sock, and shirts and then in 1918, the government cancelled all contracts. They all stopped, warehouses were full of cotton. The price had hit 80 cents a pound; cotton's not been that high since, and the price dropped to 5 or 6 cents a pound. The actual cost of cotton is 5 or 6 cents a pound. Hundreds of mills went broke." Interview, January 31, 2008.
353. W.J. Brewer, interview, July 5, 2007, and phone interview, January 7, 2008.
354. Alamance County Landfill information sheets.
355. Ibid.
356. Fred Lee and Ann Jones-Lee, "Overview of Subtitle D Landfill Design, Operation, Closure and Postclosure Care Relative to Providing Public Health and Environmental Protection for as Long as the Wastes in the Landfill will be a Threat," January 2004, http://www.fredlee.com/Landfills/LFoverviewMSW.pdf, 4–5, and citing Freeze and Cherry (1979) (accessed April 9, 2010).
357. Mike Holland makes the leacheate-sludge connection: "It's a health hazard to shuttle toxins from the basement of a landfill onto the crops that milk cows eat and into the air our children breathe." Mike Holland, list posting, sludge-bounces@lists.ibiblio.org, October 15, 2009.
358. Landfill information sheets.
359. "County Short on 3R's," *Times-News*, May 25, 2009 B1. Figures from 2007–2008 were used.
360. Lee and Jones-Lee, "Overview," 7.
361. "In the Trash," *Times-News*, April 21, 2008, B1.
362. "County Short," B1.
363. Lee and Jones-Lee, "Overview," 17.
364. Bob Brueckner, interview, May 31, 2008.
365. The website http://waterdata.usgs.gov/nwis/uv?02096500 is a source for water levels and flow at the town of Haw River.
366. Chilton, *Historical Atlas*, 29.4.
367. Troxler, *Shuttle and Plow*, 77.
368. Bulla, "History," 8.
369. Elaine Chiosso, "And It All Comes Down to the Lake," *Voice of the Haw*, Haw River Assembly, April 2008, 8.
370. "Tapped Out," *Environmental Defense Newsletter*, Environmental Defense Fund, 39, No. 4, September 2008, 10.

371. Ray J. Coomans, *Alamance County Natural Heritage Inventory* (Raleigh, NC: Department of Environment and Natural Resources, 2002), 43.
372. Stan Tekiela, *Trees of the Carolinas: Field Guide* (Cambridge, MN: Adventure Publications, 2007), 261.
373. Chilton, *Historical Atlas*, 27.3.
374. Greg Feller, OWASA Public Affairs, e-mail message, June 18, 2008.
375. Bob Brueckner notes we are .4 mile off of Ferguson.
376. Ward and Davis, *Indian Communities*, 5. They add: "This lack of abundant floodplain soils may have been a major factor contributing to the relatively small, dispersed nature of the native settlements during the Late Prehistoric and Contact periods."
377. Daniel Simpkins, Gary Petherick, Roy Dickens, "First Phase Investigations of Late Aboriginal Settlements Systems in the Eno, Haw and Dan River Drainages" (Chapel Hill: Research Laboratories of Anthropology, University of North Carolina, October 1985), 10.
378. Ibid., 106.
379. George Milner, *The Moundbuilders: Ancient Peoples of Eastern North America* (London: Thames and Hudson, 2004), 177.
380. Ward and Davis, *Indian Communities*, 5.
381. Ibid., 19 and 21.
382. Ibid., 78 and 107–108.
383. Theda Perdue, *Native Carolinians: The Indians of North Carolina* (Raleigh: NC Department of Cultural Resources, 1985), 16.
384. Dorcas Miller, *Stars of the First People* (Boulder: Pruett, 1997), 273.
385. Wade Hadley, Doris Horton and Nell Strowd, *Chatham County 1771–1971* (Lillington: Edwards Brothers, 1997), 231.
386. Grimsley Hobbs, Jr., talk at the Chapel Hill Historical Association, January 27, 2004.
387. Troxler, *Shuttle and Plow*, 201.
388. Troxler, *Shuttle and Plow*, 128, citing figures from the *Edinburgh Magazine* of January 1782.
389. Hadley et al., *Chatham*, 223.
390. Pat Shaw Bailey, *Land Grant Records of North Carolina: Orange County 1752–1885, Vol. 1* (Graham: Pic by Pat Publications, 1990), 45.
391. Stephen Hall and Marjorie Boyer, *Inventory of the Natural Areas and Wildlife Habitats of Chatham County, North Carolina*, June 2006, Triangle Land Conservancy and County of Chatham, 107–08.
392. Hobbs, talk, January 27, 2008.
393. Henry Armand London, "An Address on the Revolutionary Anniversary of Chatham County Delivered at Pittsborough NC on the Fourth Day of July 1, 1876" (Sanford: Cole Printing), 10.
394. In 1840, Chatham County's mills numbered 64 grist mills, 12 flour mills, 40 saw mills. Osborn, *Architectural*, 20.

Chapter 4

1. Barbara Clark Pugh, "The Day Chicken Bridge Got Its Name," Chatham County Historical Association, http://chathamhistory.org/archive.html (accessed July 9, 2006).
2. Cassie Wasko, *Chatham Record*, November 5, 1987.
3. Robert Martin, "Haw NC," American Whitewater, http://www.americanwhitewater.org/rivers/id/1083 (accessed December 15, 2009).

4. Hall and Boyer, *Inventory*, 147–148.
5. Herbert Poole, "Bygone Mills on the Haw River in Chatham County," *Chatham Historical Journal*, 8, No. 3, November 1995, 1–4.
6. Osborn and Selden-Sturgell, *Architectural Heritage*, 268.
7. Hadley, et al., *Chatham*, 418–9.
8. Ibid., 268 and 281.
9. Poole, "Bygone," 4.
10. Cathy Markatos, phone interview, February 4, 2007.
11. Judith Peterson, interview, January 24, 2007.
12. J. Lamont Norwood, "Mt. Pleasant Church and Pace's Mill Bridge," *Chatham Historical Journal*, September 1994, 7, No. 1, 1.
13. Beyer, *Years before Man*, 6.
14. "A Terrible Storm: Three People are Killed When House Is Blown Down," *The Chatham Record*, May 8, 1924, 1. Chilton, *Historical Atlas*, 16.8, dates the first tornado in 1892.
15. Poole, "Bygone Mills," 4.
16. Hall and Boyer, *Inventory*, 139–140.
17. Wallace Kaufman, *Coming Out of the Woods* (Cambridge, MA: Perseus, 2000), 7.
18. Ibid., 296.
19. Ward and Davis, *Indian Communities*, 161.
20. Jane Madeline McManus and Ann Marie Long with report on pottery of Alamance County by Linda Carnes, *Alamance County Archeological Survey Project of Alamance County* (Chapel Hill: Research Laboratories of Anthropology, University of North Carolina, September, 1986), 56.
21. Ward and Davis, *Indian Communities*, 55.
22. McManus and Long, *Survey*, 50–51.
23. Stephen Claggert, John Cable, and Curtis Larsen, *The Haw River Sites: Archaeological Investigations at Two Stratified Sites in the North Carolina Piedmont*, Vols. I–III (Wilmington, NC: Commonwealth Association, Inc., for the U.S. Army Corps of Engineers, 1982).
24. Kenneth E. Sassaman and David G. Anderson, eds. *Archeology of the Mid-Holocene Southeast*, Ripley Bulletin Series (Gainesville: University Press of Florida, 1997), 158.
25. Trawick Ward, "A Review of Archaeology in the North Carolina Piedmont: A Study in Change," in Mark Mathis and Jeffrey Crow (eds.), *The Prehistory of North Carolina: An Archaeological Symposium* (Raleigh: Division of Archives and History, NC Department of Cultural Resources, 1983), 65.
26. Ward and Davis, *Time Before History*, 55–58, 75.
27. Ibid., 57.
28. Ibid., 57 and 67.
29. Ibid., 55–70.
30. Ibid., 75.
31. Elaine Chiosso, interview, January 4, 2007.
32. Don Francisco quoted in "River's Cleanup Effort," C2. In a May 14, 2010 e-mail to me, Don Francisco added, "I learned that this was not terribly unusual.... It is observations like these that establish that the greatest threat to North Carolina streams and lakes is surface runoff. Waste water treatment improvements over the past 30 years have decreased the significance of this source considerably. Now as population increases, it is possible, if not likely, that the capacity for wastewater treatment will be overwhelmed by increased amounts of wastewater pollutants requiring removal. There are thermodynamic limits to how much can be removed at a cost that the public will bear."
33. "Pollution Issues in the Watershed," *Voice of the Haw*, Haw River Assembly, April 2005, 7.

34. "Bluegreen Corporation Adds 240 Acres to Chatham's Chapel Ridge," *Chatham Weekly Journal*, November 3, 2006 http://www.chathamjournal.com/weekly/business/real-estate/chapel-ridge-grows-61103.shtml (accessed April 9, 2010).
35. "Concerned Citizens Review of the Buck Mountain Golf Community Environmental Impact Assessment," Alison Weakley, ed. submitted to the Chatham County Planning Board, April 28, 2004, 19.
36. Ibid., 11, 20.
37. Ibid.
38. Ibid., 22.
39. Lew Sichelman, "It Took a Village." *Big Builder Online*. http://www.bigbuilderonline.com/industry-news.asp?sectionID=0andarticleID=183083 (accessed January 20, 2010).
40. Sally Erickson, speaking at public hearing, Pittsboro, NC, December 13, 2005. Numbers found in "Development Watch," *Voice of the Haw*, Haw River Assembly, April 2005, and on Chatham Citizens for Effective Communities' Web site, www.chathamcitizens.org/development-info/development-watch.htm (accessed January 8, 2007).
41. Sichelman, "Village."
42. Mitch Renkow, "The Cost of Community Services in Chatham County, Land Preservation Notebook." http://www.cals.ncsu.edu/wq/lpn/PDF documents/chathamCOCS.PDF (accessed February 21, 2011).
43. *Cape Fear River Basinwide Water Quality Plan*, Division of Water Quality, NC Department of Environment and Natural Resources (Raleigh: 2005), 46. Dry Creek's bio-classification was listed good-fair in 1998 and good in 1993. *Cape Fear River Basinwide Water Quality Plan*, Division of Water Quality, NC Department of Environment and Natural Resources (Raleigh: 2000), 134.
44. Jerry Markatos, phone interview, February 2, 2007.
45. *Basinwide Report*, 2005, 46. The report does note that "a new development in a tributary to Dry Creek is a potential source of sediment.... BMP [best management practices] are in place."
46. "Muddy Waters Continue to Plague the Haw," July 21, 2005, Haw River Assembly, www.hawriver.org (accessed January 12, 2007).
47. Lisa Hoppenjans, "Muddying the Waters," *News and Observer*, August 30, 2005, B1, www.newsobserver.com/new/growth/story/188520-p.2.html (accessed March 4, 2006).
48. Mark Shultz, "Project Violates Pollution Rules, *News and Observer*, August 16, 2005, B3.
49. Quoted in Hoppenjans, "Muddying." Also mentioned in the article, new developments in Wake Forest are inspected twice a month and, when "stabilized," once a month.
50. Elaine Chiosso, e-mail message, May 5, 2009.
51. "Realtors, Homebuilders Saturate General Assembly with Contributions," July 2007, *Democracy North Carolina*, www.democracy-nc.org.
52. Alison Weakley, phone interview, January 19, 2007. Fining is not currently an effective tool, the Conservation Council of North Carolina reports: "North Carolina's maximum daily penalty for sedimentation violations is $5,000 which is rarely assessed in full. From 2002 to 2005, the average assessed penalty was $31,437, while the average final penalty amount was only $8,988." "Who's Afraid of a Little Dirty Water? You Should Be," *The Carolina Conservationist*, spring 2005, 2.
53. Allison Weakley reports further problems at Briar

Chapel that impact Pokeberry Creek: "For piped stream crossings, they were supposed to bury the pipe to allow for aquatic passage — evaluations of their crossing have revealed many violations.... It's up to the state in most cases to enforce conditions, and then revise permits based on what they find, but they don't often have the time or resources to do so. It takes local oversight." Allison Weakley, e-mail message, March 15, 2010. In a report at the Haw River Assembly Annual Meeting, November 7, 2009, Catherine Deininger of HRA's Streamwatch noted that the abundant wetlands along Pokeberry will help keep the Haw clear as Briar Chapel is developed unless these wetlands are filled in by sediment. Old, inadequate septic systems are already a problem on Pokeberry Creek.

54. As Cynthia Crossan of River Watch puts it: "A lot of this is common sense — don't bulldoze the stream buffers; re-seed immediately (instead of taking the maximum time allowed); use contour plowing; use mulch, berms and other low-tech methods to keep soil in place during construction. How do you know if it works? Go out during a storm and watch where the mud is flowing!" "Muddy Waters," July 21, 2005, Haw River Assembly, www.hawriver.org (accessed September 1, 2007).

55. Lynn Featherstone, interview, July 21, 2007.
56. Chilton, *Historical Atlas*, 14.5.
57. Swain et al., *Waterpower*, 4.
58. Hadley et al., *Chatham*, 6, citing *Colonial Records*, Vol. 4, x–xi.
59. London, "Address," 7.
60. Hadley et al., *Chatham*, 4–5, citing Poe family records (La Barr, 1930) concerning a neighbor.
61. Ibid., 151.
62. Ibid., 231.
63. Ibid., 202 and caption of first illustration, no page.
64. Ibid., 406–7 and Douglas DeNatale, "History of Bynum: 1872–1979," *Chatham Journal*, March 1998, Vol. 1, No. 11, 6.
65. London, "Address," 10. He adds (in 1876): "The future historian may desire to know our present number of bridges ... and they are mentioned therefore for future information ... on Haw River are six, as follows: at Love's Mill, Pace's Mill, Bynum's Factory, Henley's Mill, Moore's Mill and at Haywood."
66. Hadley et al., *Chatham*, 152. Prices are from a bridge on the Deep River, a tributary of the Haw.
67. Ibid., 59.
68. Lawrence Fooshee London, interview with Gary Freeze, December 7, 1978, Southern Oral History Project, University of North Carolina at Chapel Hill, H-93, 4.
69. Frank Sidney Durham, interview with Douglas DeNatale, September 10 and 17, 1978, Southern Oral History Project, University of North Carolina at Chapel Hill, H-67, 45.
70. London, "Address," 10.
71. Hadley et al., *Chatham*, 125–6.
72. DeNatale, "History of Bynum," 6–8, citing Hadley, 1971: 377.
73. Lawrence Glickman, *A Living Wage: American Workers and the Making of Consumer Society* (Ithaca: Cornell University Press, 1997), 18.
74. Flossie Moore Durham, interview by Mary Frederickson and Brent Glass, September 2, 1976, #4007 H-66, Southern Oral History Program Collection, http://docsouth.unc.edu/nc/durhamf/durhamf.html (accessed May 3, 2004).
About a quarter of workers were children in 1900 in North Carolina, but change came in the next decade, as William Powell reports in *North Carolina through Four Centuries*, when 14 became the official age limit for workers and those less than 16 were confined to 60 working hours a week. 455–458.

75. Louise R. Jones, quoted in DeNatale, "History of Bynum," 6.
76. John Snipes, quoted in DeNatale, "History of Bynum," 6.
77. DeNatale, Douglas, *Traditional Culture and Community in a Piedmont Textile Mill Village*, thesis in folklore (Ph.D. dissertation, University of North Carolina at Chapel Hill, 1980), viii, citing Potwin, 1927: 29.
78. Ibid., 14, citing Clarence Poe, editor of *Progressive Farmer*.
79. Ibid., 14–16.
80. Ibid., 27.
81. Chilton, *Historical Atlas*, 13.8.
82. DeNatale, "History of Bynum," 6.
83. Anonymous interview with Douglas DeNatale, Southern Oral History Project, University of North Carolina at Chapel Hill, H-60, 3.
84. DeNatale, "History of Bynum," 6–7.
85. Douglas DeNatale, "Work and Culture in a Piedmont Mill Town," *Chatham Historical Journal*, 2, No. 2, October 1989, 4.
86. Frank Durham, interview, 47.
87. DeNatale, "History of Bynum," 7 and 8.
88. John Wesley Snipes, interview with Brent Glass, September 20 and November 20, Southern Oral History Project, University of North Carolina at Chapel Hill, H98, http://docsouth.unc.edu/nc/snipes/snipes.html, 2–13.
89. Ibid., 2–17.
90. Lawrence Fooshee London, interview, H-93/34.
91. Vernon Durham and Eula Durham, interview with James Leloudis, November 29, 1978, Southern Oral History Project, University of North Carolina at Chapel Hill, H-64, 27–28.
92. Ibid., 28.
93. Snipes, interview, 2–30.
94. Ibid., 2–29.
95. Ibid., 2–35–36.
96. Mary Gattis, quoted in DeNatale, *Traditional Culture*, 9 and 10.
97. Snipes, interview, 2–36–37.
98. Ibid., 2–38.
99. Eula Durham quoted in DeNatale, *Traditional Culture*, 10. He notes many concurring voices.
100. Louise Harris, interview with Helen Bresler, Southern Oral History Project, University of North Carolina at Chapel Hill, H 78.
101. Lint was great enough in quantity to show up in rivers with cotton oils, machine oils, hypochlorites for bleach and sulfite stain removers. Indigo and plant dyes were replaced by chemical aniline dyes. Rogers, *History and Environment*, 79.
102. Mozelle Riddle quoted in DeNatale, *Traditional Culture*, 103.
103. John Wesley Snipes, interview with Douglas DeNatale, August 22, 1979, Southern Oral History Project, University of North Carolina at Chapel Hill, H99, Tape 1, Side B.
104. "Brown Lung," Brown Lung Association, lib.unc.edu/mss/inv/Brown_Lung_Association.html. Efforts were led by Mike Szpak.
105. Anonymous worker, interview with Jim Leloudes and Mary Murphy, C-24, Southern Oral History Program Collection, UNC–Chapel Hill, 20.

106. Ibid.
107. DeNatale, *Traditional Culture*, 55.
108. Eula Durham, interview with Douglas DeNatale, March 1, 1979, Southern Oral History Project, University of North Carolina at Chapel Hill, H-65, 8.
109. Roy Eubanks quoted in DeNatale, *Traditional Culture*, 65. Wesley Snipes had his own misadventure on the Haw's banks: "I had never seen no homebrew. They'd take and get a can of yeast and five pounds of sugar, and take an old crock or five gallon jug down in the woods. They'd put that yeast and sugar in there and fill it up with water and let it work off. And it was strong! ... And when you first started to drinking it, it weren't no more than a Coca-Cola or sweet apple cider. Well, I followed them down there in the woods ... and the first thing I knew the world started turning around. And somebody run (I reckon they thought they'd get my job) and told the superintendent that I was down the river drunk. He called me to the office next morning and asked me about it. I said, 'Well, the world started turning around, and I laid down and went to sleep down there. The mill was standing part of the time and ... he said, 'Well, I'm going to have to fire you.' I come on home. That night he sent for me, and he said, 'Come on back on in the morning.' I went on back to work, and never lost another day from it." Snipes, interview with DeNatale, 47.
110. Hadley, "Fishstreams in Chatham," 1, citing the *Chatham Record* of April 14, 1887.
111. DeNatale, *Traditional Culture*, 89. Elaine Chiosso also reports hearing of runs of shad until Jordan Dam came in.
112. Louise Jones, interview with Mary Frederickson, September 20 and October 13, Southern Oral History Project, University of North Carolina at Chapel Hill, H-85, 43–45.
113. Sally Fowler quoted in DeNatale, *Traditional Culture*, 88.
114. Ibid., citing Paul and Louise Jones.
115. Ibid. 89.
116. John London, interview with Gary Freeze, Southern Oral History Program Collection, University of North Carolina at Chapel Hill, H-91, 3.
117. Anonymous worker quoted in DeNatale, *Traditional Culture*, 133.
118. DeNatale, "History of Bynum," 8. DeNatale reports that the Odell Mill closed in 1979. J.S. Osborn notes a final mill closing in 1984 in his report on cancer in Bynum.
119. DeNatale, "Work and Culture," *Chatham Historical Journal*, 2, No.1, August 1989, and 2, No. 2, October 1989.
120. Jimmy Stubbs, Allen and Sons Barbeque, conversation, June 17, 2008.
121. Jerry Partin, interview, January 13, 2004.
122. Clyde Jones, interview, January 13, 2004.
123. Louise Kessel, interview, October 19, 2002.
124. Dave Foreman, "Dave Foreman," in *Listening to the Land*, 6.
125. Christopher Manes, "Christopher Manes," in *Listening to the Land*, 20.
126. Joe Jacob, interviews, February 29, 2008, and April 17 and 25, 2007.
127. Michael Norris, coordinator of the National Stream Flow Information Program, quoted in John Schwartz, "Experts See Peril in Reduced Monitoring of Nation's Streams and Rivers," *New York Times*, April 11, 2006, D3.
128. Nelson, *Geology of the Haw River Basin*, 52.
129. Poole, "Bygone Mills," 3.
130. A recent report suggests the number of imperiled North American fish is up 92 percent. "Back Porch," *Wildlife in North Carolina*, December 2008, 36.
131. Mander, *In the Absence*, 75. One recent study hints at the dimensions of the problem with its findings of how children spend their time: 4–7 minutes average daily outdoor time; 6.5 hours average daily plugged in time. Diane Lamb, "Get Wild," *News and Record*, March 29, 2010, B1.
132. Chilton, *Historical Atlas*, 10.5.
133. Beyer, *Time Before Man*, 122–125.
134. Chilton, *Historical Atlas*, 10.2 citing Hadley, 1971.
135. Ibid., 9.4 and Poole, "Bygone Mills," 3.
136. Poole, ibid.
137. Chilton, *Historical Atlas*, 7.9.
138. Ibid., 6.0.
139. Poole, "Bygone Mills," 2.
140. Hadley, "Fish in Streams."
141. Swain et al., *Waterpower*, 151.
142. Frederick Reimers, "William Nealy, 1953 to 2001," *Paddler Magazine*, http://www.paddlermagazine.com/issues/2001_6/article_148.shtml (accessed March 1, 2008).
143. Holly Wallace, e-mail message, October 8, 2008.
144. William Nealy, map, Haw River Section IV.
145. William Nealy, *Whitewater Home Companion: Southeastern Rivers*, Vol. 1 (Birmingham: Menasha Ridge Press, 1981), 1.
146. Lynn Featherstone, *Voice of the Haw*, Haw River Assembly, December 2001, and interview, July 21, 2008.
147. Bob Sehlinger, president, Menasha Ridge Press, phone interview, June 11, 2008.
148. William Nealy, *The Nealy Way of Knowledge* (Birmingham: Menasha Ridge Press, 2000), 217.
149. Ibid.
150. Ibid., Introduction (unnumbered).
151. William Nealy, *Whitewater*, 22.
152. William Nealy, *Kayak: The Animated Manual of Intermediate and Advanced Whitewater Technique* (Birmingham: Menasha Ridge Press, 1986), 167.
153. Quoted in Kate Geis, "The One and Only William Nealy," *Riversense*, http://www.riversense.com/more_william.html (accessed June 1, 2007).
154. Hall and Boyer, *Inventory*, 105–106.
155. Catherine Deininger, interview, January 29, 2008.
156. Nancy Guthrie, interview, January 22, 2008.
157. Nancy Guthrie adds: "We helped Pittsboro upgrade their wastewater treatment plant to re-use quality, to send on to the 3M Plant; 3M pays for work at their end and splits pipe costs. So it gets a project moving."

Chapter 5

1. Curtis Larsen, Foreword in Claggert and Cable, *Haw River Sites*, Vol. 1, iii, and Vol. 3, 767 and 380.
2. Ibid., Vol. 1, vii.
3. Ibid., Vol. 2, 317.
4. Ibid., Vol. 1, iv.
5. McManus and Long, *Survey*, 8.
6. Ward and Davis, *Time Before History*, 37.
7. Claggert and Cable, *Sites*, Vol. 3, 777–778.
8. Ibid., Vol. 3, 778–780.
9. Ibid., Vol. 3, 771–772.
10. John Wilford, "Evidence Hints at Earlier Humans in Americas," *New York Times*, November 18, 2004, 1, Science section.

11. Crist Holden, interview, November 4, 2003.

12. His non-profit organization, Crist Holden Project, educates the public through the show *Shadows from the Past: The Native American Story Told in Stone*, as well as lectures and field tours.

13. Discharge figures indicated the flood occurred during September 1945.

14. Crist Holden has worked under Al Goodyear at the Topper Site on the Savannah River where carbon dating of pre–Clovis findings has put people here far earlier than 12,000 A.D. Dennis Stanford's Solutrean theory is that they came from Iberia. Holden adds: "At Meadowcraft, Pennsylvania, and Cactus Hill, Virginia, findings broke the Clovis first barrier and DNA studies show that the first migration happened between 20,000 and 30,000 years ago. I think it derived from the Solutrean cultures of Europe, and they got here by skirting the continental shelf, which was exposed."

15. George M. Horton, *The Poetical Works of George M. Horton, The Colored Bard of North Carolina*, Documents of the American South, UNC–Chapel Hill, http://docsouth.unc.edu/fpn/hortonpoem/hortonpoem.html, iii–v.

16. Ibid., v–vi.

17. Richard Walser, Introduction in George Moses Horton, *Naked Genius* (Chapel Hill: Chapel Hill Historical Society, 1982).

18. Horton, *Poetical Works*, xiii–xiv.

19. Ibid., 8.

20. Joan Sherman (ed.), *Black Bard of North Carolina: George Moses Horton and His Poetry* (Chapel Hill: North Carolina University Press, 1997), 14.

21. Walser, Introduction, *Naked Genius*.

22. Sherman, *Black Bard*, 1 and 10–11.

23. Ibid., 20–23.

24. Walser, Introduction, *Naked Genius*.

25. Horton, *Poetical Works*, 12.

26. From "On Summer," quoted in Linda Burnham, "Freedom Path," *Independent Weekly*, November 8–14, 2000, 29.

27. "Shadows from the Past: Twenty-five Years of Archeological and Historical Investigations at the B. Everett Jordan Lake Project," U.S. Army Corps of Engineers, Wilmington District, 7. The project was first named for New Hope Creek, but later changed to honor B. Everett Jordan, chair of the Senate Public Works Subcommittee on Flood Control, Rivers, and Harbors, who had run Sellers Mill in Saxapahaw.

28. *The Land Beneath the Waters: The New Hope Valley and Jordan Lake, Chatham County, North Carolina*, 1998, Chatham County Historical Association, DVD.

29. Ibid.

30. Though it is the largest tributary of the Haw, Swain, Holmes and Myers reported that New Hope Creek only turned small grist and saw mills. Swain et al., *Waterpower*, 156.

31. Troy Roberson, phone interview, March 22, 2007.

32. Corey Oakley, phone interview, March 22, 2007.

33. "Jordan Lake Nutrient Rules — And First Fish Kill of 2006," Haw River Assembly, http://www.hawriver.org/index.php?topgroupid=andgroupid=26 (accessed January 5, 2010).

34. "State of the Lake Current Indicators of Pollution Problems at Jordan Lake, 2005–2006," Haw River Assembly, http://home.earthlink.net/~jordan.lake.watch/State%20of%20the%20Lake.pdf (accessed January 4, 2010).

35. *Basinwide Water Quality Plan 2005*, 56. The 2010 draft report from the Division of Water Quality put 57 sections of streams, lake and river in the Haw watershed on the 303(d) list.

36. The first inch of rains carries this runoff to streams; in fact, that first inch brings in 90 percent of it. Rozzelle, presentation.

37. Eric Davis, interview, August 15, 2006.

38. *Basinwide Water Quality Plan 2005*, 322.

39. Mark Vander Borgh, Department of Water Quality, interview, March 20, 2007.

40. *Basinwide Water Quality Plan 2005*, 324.

41. Ibid., 321.

42. Mark Kemp, "Does the Haw Cause Cancer?" *Burlington Times-News*, February 26, 1987, 1A, and Dillon, "River's Cleanup Effort."

43. *River Currents of the Cape Fear*, Cape Fear River Watch newsletter, August 2002. According to Division of Water Quality figures for the Piedmont from 1970 to 2000, the following reductions occurred dissolved oxygen (-7 percent), turbidity (-15 percent) and fecal coliform bacteria (-19 percent). Department of Environment and Natural Resources, http://h20.enr.state.nc/esb/documents/Ambient Indicators.pdf (accessed November 21, 2009).

44. Peter Caldwell, phone interview, March 20, 2007.

45. *Basinwide Water Quality Plan 2005*, A-I-1.

46. Vander Borgh, interview, March 20, 2007.

47. Unfortunately, Mark Vander Borgh adds, bluegreens are not a preferred food. Zebra mussels, for example, will flush them out and prefer green algae or diatoms over them. This may be due to the toxins in blue-greens functioning like protection. Whatever the cause, their relative undesirability contributes to their dominance.

48. Amy Pickle, phone interview, March 15, 2007.

49. Cam McNutt, "Protection of our Watershed and Water Resources," A River Runs Through Us Conference, September 29, 2006, Elon University.

50. Michelle Woolfolk, phone interview, March 23, 2007.

51. Janet MacFall, e-mail message, September 24, 2009.

52. Grady McCallie, e-mail message, August 27, 2009.

53. Stephen Shoaf, interview, March 7, 2006. An Associated Press investigation estimates that 250 million pounds of unused pharmaceuticals are annually flushed into our waters by hospitals and long-term care facilities. While prescriptions are now likely to say "dispose of properly," take-back programs for unused drugs are still in the works; meanwhile antibiotics, birth control hormones, anti-depressants that can be potent at parts per billion or trillion show up in our water supply. Jeff Donn, Martha Mendoza, and Justin Pritchard, AP Investigation: "Scant Advice on Disposal of Meds," http://hosted.ap.org/specials/interactives/pharmawater_site/sept15a.html (accessed February 7, 2010).

54. K. Bradsher and A. Martin, "Cost and Shortages Threaten Farmers' Crucial Tool: Fertilizer," *New York Times*, April 30, 2008, A1-A8; and "Farm Runoff: Down the Mississippi to the Gulf of Mexico," in *Results: Research and Graduate Studies at North Carolina State University*, Vol. 9, No. 2, Summer 2009.

55. Henry David Thoreau, November 12, 1841 journal, *The Writings of Henry D. Thoreau* (Princeton, NJ: Princeton University Press, 1971), 1: 342 cited in France, *Thoreau on Water*, xxi.

56. "Summary of Existing Jordan Water Supply Allocations," Division of Water Resources, http://www.ncwater.org/Permits_and_Registration/Jordan_Water_Supply_Allocations (accessed March 1, 2009).

57. Clarke and King, *Water Atlas*, 52. Mike Magee says a quarter of all deaths are water-related and stresses healthy water supplies' correlation with poverty and lack of education (111). A World Health Organization report he cites suggests paybacks in health and productivity are high for investment in good water systems. Magee, *Healthy Waters*, 32.

58. Kelvin Creech, interview, March 23, 2006.

59. "The History of Drinking Water," Environmental Protection Agency, February 2000, http://www.epa.gov/safewater/consumer/pdf/hist.pdf (accessed April 9, 2010).

60. "Ozone," *Tech Brief Twelve*, December 1999, National Environmental Services Center, West Virginia University Web site http://www.nesc.wvu.edu/ndwc/pdf/OT/TB/TB12_ozone.pdf (accessed April 19, 2010).

61. "Cary/Apex Water Treatment Plant, NC, USA," NRI: Net Resources International, http://www.water-technology.net/projects/cary/cary5.html (accessed April 9, 2010). For comparison, worldwide water use is: agriculture, 70 percent; industry, 22 percent; and domestic, 8 percent. Magee, *Healthy Waters*, 41 and 55.

62. Vicki Westbrook, phone interview, April 18, 2006. A controversy occurs when it drains back to North Cary Waste Water Treatment Plant and heads not to the Haw, but to a stream that feeds the Neuse. This interbasin transfer is taken seriously downstream — especially in Fayetteville. Creech has answers to this: the new wastewater treatment planned for Cary in the next decade has the capacity to pour 18–30 million gallons a day back into the Cape Fear Basin below Buckhorn Dam. Creech also points out that water already flows the other way; Durham pipes an average of 10.5 million gallons a day from the Neuse River into New Hope Creek and then the Haw.

63. Leila Godwin, phone interview, March 28, 2006, and Division of Water Resources Web site. http://www.ncwater.org/ and http://www.ncwater.org/Permits_and_Registration/Jordan_Lake_Water_Supply_Allocation (accessed August 15, 2008).

64. "H2OUSE: Water Saver Home," California Urban Water Conservation Council, http://www.h2ouse.org/tour/index.cfm. (accessed April 19, 2010).

65. Clarke and King, *Water Atlas*, 48, 49, 22.

66. Ibid., 25.

67. Ibid., 33.

68. Ibid., 78 and 35.

69. Magee, *Healthy Waters*, 109.

70. Clarke and King, *Water Atlas*, 83.

71. "New Kerala Line: Coke Bad, Pepsi Good," *Tehelka*, April 7, 2007, http://www.tehelka.com/story_main28.zp+?filename=Ne070407New_Kerala.asp While the closing did not solve the water supply problem, it did ameliorate it.

72. Maude Barlow, "Water as Commodity — The Wrong Prescription," *Backgrounder*, Food First Institute for Food and Development Policy, Summer 2001, Vol. 7, No. 3, 3.

73. Leopold, *View*, 1.

Chapter 6

1. Chilton, *Historical Atlas*, 4.2.

2. Laurie Craft, "Paddling with Purpose," *Pendulum*, Elon University, March 8, 2007, 15.

3. John Chappell, "Robbins Paddlers Begin Journey Down Bear Creek, Deep River," *ThePilot.com*, http://www.thepilot.com/stories/20070317/news/local/20070317Paddlers.html (accessed March 27, 2007).

4. Matt Steible, e-mail message, April 8, 2007.

5. Matt Steible, "Is This the Finish or the Start?" *Voice of the Haw*, Haw River Assembly, April 2007, 7.

6. John Hairr, *From Mermaid's Point to Raccoon Falls: A Guide to the Upper Cape Fear River* (Erwin, NC: Averasboro Press, 1996), 11–12.

7. Malcolm Ross, *The Cape Fear* (New York: Holt, Rinehart and Winston, 1965), 195.

8. Hadley et al., *Chatham*, 23–24.

9. Ross, *Cape Fear*, 290.

10. Hairr, *From Mermaid's Point*, 23–24.

11. Ross, *Cape Fear*, 195–196.

12. Ibid.

13. Ibid., 272.

14. Nelson, interview.

15. Hairr, *From Mermaid's Point*, 13.

16. Environment Sciences Branch, *Basinwide Assessment Report*, June 1999, NC Department of Environment and Natural Resources, Division of Water Quality, http://h2o.enr.state.nc.us/esb/CPF1999.pdf.

17. Hairr, *From Mermaid's Point*, 9, and Smyth, *Tour*, 169.

18. William P. Cumming, *Mapping the North Carolina Coast* (Raleigh, NC: Division of Archives and History, NC Department of Cultural Resources, 1988), 86–87.

19. This is the name given by settlers of 1664. William P. Cumming, *The Southeast in Early Maps* (Chapel Hill: University of North Carolina Press), 1998, 15.

20. Herman Moll's 1708 Map in Powell, *Four Centuries*, 71. Prof. George Troxler of Elon University writes that the Earl of Clarendon, Edward Hyde (1609–1674), was one of the eight lord proprietors and the king's first minister. E-mail message, March 1, 2010.

21. From Barnwell's 1721 map, reproduced on a CD of early maps compiled by Carole Troxler.

22. Hairr, *From Mermaid's Point*, 31.

23. The ferry is on Elswell Ferry Road in Bladen County; the brief ride is well worth the experience.

24. "Cape Fear River Basin," Office of Environmental Education, North Carolina Department of Environment and Natural Resources, http://www.p2pays.org/ref/02/01216.pdf (accessed April 20, 2010).

25. "Rediscovering the River," Special Section, Fayetteville *Observer-Times*, 1999.

26. Hadley et al., *Chatham*, 157.

27. David Cecelski, "Erwin Mill No. 2 in Erwin, NC," *News and Observer*, December 9, 2001, Old West Durham Neighborhood Association, http://www.owdna.org/History/history18.htm (accessed February 7, 2010).

28. Powell, *Four Centuries*, 136.

29. "Our View: Rising Levels of C8 in Bladen County Is a Regional Concern," *Fayetteville Observer*, April 2, 2008, http://www.fayobserver.com/article?id=290185 (accessed July 2008).

30. Press release, November 29, 2005, North Carolina C8 Working Group, http://www.c8nc.org/pages/5/index.htm (accessed July 9, 2008).

31. *Basinwide Water Quality Plan 2005*, 277.

32. "Frontline Farmers and the Environmental Defense Fund Working to Turn Hog Waste into a Saleable Fertilizer," *Solutions*, Environmental Defense Fund, September 2008, 6.

33. Venita Jenkins, "Smithfield Workers Approve Union," *Fayetteville Observer*, December 12, 2008, http://www.fayobsever.com/article?id=312956 (accessed May 3, 2009).

34. William L. Murray, education and outreach with Cape Fear River Watch, phone interview, July 11, 2007.

35. The amount is for 2004. Louis Zeller, clean air campaign coordinator, BREDL, Blue Ridge Environmental Defense League, letter to Scott Miller, January 27, 2004, http://www.bredl.org/air/IP-Riegelwood012704.htm (accessed April 13, 2008).

36. Hairr, *From Mermaid's Point*, 99–100.

37. Beyer, *Time Before Man*, 177.

38. Sid Perkins, "Ice Age Ends Smashingly," *Science News*, June 2, 2007, http://www.restorationsystems.com/news/index/asp?ID=26 (accessed May 15, 2008).

39. The whole Mississippi Valley from Missouri down was once a vast ocean inlet filled in in the same way. *Surrounded by Walls*, DVD, Walter Williams, 2008, Dreamsite Productions.

40. Beyer, *Time Before Man*, 139–143.

41. Timothy B. Tyron, "Ghosts of 1898," November 17, 2006, *News and Observer*, www.Newsobserver.com/1370/v-print/story/511596.html (accessed March 1, 2007).

42. Ibid.

43. Ross, *Cape Fear*, 311.

44. Swanton (1946: 103), cited in Thomas E. Ross, *American Indians in North Carolina: Geographic Interpretations* (Southern Pines, NC: Karo Hollow Press, 1999), 26.

45. Ibid.

46. Fenn and Wood, *Natives and Newcomers*, 12–13.

47. Powell, *Four Centuries*, 81–83.

48. David McNaught and T. Edward Nickens, *Horizon 2100: Aggressive Conservation for North Carolina's Future*, Environmental Defense, http://www.edf.org/documents/2777_nchorizon2100.pdf (accessed February 7, 2010), 9.

49. Ibid., 5–31.

Bibliography

Berry, Thomas. *The Great Work.* New York: Three Rivers Press, 2000.

Besse, Dan. *See No Evil: Why Our Environmental Laws Aren't Being Enforced.* Conservation Council of North Carolina, 2002.

Beyer, Fred. *North Carolina: The Years Before Man.* Durham: Carolina Academic Press, 1991.

Bulla, Ben. "Saxapahaw Cotton Mills, The First Hundred Years. *The Cotton History Review* 2, (1961): 132–138.

Butler, Lindley S. *Rockingham County: A Brief History. Raleigh*: Division of Archives and History, NC Department of Cultural Resources, 1982.

Cape Fear River Basinwide Water Quality Plan. Raleigh: Division of Water Quality, NC Department of Environment and Natural Resources, 2000 and 2005.

Chatham Historical Journal. Pittsboro, NC: Chatham County Historical Association.

Chilton, Mark. *An Historical Atlas of the Haw River.* Carrboro, NC: Self-published, 2008.

Cook, David. *Piedmont Almanac.* Raleigh: Mystic Crow, 2001.

Coomans, Ray J. *Alamance County Natural Heritage Inventory.* Raleigh: Department of Environment and Natural Resources, 2002.

Cumming, William. *The Southeast in Early Maps.* Chapel Hill: University of North Carolina Press, 1998.

Davis, R.P. Stephen, Jr., "John Lawson and the Native Peoples of Carolina," in *Lawson's Legacy: Nature Writing and North Carolina, 1701–2001,* ed. Robert Anthony, North Caroliniana Society (in press).

DeNatale, Douglas. "History of Bynum: 1872–1979," *Chatham Historical Journal* 1 No. 11, (March 1998): 1–8.

_____. *Traditional Culture and Community in a Piedmont Textile Mill Village.* Thesis in Folklore (PhD diss., University of North Carolina at Chapel Hill, 1980).

Dickens, Roy, Trawick Ward and Stephen Davis. *The Siouan Project: Seasons I and II.* Chapel Hill: Research Laboratories of Anthropology, UNC, Monograph Series No. 1. 1987.

Ferguson, Paul, *Paddling in Eastern North Carolina.* Raleigh: Pocosin Press, 2002.

Freeland, Nan, Ted Outwater, Raelyn Grasso and the staff of Clean Water for North Carolina. *A Citizen's Toolkit for Protecting Your Environmental Rights.* Clean Water for North Carolina. No date.

Hadley, Wade. "Fishstreams in Chatham Over 100 Years Ago." *Chatham County Journal* 2 No. 2 (October 1989): 1–3.

_____. "Water-Powered Grist Mills in Chatham County, North Carolina as of 1880." *Chatham Historical Journal,* 4 #1 (January 1991), 1–3.

Hadley, Wade, Doris Horton, and Nell Strowd. *Chatham County 1771–1971.* Lillington, NC: Edwards Brothers, 1997.

Hairr, John. *From Mermaid's Point to Raccoon Falls: A Guide to the Upper Cape Fear River.* Erwin, NC: Averasboro, 1996.

Hall, J., J. Leloudis, R. Korstad, M. Murphy, L. Jones, C. Daly. *Like a Family: The Making of a Southern Cotton Mill World.* Chapel Hill: University of North Carolina Press, 1987.

Hall, Stephen and Boyer, Marjorie. *Inventory of the Natural Areas and Wildlife Habitats of Chatham County, North Carolina.* Triangle Land Conservancy and County of Chatham, June 2006.

Haw River Assembly Stream Steward Handbook, 2002.

Haw River Watch Macroinvertebrate Manual, Haw River Assembly, 1999.

Horton, J. Wright and Victor A. Zullo, eds. *The Geology of the Carolinas.* Knoxville: Geological Society 50th Annual Volume, University of Tennessee, 1991.

Hughes, Fred. *Guilford County, North Carolina: A Map Supplement.* Jamestown, NC: Custom House, 1988.

_____. *Guilford County, North Carolina Historical Documentation Map,* 1980.

Hughes, Julian. *Development of the Textile Industry in Alamance County: Evolution of Warp and Welt in Alamance.* Burlington, NC: Burlington Letter Shop, 1965.

Justice, William, and C. Ritchie Bell. *Wild Flowers of North Carolina.* Chapel Hill: University of North Carolina Press, 1979.

Knauff, Gail, and Bob Knauff. *Fabric of a Community: The Story of Haw River, North Carolina.* Haw River, NC: Haw River Historical Association, 1996.

Lawson, John. ed. Hugh Talmadge. *A New Voyage to Carolina.* Chapel Hill: University of North Carolina Press, 1967.

Lederer, John. ed. Cumming, William P., and Douglas Rights. *The Discoveries of John Lederer.* Charlottesville: University of Virginia Press, 1958.

Lefler, Hugh and Paul Wager, ed. *Orange County—1752–1952.* Chapel Hill: Orange Printshop, 1953.

Leopold, Luna. *A View of the River.* Cambridge: Harvard University Press, 1994.

McManus, Jane Madeline, and Ann Marie Long, with Report on Pottery of Alamance County by Linda Carnes. *Alamance County Archeological Survey Project, Alamance County, North Carolina.* Chapel Hill: Research Laboratories of Anthropology, 1986.

McNaught, David and T. Edwards Nickens. *Horizon 2100: Aggressive Conservation for North Carolina's Future.* Environmental Defense, 2003.

Mobley, Joe A., ed. *The Way We Lived.* Chapel Hill: University of North Carolina Press, 2003.

Nealy, William. *Kayak: The New Frontier: A Manual of Intermediate and Advanced Whitewater Technique.* Birmingham: Menasha Ridge Press, 2007.

_____. *White Water Home Companion.* Birmingham: Menasha Ridge Press, 1982.

North Carolina Conservation Network http://www.ncconservationnetwork.org/issues

North Carolina Department of the Environment and Natural Resources, www.enr.state.nc.us and http://www.ncstormwater.org/pages/stormwater_faqs page.html

Outwater, Alice. *Water: A Natural History.* New York: HarperCollins, 1996.

"A River Runs Through It." Haw River Assembly. http://www.hawriver.org/library/publications/Alice%20Outwater%20Talk%20on%20Haw%20River.pdf

Pielou, E.C. *Fresh Water.* Chicago: University of Chicago Press, 1998.

Poole, Herbert. "Bygone Mills on the Haw River in Chatham County." *Chatham Historical Journal* 8 #3 (November 1995) 1–4.

Redington, Charles B. *Plants in Wetlands.* Dubuque, IA: Kendall and Hunt, 1994.

Robinson, Blackwell, and Alexander Stoesen. *History of Guilford County North Carolina, Vol. I.* Guilford County Bicentennial Forum, 1980.

Rodenburgh, Charles Dyson, ed. *The Heritage of Rockingham County, North Carolina.* Winston-Salem: Rockingham County Historical Society, Inc., 1983.

Rogers, John J.W. *History and Environment of North Carolina's Piedmont: Evolution of a Value-Added Society.* Chapel Hill: J. Rogers, 1999.

Ross, Malcolm. *The Cape Fear.* New York: Holt, Rinehart and Winston, 1965.

The Shuttle, Newsletter of the Haw River Historical Association.

Simpkins, Daniel, Gary Petherick, and Roy Dickens. *First Phase Investigations of Late Aboriginal Settlements Systems in the Eno, Haw and Dan River Drainages.* Chapel Hill: University of North Carolina at Chapel Hill–Research Laboratories of Anthropology, October, 1985.

Simpkins, Daniel, and Gary Petherick [Roy Dickens, Primary]. *Second Phase Investigations of Late Aboriginal Systems in the Eno, Haw and Dan River Drainages, North Carolina.* Chapel Hill: University of NC at Chapel Hill: Research Laboratory Anthropology, October, 1986.

Steingraber, Sandra. *Living Downstream: A Scientist's Personal Investigation of Cancer and the Environment.* New York: Vintage Books, 1997.

Swain, Geo. F., J.A. Holmes, and E.W. Myers. *Papers on the Waterpower in North Carolina: A Preliminary Report.* Raleigh: Guy V. Barnes, Public Printer, 1899.

Tekiela, Stan. *Trees of the Carolinas: Field Guide.* Cambridge, MI: Adventure Publications, 2007.

Terres, John K. *From Laurel Hill to Siler's Bog.* Chapel Hill: University of North Carolina Press, 1993.

Textile Trail, newsletter of the Textile Heritage Museum, Glencoe, NC.

Troxler, Carole Watterson. *Pyle's Defeat: Deception at the Racepath.* Graham, NC: Alamance County Historical Assn., 2003.

Troxler, Carole Watterson, and William Vincent. *Shuttle and Plow: A History of Alamance County.* Alamance County, NC: Alamance County Historical Association, 1999.

Tullos, Allen. *The Habits of Industry: White Culture and the Transformation of the Carolina Piedmont.* Chapel Hill: University of North Carolina Press, 1989.

Voice of the River, Newsletter of the Haw River Assembly, Bynum, NC. www.hawriver.org

Ward, T. Trawick, and R.P. Stephen Davis, Jr. *Time Before History: The Archeology of North Carolina.* Chapel Hill: University of North Carolina Press, 1999.

Wildlife in North Carolina, monthly magazine, 1710 Mail Service Center, Raleigh, NC, 27699 or www.ncwildstore.com.

Index

Numbers in ***bold italics*** indicate pages with illustrations.

air pollution 82–84
Alamance, Battle of 104–105
Alamance Cotton Mill 105 ***106***
Alamance County Parks and Recreation 131, 209*n*141
Alamance Creek 95, 99, 104–106, 120, 141
algae 142, 181–188
Alston Quarter 126–127, 130
Altamahaw 43, 45, 46–50, 141
Altamahaw Cotton Mill 48–50, 106, 210*n*34
American Canoe Association 21
Andrews, Mark 62
antibiotics 70–71, 188
"Aragon Mill" 53–54
Archaic Period 141–142, 177, 178–179
Avalon Terrane 39, 40

Back Creek 82, 84–88, 132
Baker, Ashby 100, 104
Baker, Brian 79, 81, 82
Baker, Minnie 100
Baldwin Mill 133–134
baptism 14, 60, ***61***, 91
barn swallows 148
Barry, Kathy 57–58
Bartram, John and William 199
baseball 101, 102, 113, 214*n*270
basin *see* river basin
Bass Mountain *see* Cane Creek Mountains
Beaver Creek 18, 105
beavers ***6***, 8–11, ***9***, 111, 207*n*13
Beggs, Patrick 34, 124
Bellemont 106
Benaja Creek 33
Benjamin Vineyards 130
benthic system 116–118
Berry, Thomas 20–21
best management practices (BMP) 146, 183, 217*n*44
biological oxygen demand (BOD) 183, 185, 209*n*93, 220*n*43
bio-solids 67–68

birds 15; barn swallows 148; cormorants 111; eagles 45, ***137***, 165, 181; heron 11, 110, 135, 148, ***193***, 194; kingfisher ***13***, 35; osprey 45, 109, ***110***, 165; turkey vultures 11, 110
Black Creek formation 199
Black River 198–199
Blakeley, Capt. Johnston 136, 138
Bland's Mill 167
bluegreen algae 185–186, 194, 220*n*47
Bluegreen Corporation 22, 143, 147
boats 75–76, 100, ***102***, 148, 150, 196, ***197***, ***199***–200
BOD *see* biological oxygen demand
bottled water 69
bottomlands *see* floodplains
Brady, Chuck 149, 192–193
Brenner, Bob 43, 129, 167
Brewer, W.J. 126–127
Briar Chapel 143, 144, 147
bridges 9, 12, 36–37, 72, ***73***, ***76***, 82, 102, ***103***, 106, 108, 111, 115, 131, 132, 134, 136, 139, 140, 147, 153–***154***, 159–160, 167, 217*n*52, 218*n*64
Brooks Bridge 41, 43, 45
Brooks Creek 141, 152
brown lung disease 158
Brown's Mill 164
Brueckner, Bob 129–131, 133–135
Buckhorn Dam 192, 197
Buckhorn Falls 195, 196
buffalo 17, 86
Buffalo Creeks 18, 20, 21, 76, 86, 173
buffer *see* riparian buffer
Bulla, Ben 111, 113
Burch Bridge 53
Bynum 45, 70, 140, 152–163, 173; General Store 159–161, ***162***; mills 153–159
Bynum, Carney 153, 155

Bynum, Luke 153
Bynum, Luther 155
Bynum Manufacturing Mill *see* Odell Manufacturing Co.

Caldwell, Peter 184
canals ***97***, 102
cancer 69–70
Candy Creek 34
Cane Creek (river left) 27, 29, 119, 131, 173
Cane Creek (river right) 16, 20, 120, 131
Cane Creek Mountains 41, 108, 120, 122–123, 131–133
Cape Fear and Deep River Navigation Co. 195–196
Cape Fear Arch 41
Cape Fear estuary 199
Cape Fear Indians 200
Cape Fear River 7, 16, 30, ***90***, 91, 116, 196–202, ***199***, ***203***, 205
Cape Fear River Basin Water Quality Plan 145–146, 183
Cape Fear River Watch 200, 205
Cape Fear shiner 46, 143, 165, 170
Carolina bays 198–199
Carolina Canoe Club 129, 148, 150
Carolina Mill 60, ***79***, 106
Carolina Slate Belt 40
Carteret, George 71
Carteret, John 71
Cary Apex Water Treatment Plant 188–190
Catawba River 40
Catawbas 93, 94
Cedar Cliff Mill 108–109
Chapel Ridge 143–147
Chavis, William 48
Chicken Bridge 94, 129, 134, 136, 148
child labor 49, 58, 101, 218*n*74
Chilton, Mark 107, 110, 167
Chiosso, Elaine 124, 142–145, 160, ***195***
Christmas fern 23

225

Church Street *see* Sandy Cross Road
Churton, William 34
Civil War 13, 27, 98–99, 106, 154–155, 181, 196, 198, 200
Clapp's Mill 105
clean up programs 131, 205
Clean Water Act 1972 75, 82, 116, 184
Clean Water Management Trust Fund 173, 219*n*157
Clovis ***177–179***, 220*n*14
Coffin, Addison 19
Coffin, Katherine 19–20
Coffin, Levi 19–20
Coffin, Vestal 19
Collett, John 34
company store 58, 100, 106, 114–115, 157, 211*n*57
conservation (water) 30–33, 190–191
Contact Era 89–95, 179, 200, 213*n*218
Copland, James III (Jim) 62–65
Copland, James IV (Jason) 61, 65
Copland, James R. 61, 63
Copland Fabrics 61–65, **62**, 79, cormorants 111
Cornwallis, General 25, 28, 37, 51, 72, 78, 105, 195; *see also* Revolution
cotton 27, 55, 76, 100, 102, 112, 114, 133, 155, 156, 216*n*352
Cotton Mill Colic No. 3 54
Cowan, Lynn 56–57
crappie 182, 194
Creech, Kelvin 188–190
Cross Creek *see* Fayetteville
Crossen, Cynthia 116–118, 146*n*45 **195**, 218*n*53
Cunningham Road 21, 27, 29, 33

dams 41, 43–46, **47**, 53, 59, 60, 100, 102, **103**, 113, 115–116, 119, 136, **152**, 173, 181–188, 192, 197; effects on river 45–46, 102, 108, 178, 181–188, 197, 208*n*59; removal 38, 46, 78, 102, 197; uses 45–46; *see also* hydropower; safety
Dan River 7, 26–27, 40, 63, 85
Dark's Mill 136, 149
Davidson, Berry 48–49
Davis, Eric 66–69, 183
Davis, Stephen 89, 93–95, 141–142
dead zones 188
Deep River 41, 46, 82, 167, 193, 195–196, 198, 205
Deininger, Catherine ***ii***, 170, 172–173
The Depression 63, 113, 156–157
development 27, 29, 52, 109, 119–120, 124–125, 142–148, 165–166, 172, 183–184, 200; Briar Chapel 143, 144, 147; Chapel Ridge 143–147; river impact 22, 142–147, 182–184, 185; sustainable 29,

144; *see also* Jordan Lake; sediment
Dillon, Daniel 18
dinosaurs 40, 174, 199
dioxin 70, 83, 198
diseases (waterborne) 69–71, 188, 221*n*57; *see also* pollution; water treatment
Division of Water Quality 34, 182–184, 187
Dixon, Hoover 113–114, 215*n*311
Doll, Jason 34, 35
dragonflies 136, **138**
drainage area *see* river basin
drinking water *see* water treatment
drought 22, 29–33, 109
drugs 70–71, 188, 220*n*53
Drummond Creek 141
Dry Creek 139, 142–147, **146**, 151, **195**
DuBose, Howard 129, 150
Durham, Eula 157
Durham, Frank 154
Durham, Vernon 156–157
Durham, Willis 138

eagles 45, **137**, 165, 181
easements 22, 53, 107, 164, 208*n*45
ecological education 25–26, 162–163, 165, 219*n*131
ecological perspectives 11, 20–21, 25, 134–135, 163, 165–166
eddy 130, 151, 164
Elizabethtown 198–199
enforcement 145–147
Environmental Impact Assessment (EIA) 143
Environmental Protection Agency (EPA) 88, 128, 172, 198, 212*n*173; standards 70, 128, 198
Erwin 198
eutropic conditions 181–188
Eversfield Road 12, 207*n*2
excess nutrients *see* eutropic conditions; nitrogen; phosphorous

fall zone 41
Falls Neuse Manufacturing Co. 99–100
Fanning, Edmund 104–105
farm animals 51–52, 198
farming 29, 52–53, 110, 113, 126–127, 208*n*20, 208*n*59; conservation 27, 28, 52–53; river impact 26–27, 28, 29, 52–53; versus mill work 54, 101, 111, 113, 156
Farrell, Duard 102–104
Faucette, Don and Paul 57, 59
Fayetteville 16, 82, 198
Featherstone, Brenda 148, 149, 151, 152
Featherstone, Lynn 148–153, 167, 169, 207*n*1
fecal coliform 34, 88, 220*n*43
Ferguson, Paul 43, 45, 129, 130, 167
ferries 82, 130, 153, 164, 167, 197, 198, ***199***, 200

fertilizer 52, 69, 184, 188, 198
fines 83, 147, 217*n*51
fish 3, 11, 46, 49, 59, 143, 165, 201; Cape Fear shiner 46, 143, 165, 170; crappie 182, 194; migration 46, 201; stripe bass ***175***, 181–182; sturgeon 3, 167; *see also* locks and dams
fish kills 142, 182, 183, 184, 185, 187, 188
fishing 46, 59, 60, 103, 115, 133, 142, 149, 159, 160, 181–182, 197, 201; *see also* weirs
floodplains 13, 15, 17, 21, 22, 23, 26, 109, 136, 149, 178, 216*n*376
floods 8, 62, 101, 123, 127, 136, 154–155, 159, 178, 196, 197
fords 14, 28, 36–37, 51, 71, 80, 91, 101, 104–105, 108–109, 111, 153–154, 167
forest 42, 123–124, 131, 164–165, 198; historical 16, 26–27, 198; tree varieties 21, 107, 131, 141, 194
Fort Fisher 200
Fort Snug **53**
Francisco, Don 75, 142–143, 217*n*32
frontier 46, 48, 105
Fusion Party 200

Galbreath Bridge 82
Gant, Jesse 80
Gant, John Q. 48–49, 80
Gant, Roger 48–50, 51
geology 15, 26–27, 32, 38–41, 72, 139, ***140***, 164, 167, 197–198, 209*n*116, 209*n*127, 215*n*340; Avalon Terrane 39, 40; Black Creek formation 199; capture 40, 41; Carolina Slate Belt 40; Pangaea 40, 41, 167; rivers and land creation 39, 41, 199; *see also* Cape Fear Arch; Triassic basin; volcanoes
Gerringer Mill 51
Glen Raven 48–50, 80
Glencoe 45, 53–60, 81, 106, 120, 173
global warming 33, 198, 207*n*7
Goat Island 65, 78, 80
gold mining 40, 84, 98
Gomolski, Susanne 210*n*1
Granite Mill 72, ***73***, ***76***, 81, 106
Green, Holt 57
Green, Walter G. 57
Greene, Gen. Nathanael 28, 37–38, 39, 51, 105, 195
Greensboro 29–33
groundwater 7, 128, 207*n*3
Grove Winery 41, 42
Guilford Courthouse, Battle of 28, 105
Gun Creek 104
Guthrie, Nancy 173, 219*n*157

Hadley-Williams Mill 167
Harkless 160, 164
haw (shrub) 24

Haw Creek 103, 108
Haw River (town) 71–78, **76**, 81
Haw River Assembly (HRA) 24, 69, 87, 142–145, 148–150, 152, 161–163, 183–184, 192, 205; Festival 161–163; Learning Celebration 150, 163, 187; River Watch 116–118 **117**, 146–147, 162; Stream Steward Program 170, 172–173
Haw River Canoe and Kayak Co. 107, 118, 210n1
Haw River Museum 75, 81
Haw River Paddle Trail 81, 119, 120, 171
Haw River State Park 21, 22, 23–26, 194
Haw River Trail 79, 81, 109, 118, 120, 131, 164
Hawfields 82, 89, 91, 110, 213n196
Haywood 97, 194–196
Hazel, Forest 85–**86**
headwaters 5, **7**, 84–85
helical motion 14–15
Hermitage 95, 96, 98–99
heron 11, 110, 135, 148, **193**, 194
Herring, Raymond 100–102
High Falls Manufacturing 61
High Rock 33, 36–38
Hobbs, Grimsley, Jr. 134
Hobbs, Grimsley T., Sr. 38, 134
Hobbs, R.J. Mendenhall 134
Holden, Crist **177**–180, 220n14
Holland, Michael 78, 118–120, 170, 205, 216n357
Holt, Bruce 120–123, **121**, 163
Holt, Charles 74
Holt, Edwin Michael 48–49, 54, 72, 98–99, 105–**106**, 113
Holt, James 54, 79
Holt, Lawrence 106
Holt, Robert 54, 56
Holt, Thomas 72–73, 74, 81, 105
Holt, V. William 54, 79
Holt Family 78, 100, 105, 106
honeysuckle: American 25, 33; Japanese 25
hormone disruption 68, 70; *see also* dioxin
Horsepen Creek 16, 18, 19
Horton, George Moses 180–181
Hunter's Ferry 108
Husband, Herman 104
hydro power 45, 48–49, 56, 57, 58–59, 107, 113, 115–116, 156, 158, 167, 192
hydrology 14–15
hymns (of river) 13, 20, 60, 61
hyporheic zone 109

ice 104, 111, 113
impaired streams *see* 303(d) list
impervious surface 123–**124**, **125**, 143, 172; *see also* runoff
Indians *see* Native peoples
interbasin transfer 221n62
invasive plants 25
Iola Mill 51
Ironworks (Speedwell) 25, 27–28

Iroquois 93, 94
Iseley, Jane 52–53, 173

Jacob, Joe 107, 163–166, 194, **195**, 210n1
Jarrett, Richard 75–**77**
Jeanne's Falls 133
Jones, Clyde 159, 160, **161**, 162
Jones, Edward and Mary Curtis 136, 138
Jones, Louise 155
Jordan, B. Everett 111, 113–116, 220n27
Jordan, Ben 114–116, 216n352
Jordan, John 45, 114–116
Jordan Dam 45, 113, 116, 176, 177, 181, 185, 188, 192, **194**
Jordan Lake 27, 28, 29, 113, 142, 164, 174–194, **175**, **176**; history 116, 149, 174, 180–181; native sites 141, 174, 175, 177–179; rules 118, 183–184, 187–188; water quality 34, 118, 142, 149, 181–191
Juanita Mills 61

Kahn, Si 53–54
Kaufmann, Wallace 141
Kessel, Louise 163
Kinady Creek *see* Candy
kingfisher **13**, 35
Kirk-Holden War 73, 80
Knauff, Gail 71, 75, **77**
Ku Klux Klan 33, 73, 80

labor *see* mill workers; unions
Lake MacIntosh 106
landfill 119, 127–128, 130
Laws, Tony 81
Lawson, John 82, 89, 91–93, 95, 213n195; Haw River 91; native culture 91–92
leacheate 128, 216n357
Lederer, John 89–91
Lenox Castle 38
Lieberman, Max 74–75
Lindley Mill 133
Linville Road 8, 9, 10
Little River 198
Little Saxapahaw Falls 131
Little Troublesome Creek 34–35
locks and dams 195–196, 197
London, Henry Armond 134, 154
Long, Tom 80
Love's Mill 134
Lutterloh Mill 136, 149

MacFall, Janet 51, 52, 71, 109, 120, 173, 187
Magnuson, Tom 36–37, 46, 48, 105, 108–109, 111, 209n99
maps ii, **6**, **13**, 17, 33, 34, **44**, 89, **90**, **96**, **97**, 130, **137**, **175**, **193**, 197, **203**
Markatos, Cathy 138
Markatos, Jerry 138, 145–146
maroon dogwood 24
marsh *see* wetlands
Martin, Alexander 17

Martin, James 17
Martin, William 17
Mary's Creek 130
Massey, Donnie, Garland and Thelma 59–61
May apple 23
McBain Mill 133
McCain, Joseph 38
McCallie, Grady 30, 187
McCrackin, Samuel 12
McNutt, Cam 187
Meadow Creek 108
meanders 14–**15**
Mears Fork 21, 173
medical waste *see* Stericycle; drugs
mercury 82–84
Mermaid's Point 193, 196
micro-invertibrates 24, 46, 116–118, **117**, 162
migration 46, 201
Mill Creek 41
mill management: closings 54, 64–65, 101, 116, 159; management styles 57, 58, 113, 114–116; mill owner control 54, 58, 59, 73–74, 100, 101, 103, 104, **106**, 113, 155, 157–158, 211n57, 219n109; profits 49, 58, 114; *see also* The Depression; mill town/services; trade policy
mill pollution 76–77, 82, 107, 135, 149, 161, 184, 218n101
mill race 49, **50**, 79, 108, 111–112, 167
mill sites 107, 115, 167
mill towns **76**, 134; church 60, 114, 155, 158–159; community 56–61, 101, 158–160; family life 57, 59–61, 101; gardens 55, 57; housing **55**–56, 100, 103, 113, 157; recreation 57, 58, 75–76, 106, 113, 158–159; schools 58, 73, 101; services 58, 106, 113; village design 54, 55, 56, 57, 155; *see also* Altamahaw; Bynum; company store; Glencoe; Haw River; Saxapahaw; Swepsonville
mill workers 56, 60, 75, 101, 112, 155, 166; credit terms 114–115, 155; health 49, 59, 158; hours 58, 73, 100, 156, 157; rights (lack of) 59, 73, 106, 155–159; termination 115, 157, 158; wages 49, 54, 58, 60, 64, 100, 101, 106, 114, 155–157; work conditions 54, 56, 58, 101, 114, 158, 162; *see also* child labor; mill management; strikes; unions
mills **18**, 27, 28, 37–38, 42, 48–50, 53–65, 71–75, **72**, 95–96, 99–104, 105–108, 111–116, 126, 130–131, 133–134, 136, 138, 140, **154**–159, 167, 211n69, 212n116, 216n394
Minifee, George 126–127, 130
Mitchum site 89, 93–95, 148–149
mitigation 21–22
Moore, Edgar 156-7

Moore, Kevin 28–29, 34
Moore's Mill 167
Morgan Branch 110, 141
Morrow Mill 131
Mote's Creek 111, 130
Mountain to Sea Trail 27, 118–119
Murphey, Archibald 38, 95–99, *97*, 194; *see also* river transportation
mussels 46, 141, 170

Nall, Jerrie 57, 79
National Union of Textile Workers (NUTW) 74
Native peoples 48, 50, 84–86, 89–95, 108, 120–123, 131–132, 174–180, 194, 200, 213n218; cultural values 85, 91–93, 121–123, 132; food 90, 91, 92, 131–132, 141–142, 149, 174, 177–178, 200; houses 84–**86**, 95, 214n228; points 141–142, ***177–179***; pottery 93, 132, 177; settlements 85–86, 94–95, 132, 141, 149, 174, ***175–***179; spirituality 85, 120–123, 132; trade 11, 89, 93–94, 179, 213n226; *see also* Archaic; Clovis; Contact Era; Occaneechi; Paleo; Shakori; Woodland
native plants 17
nature inventories 141, 170
naval stores 198
navigation *see* river transportation
Nealy, William 129, 150, 167–***171***
Nelson, Barry 15, 39–41, 196
Neuse River 173, 183, 200, 221n62
New Hope Creek 40, 41, ***176***, 180–181, 183–184, 192
Newlin, John 111–112, 133
nitrogen 8, 34, 52, 68, 109, 118, 181–188, 207n7; *see also* eutropic conditions; Jordan Lake
non-point source 35, 123–126, 183–184, 187–188
North Carolina Coastal Federation 200
North Carolina Conservation Network 30, 187, 205
Northeast Cape Fear River 199
Northeast Park 21, 120
nuclear power 30, ***31***, 197

Oak Ridge Academy 19
Oakley, Corey 182
Occaneechi 85–86, 94
Odell Manufacturing Company, J.M. 155–156
Oenuchs 90–91
Old Mill of Guilford *see* Dillon's Mill
119 bypass 86–88
Orange Water and Sewer Authority (OWASA) 29, 131
ordinances 145–147, 172
osprey 45, 109, ***110***, 165
Ossipee Mill 48–49, 81
Outlaw, Wyatt 80

oxygen *see* biological oxygen demand

Pace's Mill 138–140
paddle access 47, 51, 53, 78, 79, 81, 82, 102, 106–107, 159, 164, 209n141
paddling 148–152, 164, 166
Paleo Era 170, 174–***175***, ***177***–180
Pangaea 40, 41, 167
Pardo, Juan 89, 122
Partin, Jerry 159–161, 162
Patrick Cunningham Mill 33
The Patriot 25, 28
paw paw 24
Pepper Road 8, 10
pesticides 70–71, 143, 211n97
Peterson, Judith 139–141
pfisteria 200
pH factor 117, 183, 186–187, 189
phosphorus 34, 52, 109, 118, 172, 181–188
Pickle, Amy 186–187
Piedmont Land Conservancy 21, 205
Pine Needle LNG Plant 12
pirates 200
plants 22–25, 108, 122–123, 133, 136, 141; Christmas fern 23; May apple 23; Haw (shrub) 24; honeysuckle American 25, 33; honeysuckle Japanese 25; invasives 25; maroon dogwood 24; native 17; paw paw 24; poison ivy 24–25; river oats 108; skunk cabbage 12; willow 108
point bars ***15***
point source 35, 69, 123, 184; *see also* Clean Water Act
poison ivy 24–25
Pokeberry Creek 147, 164, 218n52
pollution 30–32, 188, 215n311; *see also* air pollution; algae; dead zones; eutropic conditions; farm animals; farming; fecal coliform; fertilizer; leacheate; mercury; mill pollution; non-point sources; pesticides; point sources; runoff; sediment; septic systems; synthetic chemicals; trash; wastewater treatment sewers
Poole, Charlie 74, ***75***
Powell's Mill 152
Preservation of North Carolina 54, 56
put-ins *see* paddle access
Pyle's Massacre 105

Quakers *see* Society of Friends

racism 48, 80, 85, 200; *see also* segregation
railroads 13, 72, 99–100; river impact 197; wrecks 75
rain gardens 124, 172
ramblers' paths 10
rapids 78, 107, 119, 131, 136, 143, 149, 150–151, 167, 169–170

Raven Rock State Park 197
Reconstruction 80
recycling 119, 128, 131, 198
Reedy Fork Creek 16, 17, 18, 19, 33, 37, 48, 51, 82, 120
Reeves, Jimmy, George, Royce 95, 178
Regulators 104–105, 111, 130
Reidsville Lake reservoir 29
religion on rivers 14
rescue 43, 45, 129, 141
reservoir 29, 32, 80, 131; *see also* Jordan Lake
restoration 21–22, 29
retrofits 123–126
re-use 131, 144, 219n157
Revolution, American 17, 18, 28, 37, 72, ***96***, 104–105, 153; *see also* Cornwallis; Greene
riparian buffer 16, 28–29, ***124***, ***125***, ***132***, 143, 147
river basin *ii*, 5, 17, 18, ***193***, 202–***203***
river guides 210n1
River Landing Inn 130
river oats 108
river transportation 16, 38, 41, 75, 76, 100, ***102***, 195–196, ***197***, ***199***; obstacles 16, 153, ***168***, 195–196
Roberson, Troy 182
Robeson Creek 165, 167, 170, 172–174
Rock Rest 136, 138–140, 149
Rocky River 41, 196
Rodgers, Isaac 99
Rose Creek 34
Rosenthal, George 99–100
Route 68 8, 10, 12
Royall, Kathryn 22, 27
Ruffin, Thomas 98–99, 112
runoff 28, 29, 109, 118, 123–126, 135, 142–147, 184, 217n32, 220n36; *see also* impervious surface; solutions; trash

safety 4, 41, 43–45, 129–130, 152
Sandy Cross Road 14, 17, 21, 22
Saponi *see* Occaneechi
Saxapahaw (town) 111–118; lake 45, 110, 165; mills 111–112; *see also* Sellers Manufacturing; Sissipahaw
Scott, Ralph 82
Scott, Robert 81–82
Scott, W. Kerr 81–82
sediment 24, 26–27, 107, 109, ***124***, 142–148, ***146***, 182–184, 217n51; control 144–147, 183; quantity 142–143; *see also* fines; runoff
seeps 7
segregation 73, 80, 102, 159, 198–199, 212n116; *see also* racism
Sehlinger, Bob 169
Sell, Clarence 66, 68–69
Sellers, Charles V. 113–114
Sellers Manufacturing Co. 111, 113–116
septic systems 21, 86–88, 172

Index

settlers (European) 7, 16–17, 153, 207*n*14; claims 110; early coastal 199–200; housing 16–17; food 17; farming practices 26, 27
sewers 86–88, 101, 144; *see also* wastewater treatment
Shakori 90–91, 94
Shallow Ford 51
Sharp, Jane 149, 163
Shearon Harris Nuclear Power Plant 30, **31**, 197
Shoaf, Steve 69, 183, 184, 188
Shoffner, John 106
Sissipahaw 85, 89–95, 111
skunk cabbage 12
slavery 19–20, 26, 48, 72, 96, 98, 112, 180–181, 210*n*24, 25
sludge 67, 68
Smilie's Falls 195
Smyth, J.F.D. 16, 17, 72, 82
Snipes, John Wesley 155–158
Snow Camp 133
Society of Friends 19–20, 131, 133; *see also* Underground Railroad
solutions 28–29, 187, 201–202, 211*n*83, 218*n*53; *see also* conservation; enforcement; EPA; rain gardens; recycling; retrofits; reuse; riparian buffer; sediment; stormwater control; water use
source *see* headwaters
Spirit Island 107
spring peepers 10
springs 7, 158
stage coach 101, 103
Stagg Creek 84–86
Steible, Matt 192–194, **195**, 196
Stericycle 70, 82–84, 212–213*n*174, 177
Stewart, Ray 83–84
Stigall Road 7, 8
Stoneman's Raid 49
Stony Creek 79–**80**
stormwater control 68, 173, 183, 187; *see also* runoff
strainers 43, 130
stream address 19, 133
stream boundaries 17, 154–155
streams: and land value 19
strikes 48, 73–74, 100, 156–158
stripe bass **175**, 181–182
Strudwick, Samuel 110–111
sturgeon 3, 167
sustainable development 29, 144
swamp *see* wetlands
sweeper 43
Swepson, George W. 99–100
Swepsonville 45, 95–104, **103**, 106–107, 120
Swepsonville River Park 102, 173
synthetic chemicals 69–71, 123

Taoism 36
Tar Heel, NC 198
Taylor-Henley Mill 166
Tenaglia, Dan 22–26, 27
Terrells Creek, river left 133, 136, 149

Terrells Creek, river right 133–134
Terres, John 110
Textile Heritage Museum 57–58
Thomas, Greg 128
Thompson, William 195–196
Thoreau, Henry David 36, 71, 141, 188
303(d) list 34, 145, 182, 183, 184, 187, 207*n*2
Tickle Creek **15**
Tigris and Euphrates River 30, 32, 142
TMDL *see* total maximum daily load
tornadoes 139–140
total maximum daily load (TMDL) 183–184
Town and County Nature Park 80, 81
trade policy 64, 65
trading path 36, 48, 51, 89, 103, 104, 108–109
Trading Path Association 36, 108–109
transcendentalism 36
transpiration 123
trash 79, 127–128, 130–131
Travis Creek 51
tree varieties 21, 107, 131, 141, 194
trespassing 10
Triassic basin 40, 167, **168**, 170, 174, 192, 209*n*127
Trollinger, Adam 71–72
Trollinger, Ben 72, 81
Trollinger, Henry 72
Trollinger, Jacob 72, 78
Trollinger, John 61, 72
Trollinger, Mike 77–78
Trollinger, Spencer 77–78
Trollinger, William 72
Troublesome Creek 20, 25, 27–29, 34, 38
Troxler Mill 41
Tryon, Gov. William 104–105, 130
turbidity 26–27, 118, 145, 220*n*43
turkey vultures 11, 110

Underground Railroad 19–20; *see also* Coffin; Troublesome Creek
Union School 130
unions 49, 55, 57, 59, 60, 73–74, 113, 114, 157; *see also* mill workers
University of North Carolina, Chapel Hill 180, 195, 214*n*232

valley *see* river basin
Vander Borgh, Mark 184–186
Varnal's Creek 108
Verrazanno, Giovanni de 200
Virginia Mills 63, 81, 100–103
volcanoes 39, 40, 72, 84, 139–140, 164

wages *see* mill workers
Wallace, Holly 169–170
Walls, John Freeman 20
Walton, Helen 54–56
wastewater treatment 21, 23, 34–35, 52, 63, 65–69, **67**, 82, 86–88, 101, 144, 145, 147, 183–184, 211*n*83, 215*n*311, 217*n*32
water conservation *see* conservation
water gauges 129, 134–135, 164, 166
water power *see* hydropower
water quality 27, 34, 76, 131, 220*n*43; *see also* Jordan Lake; pollution
water rights 10, 30, 156
water supply 22, 30–33, 144, 189–191; Greensboro 29–33, 41; policies 30; pricing 32–33; storage 32; worldwide 190–191; *see also* developments; droughts; water use
water treatment 69–70, 88, 187, 188–191, 209*n*85, 217*n*32; bottled water 69, 190–191; Cary Apex Water Treatment Plant 188–190; history 69, 188–189; *see also* conservation; disease
water use 30–33, 63, 189–190, 209*n*80
Watermelon Island 136, 149
watershed *see* river basin
Weakley, Alison 147, 217*n*51, 52
Weaver Creek 179
Webster Site 141–142, 148–149
weirs 78, 130, 149
Weitzel's Mill 37
West End Revitalization Association (WERA) 86–88
wetlands 8–12, 21–22, 24, 33, 40, 174, 194, 198, 207*n*7, 218*n*52
White, James 113
Whitehead Creek 109
Whitesell, Anna 73
whitewater 40, 46, 77–78, 129–131, 147–152, 160, 167–170, **171**; *see also* Brueckner; Featherstone; Nealy; paddling
wild and scenic 149
wilderness 38
Wildfire, Nolan 192–194, **195**, 196
Wildlife Resource Commission 76, 143, 182
wildness 38, 141
Williams, Allan 32–33, 209*n*85
Williamson, James 48–49, 61, 113
willow 108
Wilmington **199**, 200
Wilson, Omega 86–88
Witty, John 17
Woodland Period 131–132, 179
Woody's Ferry 130
Woolfolk, Michelle 187
workers *see* mill workers

xenoestrogens 70

Yadkin River 6, 7, 26, 40, 91
Yee Haw Paddle 79–82

zoning 143–144

www.ingramcontent.com/pod-product-compliance
Ingram Content Group UK Ltd.
Pitfield, Milton Keynes, MK11 3LW, UK
UKHW050531150426
5217IPUK00026B/1891